Guidelines for Using Activated Sludge Models

Scientific and Technical Report Series: Report No. 22

Guidelines for Using Activated Sludge Models

Authored by the IWA Task Group on
Good Modelling Practice – Guidelines for Use of Activated Sludge Models

Leiv Rieger (chair)
Sylvie Gillot
Günter Langergraber
Takayuki Ohtsuki
Andrew Shaw
Imre Takács
Stefan Winkler

IWA Publishing
London • New York

Published by IWA Publishing
Alliance House
12 Caxton Street
London SW1H 0QS, UK
Telephone: +44 (0)20 7654 5500
Fax: +44 (0)20 7654 5555
Email: publications@iwap.co.uk
Web: www.iwapublishing.com

First published 2013
Reprinted 2014

British Library Cataloguing in Publication Data
A CIP catalogue record for this book is available from the British Library

ISBN 9781843391746 (Paperback)
ISBN 9781780401164 (eBook)

Contents

GMP ON IWA WATERWIKI WEBSITE

http://www.iwawaterwiki.org/xwiki/bin/view/Articles/GuidelinesforUsingActivatedSludgeModels

The content of this website may change or be extended!

WaterWiki 1	The GMP Application Matrix
WaterWiki 2	Gujer Matrices for 7 published models
WaterWiki 3	Spread-sheet to estimate number of tanks in series
WaterWiki 4	Spread-sheet to create dynamic influent (HSG Influent Generator)
WaterWiki 5	Reprint of paper: *Activated sludge modelling in practice – An international survey* by Hauduc *et al.* 2009
WaterWiki 6	MS Word 2003/2007 macros to translate old into new Notation

Acknowledgements

We owe a lot to our home institutions and their kind support during all these years is gratefully acknowledged.

A special "Thank You" goes to Hélène Hauduc, the most dedicated Ph.D. student ever, who dedicated so much of her time to help with the GMP work.

The work of the Task Group would not have been possible without the crucial support from our two sponsors: **Hydromantis Environmental Software Solutions, Inc**. (1 James Street South, Suite 160, Hamilton, Ontario, Canada, L8P 4R5. www.hydromantis.com) and **MOSTforWATER**.

The GMP Task Group was sponsored by the International Water Association (IWA) and the GMP members would like to express their sincere gratitude to Hong Li (our main contact at IWA head quarters) and Paul Reiter (IWA Executive Director). David Burns, Maggie Smith, Michelle Jones, and Chloe Parker guided us during the publishing process.

We thank all the organisations and individuals who helped to organise our meetings. Especially Professor Krishna Pagilla should be mentioned who invited us to his house and made sure that we were not running out of substrates. Cemagref (now Irstea) holds the record for hosting three of our meetings.

Last but not least we would like to thank all the reviewers for their invaluable input to this work:

Youri Amerlinck, Belgium
James Barnard, USA
Damian Batstone, Australia
Evangelia Belia, Canada
Marie Burbano, USA
Scott Bury, USA
John Copp, Canada
Lluis Corominas, Spain
George Ekama, South Africa
Güçlü Insel, Turkey
Bruce Johnson, USA
Dave Kinnear, USA
Henryk Melcer, USA
Eberhard Morgenroth, Switzerland
Ingmar Nopens, Belgium
Paul Roeleveld, The Netherlands
Oliver Schraa, Canada
Kim Soerensen, France
Ana-Julia Tijerina, USA
Peter Vanrolleghem, Canada

List of GMP Task Group Members

Leiv Rieger (chair)
EnviroSim Associates Ltd. (2009–2012), Canada
Université Laval (2006–2009), Canada
École Polytechnique de Montréal (2005–2006), Canada
Eawag (2004–2005), Switzerland

Sylvie Gillot
Irstea (formerly Cemagref), France

Günter Langergraber
University of Natural Resources and Life Sciences Vienna (BOKU), Austria

Takayuki Ohtsuki
Kurita Water Industries Ltd., Japan

Andrew Shaw (2006–2012)
Black & Veatch, USA

Imre Takács
EnviroSim Associates Ltd. (2004–2008), Canada
EnviroSim Europe (2008–2010), France
Dynamita (2010–2012), France

Stefan Winkler
Vienna Technical University, Austria

Paul Roeleveld (2004–2006)
Grontmij Water and Waste Management, The Netherlands

Preface

THE GOOD MODELLING PRACTICE (GMP) TASK GROUP

It all started in Marrakech at the IWA World Water Congress in 2004: four groups joined forces to organize a workshop on "*Guidelines for the Application of Wastewater Treatment Plant Models*" with the ultimate goal to start a new IWA task group. The four groups were: (1) Hochschulgruppe Simulation (HSG), a group of German speaking Ph.D. students and researchers, (2) several modellers involved in the development of the Water Environment Research Foundation (WERF) study on "*Methods for Wastewater Characterization in Activated Sludge Modeling*", (3) STOWA from The Netherlands, and 4) the BIOMATH research group from Ghent University, Belgium.

A small team of modellers with international experience was formed with a mandate to collect experience and knowledge on activated sludge modelling with a clear aim to provide guidance to practitioners. Stefan Winkler (Austria), Imre Takács (Canada, later France), Paul Roeleveld (The Netherlands), and Sylvie Gillot (France) were appointed from the involved groups and Leiv Rieger (Switzerland, later Canada) was elected as chairman. To include more international experience Takayuki Ohtsuki (Japan) was invited to join the group and Günter Langergraber (Austria) was asked to take over the task of event coordinator for the group. In 2006, Paul Roeleveld left the group due to time pressures and Andrew Shaw (USA) stepped in.

THE ACTIVITIES

Mogens Henze, the chairman of the ASM Task Group, suggested that the new Task Group should not only sit down and write a guideline but should actively promote "Good Modelling Practice". We took this advice and solicited input from model users from around the globe. The Task Group initiated or linked to existing expert groups (e.g. Modeling Expert Group of the Americas (MEGA), Groupe ASTEE Modélisation as well as HSG), sent out questionnaires, and interviewed several distinguished modellers (Peter Dold, George Ekama, Willi Gujer, Mogens Henze, Mark van Loosdrecht). To discuss the work, several workshops, meetings, and courses on activated sludge modelling were organised. For instance, the Task Group played a significant role in the initial organisation of the on-going Wastewater Treatment Modelling Seminars (WWTmod). The Task Group was involved in the development of a new IWA Model Notation System (Corominas *et al.* 2010) and a reprint of that paper can be found in Appendix D.

THE WORK

From heated all night long discussions on *How to set up and calibrate models for specific objectives?* to *What is calibration in the first place?*, we learned a lot from each other and we think the discussions that ensued from questions like these helped bridge the gap between North American and European modellers.

Most productive were the long meetings we had. It was hard work but good spirits (and food) helped keeping us on track. We still remember the raw horse meat in Japan and the oysters in France. Site visits were always part of a long meeting and often triggered discussions on better ways to use models to operate or design plants. A special debt of gratitude is extended to all the plants that entertained the Task Group and answered all the strange questions.

THE SCIENTIFIC & TECHNICAL REPORT

When the work was initiated, it was our intention to write a concise guideline for activated sludge modelling. However, the discussions quickly revealed that a lot of essential information was not easily available. It was clear that all the information could not be written into a single report. We struggled between writing a reference book and coming up with short and concise guidelines. The external reviewers and especially John Copp brought us on the right track suggesting to split the work into essentially two parts: a formal *Guidelines* section and a second section with various *Examples*.

Another advice we got from Mark van Loosdrecht was to dare to suggest typical values for often used ratios, variables, and parameters in wastewater modelling. We agreed and what you will find in this report are starting points for these quantities and even though the reported values are the result of intense group discussions, it is fully expected that these will be refined with increasing experience.

In the spring of 2011 we finally had a first complete version including appendices and additional material for a dedicated website. Convinced that this would be more or less the final version, we invited more than 20 experts to review the report. The feedback received was invaluable and the Task Group is indebted to those experts for taking the time to read the original draft. However, following their advice, we decided to completely restructure the STR.

Instrumental in this last effort was Dr. John Copp who agreed to serve as the final editor. His careful edits have improved the organisation of the written material and made the report readable. The Task Group thanks him for his time and diligence.

WHAT STAYS?

In the end we do not know how much of this report will be valid after 10 years, but wastewater modelling progresses rapidly and we recommend that these guidelines be re-evaluated after a certain period of time.

However, what will not have to be re-evaluated is the friendship between the GMP Task Group members.

We hope you will enjoy reading this Scientific and Technical Report!

– Modelling is fun! –

Leiv Rieger
Sylvie Gillot
Günter Langergraber
Takayuki Ohtsuki
Andrew Shaw
Imre Takács

Chapter 1
Introduction

1.1 RATIONALE

Mathematical modelling of activated sludge systems has become a widely accepted tool for plant design and operation, training of process engineers and operators, and as a research tool. However, model results are only useful in practice if the model predictions are reliable.

The level of detail and quality required for simulation studies varies greatly depending on the project objectives, the resources committed to the project and the available expertise. Inconsistent approaches and insufficient documentation make quality assessment and comparison of simulation results difficult.

A questionnaire sent out by the IWA Task Group on *Good Modelling Practice – Guidelines for Use of Activated Sludge Models* identified the following main obstacles for promoting the wider use of activated sludge models by practitioners (Hauduc *et al.* 2009, see GMP IWA WaterWiki website for a reprint):

- *Cost and time*
- *Model structure:*
  - Missing information on model limitations
  - Need for a comparison of available models
  - Lack of information on model reliability
  - Model complexity too high and inadequate models for specific objectives
- *Model application:*
  - Available models not adapted to objectives or operational conditions
  - Need for better knowledge and experience transfer
  - Need for software improvements
- *Modelling procedure:*
  - Data collection and reconciliation: time consuming, lack of standardised procedures for specific measurements (e.g. influent characterisation) and data reconciliation procedures.
  - Calibration/validation: standardisation of modelling procedures lacking

The Task Group worked to overcome these obstacles by compiling internationally available knowledge and experience in this field through expert groups, workshops, questionnaires, modelling courses and an internet discussion forum. All this experience and knowledge was then processed and condensed and is presented in this IWA Scientific and Technical Report (STR).

1.2 SCOPE OF REPORT

This Scientific and Technical Report (STR) is intended for practitioners. It gives guidance in the process of planning and conducting simulation projects and can be used as an introductory book to learn about Good Modelling Practice (GMP) in activated sludge modelling and will be of special interest to process engineers who have no knowledge of modelling.

The STR presents a framework to deal with the practical application of commonly used process models such as the activated sludge models (ASMs) developed by the International Water Association – the so-called "ASM-type" models (Henze *et al.* 2000) – as well as models that use a similar structure (e.g. Barker & Dold model (Barker & Dold, 1997); ASM3 + BioP (Rieger *et al.* 2001); ASM2d + TUD (Meijer, 2004); and UCTPHO+ (Hu *et al.* 2007).

This framework is intended to make modelling straightforward and more systematic for practitioners. Additionally, an Application Matrix approach is presented that will help to define quality levels for simulation results. This approach provides a procedure to assess this quality and a means to assist with the proper use of the models. Examples of different model application goals are identified, classified and linked to the required effort to obtain results with sufficient quality to meet the objectives of the simulation study.

The STR describes a methodology for a goal-oriented application of activated sludge models using a concise simulation project procedure and some illustrative case studies. The case studies give examples of the required data quality and quantity and the effort required for calibration/validation with respect to a defined goal.

1.3 STRUCTURE OF THE STR

After this ***Introduction*** chapter, the STR continues with:

Chapter 2 *State-of-the-Art of Activated Sludge Modelling*, which contains a brief history of wastewater treatment modelling, a review of current practice, and perspectives on the future application of models.
Chapter 3 *Available Protocols* discusses available simulation protocols.
Chapter 4 *The GMP Unified Protocol* describes the development of this protocol.
Chapter 5 *Unified Protocol Steps* provides detailed descriptions of each of the five protocol steps and the required procedures to carry out a modelling project.
Chapter 6 *The GMP Application Matrix* introduces the matrix concept. The Application Matrix use is illustrated through 14 typical modelling case studies where different steps are linked to a required level of effort.
Chapter 7 *Examples: Walking through the GMP Unified Protocol* provides detailed examples for the application of the GMP Unified Protocol.
Chapter 8 *Use of Activated Sludge Models for Industrial Wastewater* focuses on industrial applications and the modifications required to model them.
Chapter 9 *Frequently Asked Questions* answers typical questions related to modelling activated sludge systems.
Glossary provides definitions of technical terms used in this report
References includes a complete list of references used in the text,
Index provides help to find specific topics.

The **Appendix** contains additional information including reprints of selected Task Group papers, a list of typical sources of measurement errors, and sources of uncertainties.

Appendix A provides detailed descriptions of selected sub-models.
Appendix B introduces the matrix notation for bio-kinetic models (Gujer Matrix).
Appendix C gives some useful tips on solver settings.
Appendix D presents an IWA naming notation for model state variables and parameters.
Appendix E is a reprint of a paper on model verifications for seven published models.
Appendix F is a reprint of a paper discussing model parameters.
Appendix G lists typical sources of measurement errors.
Appendix H lists sources of uncertainties as identified by the IWA DOUT Task Group.
Appendix I discusses mass balancing as a tool for fault identification.

A dedicated website has been set up to provide additional material. Information on that site includes a set of case studies, spreadsheets with error-checked Gujer Matrices and various modelling tools in spreadsheet format: **http://www.iwawaterwiki.org/xwiki/bin/view/Articles/GuidelinesforUsingActivatedSludge Models**.

1.4 MODEL NOTATION, NOMENCLATURE, UNITS

The STR uses the newly developed **notation for biokinetic models** according to Corominas *et al.* (2010, reprint Appendix D). This new notation uses systematic naming rules and is an attempt to overcome several pitfalls and shortcomings of the notations used in the literature previously. For readers familiar with the old IWA notation a translation table is provided on the IWA WaterWiki website together with corrected Gujer Matrices for 7 published activated sludge models.

A clear distinction between measured and model variables is provided by using a separate *nomenclature for measured variables.* Table 1.1 lists selected measured variables with their long names, units and the names used for the measured and model variables, respectively. The unit format used in the STR gives priority to units commonly used in practical reports.

Table 1.1 Naming rules for selected variables.

Variable	Unit	Measured	Model
Total COD	mg COD/L	COD_{tot}	T_{COD}
Filtered COD	mg COD/L	COD_{filt}	SC_{COD}
Soluble COD (after flocculation and filtration)	mg COD/L	COD_{sol}	S_{COD}
Total nitrogen	mg N/L	N_{tot}	T_N
Total Kjeldahl nitrogen	mg N/L	TKN	T_{KN}
Ammonia nitrogen	mg N/L	$NH_x\text{-}N = NH_4\text{-}N + NH_3\text{-}N$	S_{NHx}
Nitrate/nitrite nitrogen	mg N/L	$NO_x\text{-}N = NO_2\text{-}N + NO_3\text{-}N$	S_{NOx}
Total phosphorus	mg P/L	P_{tot}	T_P
Phosphates	mg P/L	$PO_4\text{-}P$	S_{PO4}

1.5 SUGGESTED READING

In an attempt to help readers navigate the STR, the following table has been compiled. Although it is hoped that readers will read the STR in its entirety, Table 1.2 is meant to help focus readers with specific interests to specific chapters.

Table 1.2 Suggested reading for selective topics.

	Ch 1: Introduction	Ch 2: State-of-the-Art of AS Modelling	Ch 3: Available Protocols	Ch 4: The GMP Unified Protocol	Ch 5.1: Step 1 Project Definition	Ch 5.2: Step 2 Data Collection and Reconciliation	Ch 5.3: Step 3 Plant Model Set-up	Ch 5.4: Step 4 Calibration and Validation	Ch 5.5: Step 5 Simulation and Result Interpretation	Ch 6: The GMP Application Matrix	Ch 7: Using the GMP UP by Example	Ch 8: Use of ASMs for Industrial Wastewater	Ch 9: FAQs	Glossary
Professionals directly involved in modelling details														
Consultant														
Beginner	X	X	X	X	X	X	X	X	X	X	X	O	X	X
Intermediate	X			X	X	X		X	X	X	X	O	O	O
Expert	X			X	X			X				O		
Researcher	X	X		X	X	X	X					O		
Instructor	X	X		X		X	X		X			O		X
Software developer	X	X	X	X		X	X	X	X			O		
Professionals involved in higher process level tasks														
Project manager	X			X	X	X								
WWTP staff														
Operator	X			O		X						O		
Plant manager	X			X	X									X
Municipality / EPA / regulator / water board	X		X	X	X							O		X

X: Read whole chapter

O: Read if of interest

Chapter 2

State-of-the-art in activated sludge modelling

Short summary

This chapter provides an overview of activated sludge (AS) modelling practice. The section *Representing Reality in Models* is a theoretical view of mathematical modelling defining major terms and structures. The sections *History of Activated Sludge Modelling*, *Review of Practice* and *Future of Activated Sludge Modelling* describe the past, the current situation and the potential future of activated sludge modelling.

2.1 REPRESENTING REALITY IN MODELS

A model is a simplified representation of a real object, state of affairs, or process. Typically, a behaviour or a few key characteristics of a selected system are represented in the model. A *numerical* model is an attempt to represent a real-life situation using mathematical equations (typically on a computer). The act of simulation describes the use of a numerical model within a software package (called a "simulator").

Typically, numerical models have to be calibrated to one or more data sets before they can be applied. Ideally, this is then followed by a validation step which ensures that the model can be used to predict the behaviour of the system under different conditions. Numerical models can be used for *prognostic*, that is predicting the future, *diagnostic*, that is understanding mechanisms or processes, and *educational purposes*, that is for communicating between experts and non-experts or for training of process engineers or operators. These different uses are illustrated in Figure 2.1.

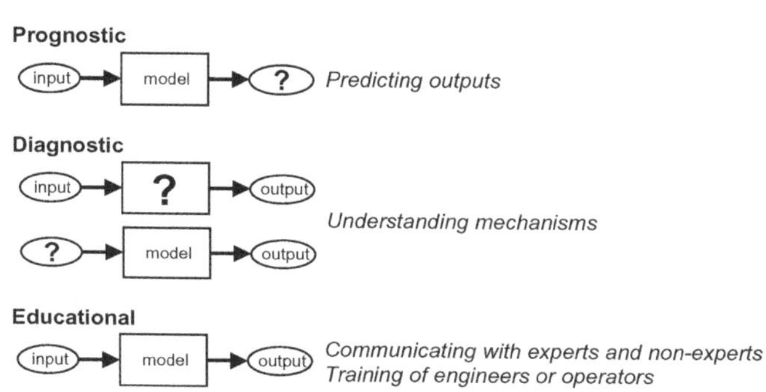

Figure 2.1 Purpose of modelling (adapted from Hug *et al.* 2009).

The key issues in developing a successful model of a wastewater system include obtaining reliable measurements (observations), the selection of key characteristics and behaviour, the use of simplifying approximations and assumptions, the accuracy of the simulation output (calibration/validation) and the reliability of the predictions.

Figure 2.2 shows how real world observations are translated into a plant model. Plant models require a number of inputs, which are then used to produce a prediction (an output). Often, the model requires inputs that must be generated from various observations obtained on the real system. Model inputs are rarely the same for different models and vary based on the scope of the model being developed. That is, the selected system boundaries may include just the plant, or maybe the whole city or could even include the entire watershed as depicted in Figure 2.2.

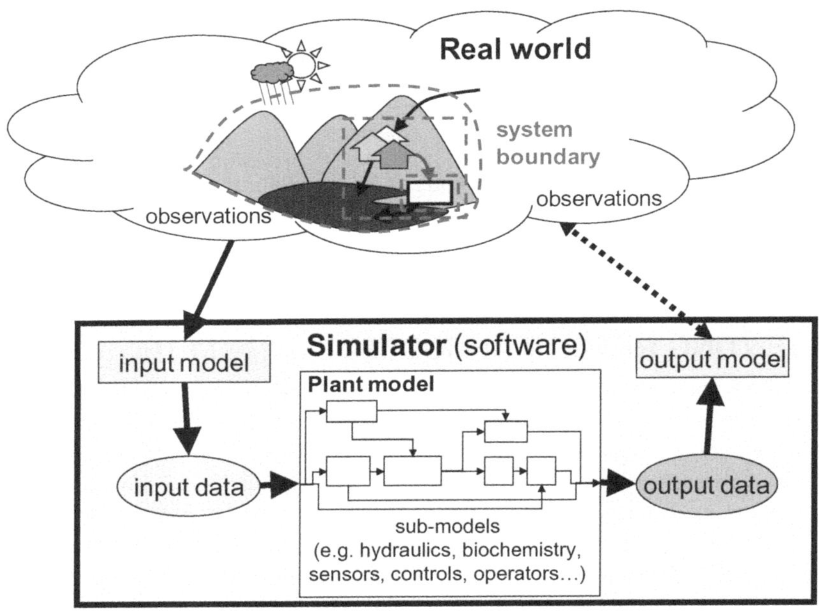

Figure 2.2 Connection between real world observations and a plant model including several sub models implemented in simulation software (adapted from Hug, 2007).

Irrespective of the system boundary, the model input is generated by translating measurements into appropriate model input variables which the model uses during simulation. For AS models these inputs are the variables describing the input concentrations of the modelled compounds and other critical components like the input flow. It is often the case that the system model will require sub-models that describe specific processes (e.g. biokinetic transformation, settling, etc.) or operational units (e.g. pump, flow splitter, etc.) and these sub-models may require their own set of inputs. Similarly, the model outputs are often converted into variables which correspond to the measured observations of the real system so that comparisons can be made between the calculated model predicted output and the measured quantities. These comparisons are used as a guide to judge the suitability of the model to a given task.

Numerical activated sludge models are based on the development of a set of differential equations that calculate the accumulation of a number of state variables (represented by C_x) within a time step, dt, in a compartment with a volume, V, under the consideration of an influent and effluent flow and biokinetic conversions represented by r (see Equation 2.1).

$$\underbrace{\frac{dM}{dt} = \frac{d(v \cdot C_x)}{dt}}_{\text{Accumulation}} = \underbrace{r \cdot V}_{\substack{\text{Conversion term} \\ \text{(biokinetic model)}}} + \underbrace{Q_{\text{in}} \cdot C_{\text{x,in}} - Q_{\text{out}} \cdot C_{\text{x,out}}}_{\substack{\text{Transport term} \\ \text{(transport model)}}} \tag{2.1}$$

The last two terms on the right-hand side of Equation 2.1 describe the transport of compounds (input and output). Through linking of several reactors in series, it becomes possible to reflect the hydraulic behaviour of the system (i.e. plug flow *versus* completely mixed). The first term on the right-hand side of Equation 2.1 describes the biokinetic reactions that are modelled by a biokinetic model (e.g. ASM1). Other terms can be added to this equation to describe other transformations (e.g. precipitation) and transport (e.g. gas transfer).

There are different biokinetic models available for activated sludge systems. A list of the biokinetic models covered in this STR is provided in Section 2.3.2 *Discussed Biokinetic Models* below. A detailed description of their notation is provided in Appendix B and error-checked matrices are provided on a dedicated website: **http://www.iwawaterwiki.org/xwiki/bin/view/Articles/GuidelinesforUsingActivatedSludgeModels.**

2.2 A BRIEF HISTORY OF ACTIVATED SLUDGE MODELLING

This section provides a brief historical overview of activated sludge modelling. For interested readers who would like to develop a deeper understanding of the development of AS models, this section references a few recommended milestone publications. The timeline presented in Figure 2.3 describes the historic development of activated sludge modelling.

Early years

In the early 20th century, Penfold and Norris (1912) showed that the generation time of bacteria was inversely proportional to substrate concentration and in 1914 Ardern and Lockett announced the invention of the "activated sludge" process. These early publications provided a springboard for this new process, but it was not until several decades later that its kinetic basis was fully appreciated.

Development of activated sludge kinetics

Monod (1942) showed that the growth of pure bacterial strains could be described using the equation that now carries his name and is widely used in many modern day models. Garrett and Sawyer (1952) stated that Monod's kinetics could also be used to describe the behaviour of the mixed bacterial cultures present in activated sludge. Herbert (1958) was arguably the first to introduce the concept of endogenous respiration which improved kinetic model predictions in systems with longer sludge retention times. In the late 50's and through the 1960's Eckenfelder (1958) and McKinney (1962) described the stoichiometry of sludge oxidation and developed mathematical expressions to represent the kinetics of activated sludge in completely mixed reactors. Throughout this period, the kinetics of nitrification were investigated widely and Downing *et al.* (1964) presented a thorough synopsis of the process and kinetics at the time. Gujer (2010) provided a more recent review of current nitrification understanding with some historical perspectives.

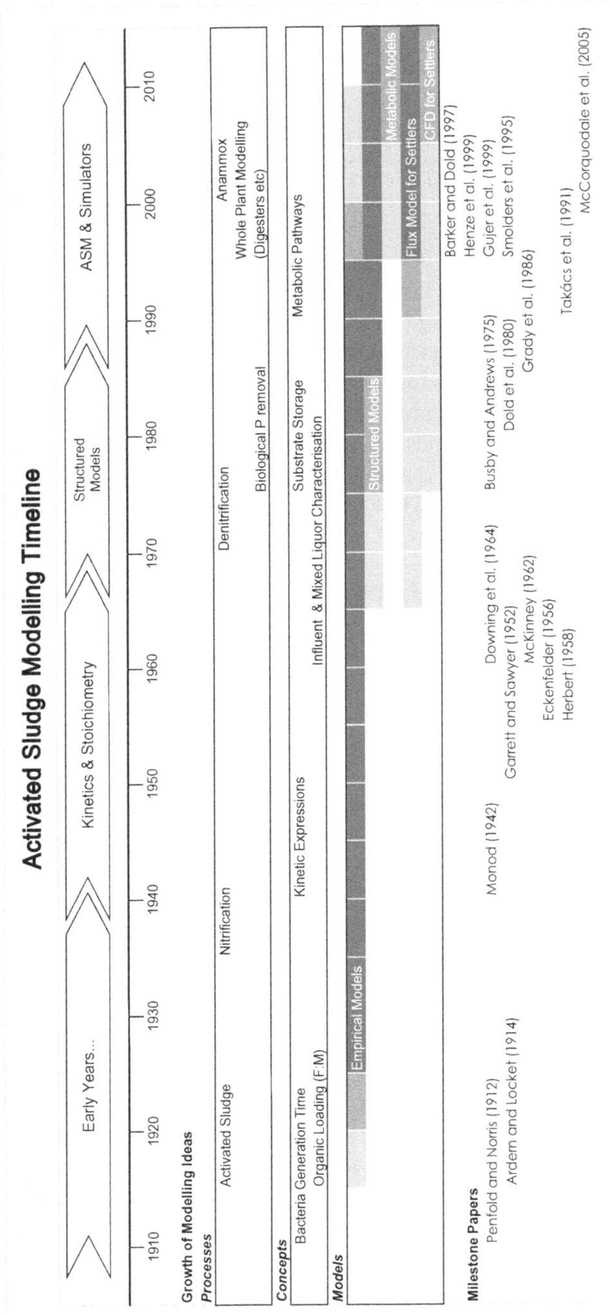

Figure 2.3 A Timeline for Activated Sludge Modelling.

Activated Sludge Models (ASMs) and beyond

With the development of increasing computer processing power came a revolution in process modelling. One of the earliest process models implemented in a computer was that of Busby and Andrews (1975). This implementation was a dynamic model using the "Continuous System Modeling Program" (CSMP) which included stored mass, active mass and a settler model. Around the same time, Dold *et al.* (1980) developed a structured dynamic activated sludge model that included both biological nitrification and denitrification.

In the early 1980's, the International Association on Water Pollution Research and Control (IAWPRC, now IWA) set up a Task Group to synthesise different modelling concepts into a "consensus" model for activated sludge. Members of the Task Group, Grady, Gujer, Henze, Marais and Matsuo, developed (Grady *et al.* 1986) what is now known as Activated Sludge Model No. 1 (ASM1). This model and its presentation format (the Gujer Matrix, Appendix B) has become the basis and framework for most model research and development since.

The initial publication of ASM1 was followed by the development of several other process models by the IWA Task Group and other researchers. These include ASM2 and ASM2d (Henze *et al.* 1995, 1999) and the Barker and Dold model (Barker & Dold, 1997) covering biological phosphorus removal, as well as ASM3 (Gujer *et al.* 1999 later updated in Henze *et al.* 2000). Other notable variants include a biological phosphorus removal extension to ASM3 (Rieger *et al.* 2001), the UCTPHO+ model (Hu *et al.* 2007) and the ASM2d + TUD model (Meijer, 2004). Secondary clarifier models such as Takács *et al.* (1991) have been widely used in conjunction with ASM1 and subsequent models to simulate complete activated sludge systems. Many of these models are implemented in the commercial simulators used today. Most simulators also contain models for other unit processes such as anaerobic digestion, thickening and dewatering which enable them to simulate the interaction between processes in the whole plant.

In addition to the structured models based on Monod kinetics, a different approach, known as metabolic modelling (e.g. Smolders *et al.* 1995), is increasingly being used to establish the stoichiometry of transformations.

Kollekolle and WWTmod

Many of the ideas for the ASM-type models were presented and refined in the 1980s and 1990s through a series of modelling seminars in Kollekolle, Denmark. In 2008 IWA and the Water Environment Federation (WEF) jointly started a new biennial modelling seminar series, WWTmod in Mont Saint Anne, Quebec. The model*EAU* research group at the Universitié Laval in cooperation with the GMP Task Group played a seminal role in this initiative. In the spirit of the Kollekolle seminars, the purpose of the WWTmod seminars is to bring together modellers from diverse backgrounds to discuss the current state-of-the-art and future perspectives of wastewater treatment modelling.

2.3 REVIEW OF PRACTICE

2.3.1 Overview of current practice

Activated sludge models have gained widespread acceptance since their initial introduction in the 1980's. This is reflected in the increasing number of papers (Figure 2.4), as well as the birth and rapid growth of the wastewater modelling industry through engineering/software companies that provide modelling products and/or services. The use of modelling is slowly changing data gathering practices in wastewater treatment plants. Variables not directly related to plant effluent limits or operational objectives are increasingly measured as they are needed for modelling-based process evaluation, mass balancing and dynamic performance assessment. Previously neglected COD measurements, nitrogen and phosphorus contents of the solids and respiration rates are just a few examples of these additional variables now being measured.

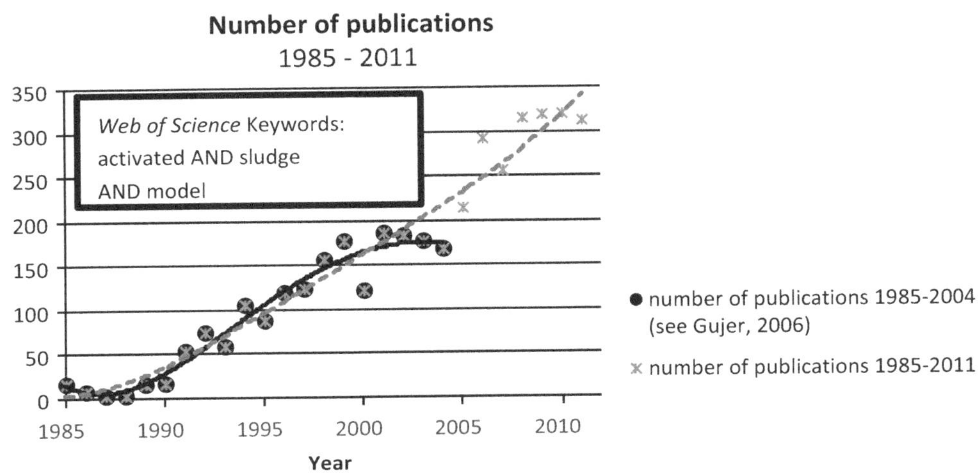

Figure 2.4 Interest in activated sludge modelling by number of publications from Web of Science® between 1985 and 2011 (adapted from Gujer, 2006).

Traditionally, activated sludge models were used to estimate required aeration capacity, the potential for denitrification in a large number of process configurations and to predict solids production for the sizing of sludge handling equipment (Henze *et al.* 1987). This new tool later found its use in improving plant control and operation. The models in use today are still based on carbonaceous material removal, nitrification and denitrification, but many additions and improvements have been proposed including: biological and chemical phosphorus removal as well as solids separation.

The same model concepts, structure and processes have been extended to various biofilm processes, reactive settler modelling, anaerobic and aerobic digestion, sidestream treatment and many other ancillary processes. The application of activated sludge modelling is more and more connected to other processes in the wastewater treatment plant and the use of models has progressed from simply modelling individual processes to plant based modelling. This approach is called "whole plant" or "plant-wide" modelling. As the scope of this STR is limited to activated sludge units, we will only consider examples related to activated sludge; however, it must be kept in mind that this is just one (although fundamental) model used in the wastewater process industry.

A mathematical model is a structured and rigorous way of storing process knowledge. The model can be used to enhance (not replace!) expert knowledge and the requirement to thoroughly understand key and more subtle characteristics of unit processes and their interactions. The model helps to consistently track interlinked mass flows, to calculate conversion rates in unit processes and to estimate likely performance regarding effluent quality, sludge production, oxygen and chemical demand and other key indicators.

2.3.2 Discussed biokinetic models

Several biokinetic models have been published since ASM1. This STR refers to the following biokinetic models:

- ASM1 (Henze *et al.* 1987)
- Barker and Dold (Barker & Dold, 1997)
- ASM2d (Henze *et al.* 1999)
- ASM3 (Gujer *et al.* 1999)
- ASM3 + BioP (Rieger *et al.* 2001)
- ASM2d + TUD (Meijer, 2004)
- UCTPHO+ (Hu *et al.* 2007)

More details on these models are given in Chapter 5.3 *Plant Model Set-up*. The continuity and correctness of the models has been carefully checked and some typos from the original publications were identified during this process. Appendix E contains a detailed report on the model verification as published in Hauduc *et al.* (2010). Corrected Gujer Matrices in spread-sheet format are provided on the following website: **http://www.iwawaterwiki.org/xwiki/bin/view/Articles/GuidelinesforUsingActivatedSludgeModels**.

2.3.3 Stakeholders in modelling projects

People involved in a modelling project – the stakeholders – can be split into modellers and non-modellers. Modellers are defined as a group that have a thorough understanding of the models and their use, whereas non-modellers might not know about the details of the models and how to simulate. However, non-modellers need to understand what answers the model can provide, the level of effort for a specific simulation project (including data needs) and what accuracy they can expect from the simulation results. Important considerations for modellers are to set realistic expectations for the modelling (Chapter 5.1) and to ensure that model outputs are communicated clearly (Chapter 5.5), and in such a manner that non-modellers are properly engaged and informed in the modelling project objectives and scope. In some countries, regulatory organisations check whether designs comply with technical standards. Some regulators approve the use of modelling and so these regulators have to be able to evaluate simulation projects which may be used to supplement regulatory filings. This is why this STR should be helpful for both groups.

2.3.4 The role of models in the life cycle of a plant

It is useful to consider process models in the context of the life cycle of a treatment facility and the fact that it is impossible to predict, with a high degree of certainty, the future for a treatment facility. Gujer (2011) gives the example of the Zurich-Werdhölzli wastewater treatment plant as it developed over a period of 20 years. During this period pilot studies, construction of the original activated sludge plant and significant modifications were carried out. The plant needed to respond to increasing levels of nutrient removal requirements driven by new regulations, aging infrastructure, new processes and changing influent characteristics (particularly when significant industries moved out of town). In developing and using process models, not only are plants able to optimise an existing facility, they are better able to respond to the uncertainties about future loads and treatment requirements. A well calibrated model can be adapted and modified over time to keep track of the current status of the plant (Figure 2.5) thus providing a reasonably stable basis for decision making (Phillips *et al.* 2010).

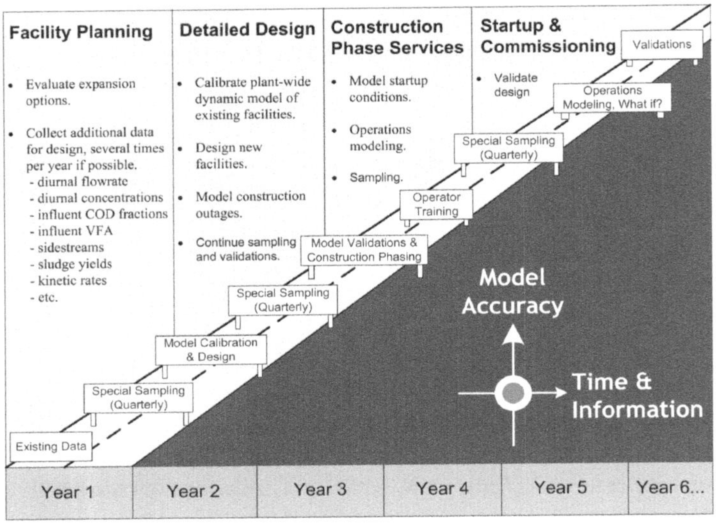

Figure 2.5 Life cycle of a plant model used in different stages of a plant upgrade (Phillips, 2010).

2.4 THE FUTURE OF ACTIVATED SLUDGE MODELLING

2.4.1 Driving forces in wastewater treatment

There are many challenges in the wastewater treatment industry that affect activated sludge modelling. Table 2.1 lists some of the drivers in wastewater treatment and their impact on modelling.

Table 2.1 Drivers in wastewater treatment and their impact on modelling.

Driving forces in wastewater treatment	Impact on activated sludge modelling
Financial drivers	
Operating costs (energy, chemical dosage, maintenance, legal costs, etc.)	Use models for operation/control optimisation
Political demand to cover full treatment costs by user fees (no subsidies)	Strong incentive for operational optimisations to keep industries in the region. Use models for operation/control optimisation. Use models for prediction of future costs and planning.
Limited funding for expansions	Use models for design optimisation, eating into safety factors, moving away from empirical design, risk based evaluation, dealing with uncertainty, investigation of alternative processes
Space limitations	Innovative process design, high-rate processes
Tightening effluent limits	
New nutrient requirements from receiving water quality issues (N&P)	Models need to describe nutrient removal processes.

(Continued)

Table 2.1 Drivers in wastewater treatment and their impact on modelling (*Continued*).

Driving forces in wastewater treatment	Impact on activated sludge modelling
Low effluent limits	Influent characterisation and state variables, new models, half-saturation coefficient values. Need to deal with uncertainties.
Microconstituents	New models needed, analytical issues, fate of specific compounds
Climate change	
Reduced water use	Change in influent characteristics. Influent strength increases and quantity decreases.
Wet weather issues	Storm event modelling, historical approach to flows and loads not applicable
Increased water reuse	New processes, increased dissolved solids (e.g. chloride) issues
Concern about greenhouse gas emissions	Energy optimisation, new models for nitrous oxide
Social drivers	
Increased accountability to stakeholders	Models used for communication and education

2.4.2 Trends and research needs

Several important trends have been observed in the use of process modelling in recent years. Some of the most important amongst these include:

- The increased use of whole plant models that enable process engineers to investigate interactions between unit processes;
- Models integrated with upstream and downstream water systems (integration of the sewer, WWTP and receiving water) to further understanding of interactions outside the plant fence-line;
- Basin scale modelling which pulls together models from all parts of the water cycle.

Increasingly dynamic simulations are being used for optimisation and control design. The use of improved online instruments provides the data necessary for dynamic simulations and this trend will continue.

Model developers and researchers continue to strengthen the microbiological foundations of the models (including metabolic modelling, population dynamics, etc.) and to develop models for specific compounds such as greenhouse gases (e.g. N_2O), heavy metals and micro-constituents. New tools and methods are being developed such as elemental balancing and specific influent characterisation methods. Models are being developed that consider physical-chemical processes in combination with biological processes and still others incorporate equilibrium chemistry to account for fundamental speciation. Increased processor power and model development are starting to make it feasible to combine computational fluid dynamics (CFD) with biological models. Researchers are investigating the use of probabilistic modelling to handle uncertainties and others are using models as part of decision support systems (DSS).

The more widespread modelling becomes, the more it highlights gaps in current knowledge and areas where models could be enhanced or improved. It is beyond the scope of this STR to describe all research needs; however, Table 2.2 presents a list of areas where, according to the authors, further research is needed. It should be recognised that establishing *Good Modelling Practice* now is important to work on these research needs so that we might be able to solve tomorrow's problems.

Table 2.2 List of further research needs in wastewater treatment plant modelling.

Area	Need for modelling research
Nutrient removal	More detailed influent characterisation, particularly for phosphorus and nitrogen compounds to determine their degradability Nutrient removal to very low levels (are our models still valid in these concentration ranges?)
Biological and kinetic fundamentals	Identification and quantification of microorganisms Hydrolysis kinetics Treatability database (kinetic and stoichiometric data for specific chemicals)
Aeration system modelling	Improved oxygen transfer models including variable alpha factors and extended aeration system models including system elements like actuators or blowers
Impact of process hydraulics	Adsorption/mixing impacts Detailed hydraulics incorporating CFD Improved settler models
Fouling	Fouling diffusers and membranes Modelling oil and grease impacts
Foaming and bulking	Floc formation and morphology with regards to foaming/bulking or also for MBR etc.? Diffusion limitations, settling tank impacts.
New processes/models	Incorporating sulphur chemistry and biology Modelling sludge reduction technologies Microconstituents (fate and treatment) Nitrous oxide models for greenhouse gas emissions Biofilm modelling and integration with activated sludge
Analysis tools	Structured data analysis tools for dynamic simulations Methods for dealing with uncertainties

FURTHER READING

Gujer, W. (2008). Systems Analysis for Water Technology. Springer, Berlin, Germany, ISBN: 978-3-540-77277-4.

Henze, M., van Loosdrecht, M.C.M., Ekama, G.A. and Brdjanovic, D. (2008). Biological Wastewater Treatment: Principles, Modelling and Design. IWA Publishing, London, UK, ISBN: 9781843391883.

WEF MOP 31 (2009). An Introduction to Process Modeling for Designers. Manual of practice No. 31. Water Environment Federation, Alexandria, VA, USA.

Proceedings from the wastewater modelling seminars in Kollekolle and WWTmod.

Chapter 3

Available protocols

Short summary

This chapter discusses benefits of modelling guidelines, gives an overview of existing protocols and analyses them according to specified criteria. The main features of each protocol are listed and discussed. The results are summarised in a tabular form.

3.1 INTRODUCTION

The great variety of approaches combined with insufficient documentation make quality assessment and comparability of simulation results difficult or almost impossible. A general framework for the application of activated sludge models is needed in order to overcome this obstacle.

Several guidelines that focus on different aspects of simulation projects have been developed recently. This chapter analyses existing protocols and guidelines and synthesises the available information into a unified protocol for activated sludge modelling.

3.1.1 Benefits of modelling guidelines

It is the authors' belief that the development of a simulation protocol can and will improve the quality of the results and may reduce the required effort. In addition, other spin-off benefits like improved data quality for operation and design may result. Some of the main benefits, however, are listed below:

- As a standardisation process, a protocol often leads to results that are typically more comparable, reproducible and transferable;
- A protocol can provide guidance that clearly defines requirements, limitations and expectations;
- Standardised procedures should lead to improved quality assurance/quality control (Shaw *et al.* 2011);
- Standardised steps will aid inexperienced modellers and clients through the project.

A GMP survey (Hauduc *et al.* 2009) found that the majority of model users had never received organised training in process modelling. This conclusion is problematic for the industry and explains why it is not uncommon to come across different approaches in the literature. A standardised protocol that leads modellers through a series of defined steps should consolidate these approaches over time. This, in turn, should reduce the time for engineer training. Following the protocol might highlight typical pitfalls and therefore save time while improving model quality. Standardised procedures should lead to better quality assurance/quality control and confidence in the model output. Guidance for the interaction between the modeller and client will help define responsibilities and the definition of clear objectives. Quality simulation projects will highlight existing (often undetected) plant problems and data inconsistencies. Standardised reporting will speed up documentation and communication with the client, which will aid the knowledge transfer. And finally, the standardisation of a modelling protocol will benefit all the stakeholders involved in any modelling project be they practitioners, operators or regulators because the procedures undertaken to arrive at a conclusion will be understood.

A simulation protocol should lead to a continuous increase in efficiency of simulation projects. As such, the goal should be to regularly re-assess the suggested methods and add improvements if commonly accepted and standardised. Newly developed technologies might require new models and a state-of-the-art simulation protocol should not exclude better suited models.

3.1.2 Potential risks of standardisation

A standardised procedure is not without risk however. One of the disadvantages of using a standardised protocol is that it may block innovative and more cost-effective solutions. A structure common for all modelling projects is suggested, but the modeller should feel free to decide on the best methods and models available for the project's specific objectives.

Furthermore, the modeller should not follow a protocol blindly. The modeller must take into account the defined objectives and case characteristics and avoid unnecessary, complex, and where possible, expensive steps.

3.2 EXISTING GUIDELINES

Several modelling protocols and guidelines have been proposed in recent years. Many are specific to wastewater whilst others were developed to help perform modelling projects in general.

3.2.1 General modelling guidelines

In fields other than wastewater treatment, promoting the correct use of models and assuring quality and modelling efficiency were identified as problematic issues (Scholten *et al.* 2000; Refsgaard *et al.* 2005; US EPA, 2009). It was identified that model quality needed to be improved, reproducibility was lacking and standardised procedures were not available. A study group of experts in water management in The Netherlands addressed these concerns by producing a Good Modelling Practice Handbook (Van Waveren *et al.* 2000). Enhancing model credibility was also one of the objectives of the European Union funded Project HarmoniQuA (http://harmoniqua.wau.nl/), which led to the development of quality assurance guidelines and a modelling support tool (MoST, 2006) to provide guidance and a quality assurance framework in water management.

3.2.2 Wastewater-oriented guidelines

The following four wastewater-oriented protocols have been published and are widely referenced:

- *STOWA* (Hulsbeek *et al.* 2002; Roeleveld & van Loosdrecht, 2002)
- *BIOMATH* (Petersen *et al.* 2002; Vanrolleghem *et al.* 2003)
- *WERF* (Melcer *et al.* 2003)
- *HSG* (Langergraber *et al.* 2004)

A comparison of these four protocols can be found in Sin *et al.* (2005) and Corominas (2006). Protocols with less international scope or coverage have been proposed by Frank (2006), and the Japan Sewage Works Agency (2006, in Japanese with an English summary, Itokawa *et al.* 2008). Guidance for WWTP modelling can also be found in many publications presenting case studies (e.g. Brdjanovic *et al.* 2000; Meijer *et al.* 2002; Third *et al.* 2007), different books (e.g. Gujer, 2008; Henze *et al.* 2008; Makinia, 2010), simulator manuals, or in proprietary company guidelines.

3.2.3 Analysis of existing guidelines

The following section highlights the main features of the different guidelines to identify the specific items that should be included in a unified protocol. Detailed analyses of these guidelines are presented in Tables 3.1 to 3.4. The strengths of the 4 main protocols are included in Table 3.1 and 3.2. The analysis of the protocols proposed by Frank (2006) and Itokawa *et al.* (2008) can be found in Table 3.3 and the general modelling guidelines (Van Waveren *et al.* 2000; Refsgaard *et al.* 2005) are compared in Table 3.4.

The four main activated sludge modelling guidelines (STOWA, WERF, BIOMATH, HSG) can be classified as Type 2: Public Technical Guidelines as defined by the the HarmoniQuA project (Refsgaard *et al.* 2005). The other, lesser known regional protocols, are defined as Type 1 documents (Internal Technical Guidelines Developed and Used Internally). Type 3 guidelines additionally include guidance on the interaction between the modeller and the client. Types 2 and 3 are developed in a public consensus building process. Other characterisation criteria used in the tables refer to:

- The scope of the protocols
- The background of the protocol developers
- The status of the work and the level of application

3.2.3.1 Short description of guidelines

STOWA's main emphasis was to help end-users model their nitrogen removal plants with ASM1 in a systematic and standardised way. An essential part of this protocol is an easy-to-use wastewater characterisation procedure. As part of the development, user groups were set up and the protocol was the result of an extensive consensus building process. Unfortunately, only a summary is available in English (Hulsbeek *et al.* 2002; Roeleveld & van Loosdrecht, 2002).

The *WERF* (Water Environment Research Foundation) guidelines (Melcer *et al.* 2003) are based on mainly North American experience with the development consisting of research on wastewater characterisation methods and a consensus building process involving a large international reviewer group. The final report includes an extensive overview of knowledge, experience and data and has become a standard reference for wastewater characterisation and simulation procedures.

The *BIOMATH* group at Ghent University (Belgium) proposed a generic calibration procedure (Vanrolleghem *et al.* 2003), using state-of-the-art parameter estimation methods. The protocol, which focuses on the biokinetic model (with sections on settling, hydraulics and aeration) requires extensive experimentation and takes advantage of systems analysis tools.

The *HSG* protocol is a generic procedure to guide modellers through all steps of a modelling project. The HSG protocol is the culmination of experience from specialised researchers in German speaking countries. An objective-oriented approach is encouraged in this protocol but deviations from the full procedure need to be explained and documented. The importance of data quality is highlighted. An 8 page journal publication in English is publicly available (Langergraber *et al.* 2004).

Table 3.1 Main features and specific features of wastewater oriented guidelines (STOWA, BIOMATH) to be included in the unified protocol.

Name	Background info	Objectives/main specificities of the procedure	Main steps of protocol	Topics to be included in the GMP unified protocol
STOWA (The Netherlands) (Hulsbeek *et al.* 2002; Roeleveld & van Loosdrecht, 2002)	*Issued by:* Dutch Foundation for Applied Water Research *Background of authors:* Round table of practitioners coordinated by STOWA *Targeted users:* Process engineers using dynamic models in their daily work. New modellers *Status of work:* Complete report only available in Dutch. 2 publications in English (2 × 8 pages)	*Main focus:* Straightforward and practical protocols for practitioners • Stimulating the practical use of dynamic simulation of AS systems, for optimisation studies and different aspects of a design phase • Standardisation of model use • Wastewater characterisation procedure • Quality control • Adapted to less experienced users as only few choices need to be made	(1) Definition of the objectives (2) Process description (3) Data collection and validation (4) Model structure selection (5) Characterisation of the main flows (6) Calibration/validation (7) Detailed characterisation (8) Validation of the project (9) Model application	• Structured overview of the protocol, including feedback loops • Detailed settling and biological (sludge)characterisation • Influent characterisation • Time estimate • Detailed data quality checks • Step-wise calibration of biological process parameters (gives few parameters to calibrate and acceptable parameter ranges) • Practical experimental methods • Adapted to consultants and new modellers
BIOMATH (Belgium) (Vanrolleghem *et al.* 2003)	*Issued by:* BIOMATH research group (Ghent University) *Background of authors:* Researchers, PhD students with background in biotechnology *Targeted users:* Researchers and other modellers dealing with advanced modelling issues *Status of work:* Extensive publication on model calibration incl. case study (28 pages)	*Main focus:* Calibration procedure for biokinetic model • Proposing a comprehensive model calibration procedure • Use of state-of-the-art methods to calibrate models (Optimal Experimental Design methodology) • Mainly dedicated to experienced modellers • Requires a high level of experimental results and expert knowledge (system analysis background)	(1) Definition of the objectives and requirements (2) Plant survey and characterisation (model selection); Additional data collection based on OED (3) Steady-state calibration (4) Dynamic calibration and evaluation of calibration results	• Structured overview of the protocol, including feedback loops • Detailed characterisation of: influent, biomass, settling, hydraulic and biological models • Use of optimal experimental design for measuring campaigns • Sensitivity analysis/parameter selection • Examples are provided • Generally applicable

Table 3.2 Main features and specific features of wastewater oriented guidelines (WERF, HSGSim) to be included in the unified protocol.

Name	Background info	Objectives/main specificities of the procedure	Main steps of protocol	Topics to be included in the GMP unified protocol
WERF (North America) (Melcer *et al.* 2003)	*Issued by:* Water Environment Research Foundation *Background of authors:* Modelling experts from consulting firms, simulator companies, chemical industries and universities *Targeted users:* Municipalities and consulting engineering companies carrying out municipal or industrial wastewater simulation studies. New modellers *Status of work:* Final report of major research initiative (575 pages)	*Main focus:* Review and suggestion of methods to characterise influent and sludge and to measure nitrifier growth rate • Methods for measuring AS model parameters and to characterise influent • Guidance to calibrate and use models • Reference manual including extensive overview of knowledge, experience and data	Not a protocol as such (no structured overview). For the calibration: (1) Plant configuration set up (2) Additional data collection (3) Calibration itself, with 4 different levels depending on the objectives - Level 1: Based on defaults and assumptions - Level 2: Based on historical plant data Level 3: On site full scale testing and gathering data - Level 4: Supplementing level 3 with direct parameter measurements (4) Validation	• Detailed procedures for influent and biomass characterisation and measurement of some key parameters (nitrification) • Sensitivity analysis/parameter selection • Detailed data quality check • Tiered approach for calibration • Several examples of case studies Practical experimental methods Adapted to consultants and new modellers
HSGSim (DE, AU, CH) (Langergraber *et al.* 2004)	*Issued by:* Hochschulgruppe Simulation (HSGSim) *Background of authors:* University researchers from German speaking countries *Targeted users:* Modellers in consulting firms, water boards and municipalities *Status of work:* Protocol structure and outline of future work (8 pages)	*Main focus:* Providing standardised structure to carry out complete simulation studies • Definition of a quality level for highest requirements as reference case • For objectives with lower requirements the changes to the reference case have to be explained • Goal is comprehensible and comparable simulation results • Importance of data validation • Applicable to any biokinetic model • Detailed documentation	(1) Definition of the objectives (2) Data collection and model selection (3) Data quality control (4) Evaluation of model structure and experimental design (5) Additional data collection (6) Calibration/validation (7) Study, evaluation and documentation	• Guidelines to support modellers in carrying out a complete simulation study • Structured overview of the protocol • Detailed hydraulic and biological characterisation • Design of measuring campaign • Data quality check • Documentation • Generally applicable

Table 3.3 Main features and specific features of wastewater oriented guidelines (Frank, Japan Sewage Works) to be included in the unified protocol.

Name	Background info	Objectives/main specificities of the procedure	Main steps of protocol	Topics to be included in the GMP unified protocol
Frank (USA) (Frank, 2006)	*Issued by:* Gannett Fleming, Inc *Background of authors:* Process engineer working in a consulting company *Targeted users:* Consulting engineers using dyn. models in their daily work *Status of work:* Applied protocol used in several studies (31 pages)	*Main focus:* Straight forward protocol for consulting engineers • Proposing a practical stepwise procedure used in process design and prediction of process performance • No mention of settling processes	(1) Identification of goals (2) Historical data analysis (3) Data reconciliation (4) Intensive sampling campaign (5) Influent WW characterisation (6) Model construction and calibration (7) Model validation (8) Influent scenario development (9) Plant performance simulations	• Structured overview of the protocol, including feedback loops • Detailed influent, hydraulic and biological characterisation • Detailed data quality check • Typical ratios of WW characteristics provided • Illustration of the procedure through 3 examples • Illustration of scenario analysis
Japan Sewage Works Agency (Japan) (Japan Sewage Works, 2006, Itokawa *et al.* 2008)	*Issued by:* Japan Sewage Works *Background of authors:* Round table of practitioners coordinated by Japan Sewage Works *Targeted users:* Consulting engineers, water boards, municipalities *Status of work:* Complete report only available in Japanese. 1 publication in English (8 pages)	*Main focus:* Practical protocols for practitioners, particularly for beginners • Promoting the use of AS models in design and operation of municipal WWTPs in Japan • Focused on main biological processes (ASM2d), excluding accompanied models such as sedimentation • Protocols based on the combination of existing ones	(1) Project definition (2) Collection of existing data (3) Additional measurements (4) Process analysis (5) Process modelling (6) Calibration (7) Simulation and evaluation	• Application example with detailed information and sensitivity analysis • Potential target list of kinetic parameters to calibrate • Review on published application examples • Typical ratios for characterisation based on intensive investigation on Japanese municipal wastewater • Description of typical pitfalls

Table 3.4 Main features and specific features of existing guidelines to be included in the unified protocol (general modelling guidelines).

Name	Background info	Objectives/main specificities of the procedure	Main steps of protocol	Topics to be included in the GMP unified protocol
GMP-Handbook (The Netherlands) 1999 (Van Waveren *et al.* 2000)	*Issued by:* Modellers in water management Main players in Dutch water management *Background of authors:* Project group from Dutch water boards, universities, consulting companies and national agencies led by STOWA *Targeted users:* Modellers *Status of work:* Finalised report available (165 pages)	• Improving the efficiency and quality of modelling in water management in The Netherlands • on the basis of a general improvement/agreement on the modelling and simulation process by users and a shared knowledge with multidisciplinary teams • General modelling procedure, not straightforward to apply for WWTP modelling in practice	(1) Starting a logbook (2) Definition of the project (3) Model set up (4) Model analysis (sensitivity analysis, calibration, validation) (5) Model use (6) Result interpretation (7) Reporting and archiving	• Structured overview of the protocol, including feedback loops Encouragement to start with a logbook • Including a glossary to help unifying the terms used in AS modelling • Providing a check-list to ensure that each crucial step of the protocol has been performed • Systematise the documentation and reporting • Including a list of usual pitfalls • Protocol tested by both experienced and inexperienced modellers
HarmoniQuA (European project) 2002–2005 (http:/f in mathharmoniqua.wau.nl/) (Refsgaard *et al.* 2005)	*Issued by:* 12 partners, coordinated by H. Scholten (Wageningen University, The Netherlands) *Background of authors:* Partners from universities, national agencies and software companies working in river basin management *Targeted users:* Modellers, water managers, auditors, stakeholder and general public	• Overcoming methodological difficulties in modelling projects • Developing Good Modelling Practice Handbooks in 7 domains of water management for different geographical regions • Developing a modelling support tool (MoST) to provide a guidance and quality assurance framework, which allow teams of different disciplines to work on the same project. • Including informing and	(1) Model study plan (2) Data collection and conceptualisation (3) Model set up (4) Calibration and validation (5) Simulation and evaluation	• Structured overview of the protocol, including feedback loops • during the definition of the objectives, stop criteria are defined for different steps (calibration, validation) • Uncertainty analysis • Guidance on the organisation of the interaction between modellers and clients • Each step includes a report that has to be evaluated by

(*Continued*)

Table 3.4 Main features and specific features of existing guidelines to be included in the unified protocol (general modelling guidelines) (*Continued*).

Name	Background info	Objectives/main specificities of the procedure	Main steps of protocol	Topics to be included in the GMP unified protocol
	Status of work: EU project finished in 2005. Work will continue as Post HarmoniQuA Collaboration Framework	consulting stakeholders and public • General modelling procedure, not straightforward to apply for WWTP modelling in practice		the client *Modelling Support Tool (MoST)* • A general structure implemented in a computer tool to provide guidance and quality assurance • Guiding users during the modelling project • Guidance is adapted to the domain and to different users: (a) modellers, (b) water managers, (c) auditors, (d) stakeholders and (e) general public

Regional protocols (e.g. JS protocol by Japan Sewage Works Agency) often focus on specific issues and constraints and may not lend themselves to generalisation. *Company protocols* (e.g. Frank, 2006) are often proprietary and not easily accessible. The focus of both types of protocols is typically on practical use. *Software manuals*, although focused on a particular software package, often provide a comprehensive source of information on how to apply models. *Published case studies* (e.g. Meijer *et al.* 2002; Third *et al.* 2007) can be used as another source of guidance but are often too specific to be used as general guidance on activated sludge modelling.

Chapter 4

The GMP Unified Protocol

Short summary

This chapter gives an overview of the *GMP Unified Protocol* including a flow diagram with short outlines of every step. The Unified Protocol includes technical guidelines for modelling projects through a stepwise procedure, as well as guidance for organising the interaction between the modeller and the stakeholders. More detailed information on every step is provided in Chapter 5.

4.1 TOWARDS A UNIFIED PROTOCOL

The *GMP Unified Protocol* is based on the wastewater oriented protocols described in Chapter 3, extended with some key elements provided in general guidelines, especially the interaction between modellers and other stakeholders.

The similarities in the existing protocols outnumber the differences and where differences exist they are mostly at the level of detail and focus. The main differences of the discussed protocols are:

(i) The design of measuring campaigns;
(ii) The experimental methods used to characterise the influent, hydraulics, settling, aeration, and stoichiometric/kinetic parameters; and,
(iii) The procedure to calibrate and validate the plant model.

These differences are undoubtedly linked to the authors' objectives, their background (e.g. researchers, consulting engineers, round tables, and the field of their expertise: process engineering or water management) and the targeted users. Differences may also be linked to the fact that models are used for different purposes in different locations that is, mainly for design/redesign purposes in North America, and for optimisation or control studies that require more dynamic simulations in Europe (Hauduc *et al.* 2009).

The last columns in Tables 3.1 to 3.4 in Chapter 3 suggest several key elements that should be considered for inclusion in a unified protocol:

- A structured overview of the practical protocol, including feedback loops;
- Level of effort (e.g. data requirements, calibration levels, time estimates) as a function of the objectives;

- Data collection and reconciliation strategies;
- Practical methodologies to perform detailed characterisation of influent wastewater, biomass, settling, hydraulics and biological processes;
- Guidance for model selection and set-up;
- A plant model calibration/validation procedure that includes parameter selection;
- Uncertainty analysis;
- Guidance for the interaction between modellers and end-users (establishment of performance criteria, reviews of the main steps of the protocol);
- Guidance on documentation;
- Examples, including typical pitfalls.

4.2 THE GMP UNIFIED PROTOCOL – STRUCTURED OVERVIEW

The Unified Protocol combines the key aspects and follows the general structure of MoST (2006). This structure was chosen because it is in accordance with general modelling concepts (e.g. Van Impe *et al.* 1998).

The aim of the GMP Task Group work was to develop a Type 3 (Public Interactive Guideline) protocol as defined by Refsgaard *et al.* (2005) including the interaction between modellers and stakeholders. The proposed protocol is illustrated in Figure 4.1 and comprises five steps:

Step 1. Project Definition
Step 2. Data Collection and Reconciliation
Step 3. Plant Model Set-up
Step 4. Calibration and Validation
Step 5. Simulation and Result Interpretation

Each step has to be reviewed and agreed upon with the stakeholders before the next step is carried out (decision boxes in black). This review is based on reports and other documents that are generated during the last task of a step. The protocol calls for a final report that encapsulates the outputs from all modelling steps, usually in summary form with the detailed reports from each step in appendices.

The 5 steps of the Unified Protocol are described briefly in the following sections. The procedures and methods required to perform these steps are detailed in Chapter 5.

Step 1: Project definition

In the project definition step, the objectives of the modelling project are defined, stakeholders and their responsibilities are identified and budget and schedule constraints are agreed upon.

Deliverables

Agreement is reached on technical conditions of the modelling project and on the budget. A project definition document is produced that defines the project. Ideally allowance should be made to modify this document during the review task of each step that is, it should be a "living" or "dynamic" document.

Step 2: Data collection and reconciliation

This step aims at collecting, assessing and if necessary reconciling data sets for simulation projects. A stepwise procedure to analyse collected data is provided including dedicated methods based on statistical analysis, engineering expertise and mass balancing.

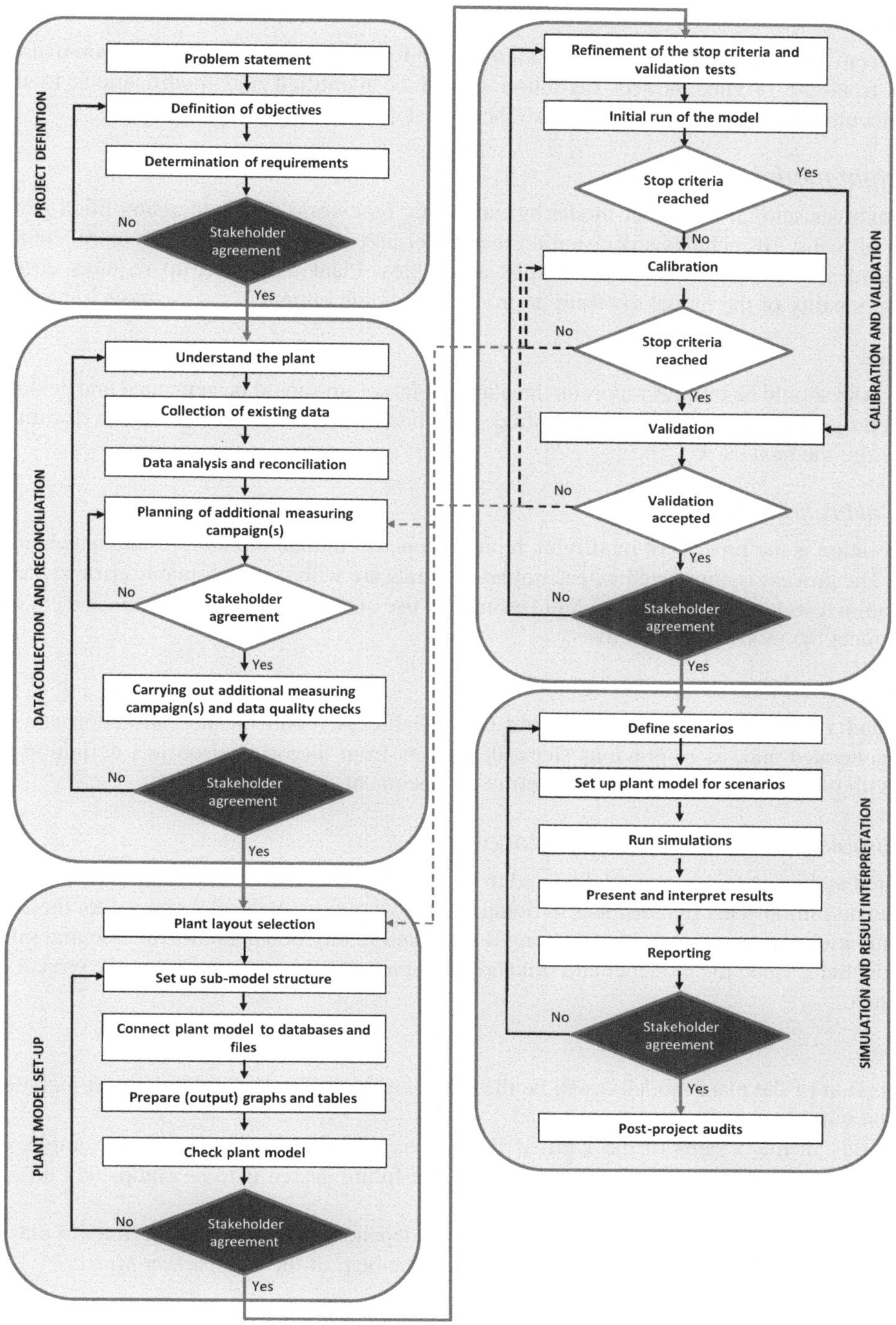

Figure 4.1 The proposed GMP Unified Protocol.

Deliverables

The output from this step will be reconciled data sets used for all subsequent steps of the simulation project. Deviations from the original project definition should be identified and modifications to the project definition document should be finalised before the next step.

Step 3: Plant model set-up

This step involves setting up a plant model by translating real world data into a simplified mathematical description of reality. It includes a decision on the model layout, the sub-model structure, connections to databases, and setting up model output graphs and tables. Plant model set-up requires checks of the general functionality of the model to ensure it produces sensible outputs.

Deliverables

The plant model should be built. A report on the plant model set-up should be generated and deviations from the original project definition should be identified. Modifications to the project definition document should be made before the next step.

Step 4: Calibration and validation

Model calibration is the process of modifying input parameters until simulation results match an observed set of data. The process is completed when simulated results are within an acceptable error to the measured data. Validation tests are performed in order to ensure the use of the plant model with the level of confidence required to meet the modelling objectives.

Deliverables

Calibrated and validated parameter sets should be available. A report on the calibration and validation should be generated and, as in previous steps, deviations from the original project definition should be identified with modifications to the project definition document.

Step 5: Simulation and result interpretation

The calibrated and validated plant model is used to run simulations in order to meet the objective(s) laid out in the project definition. This step includes defining scenarios; setting up the plant model for these scenarios; running simulations; presenting and interpreting results; and finally documenting all essential information. This step concludes when the modeller and stakeholders agree that the expectations of the project definition have been met.

Deliverables

The final version of the plant model(s) will be the outcome from this step. A final report including model interpretation will be generated.

Carrying out all the 5 steps of the Unified Protocol is not always required. For example, collecting existing data for a design project may be irrelevant if the future design is to be completely different from the existing plant.

The required level of effort associated with these steps depends mainly on the objectives of the modelling project. Modelling effort is discussed in Chapter 6 with the help of the *Application Matrix*.

Chapter 5

Unified Protocol steps

Short summary

This chapter describes the Unified Protocol steps in detail. The structure of the chapter follows that of the Unified Protocol. The first section *Project Definition* describes the initial step of the protocol which includes defining the objectives of the modelling project, identifying responsibilities and agreeing upon budget and schedule. The second step of the protocol, *Data Collection and Reconciliation* provides a methodology for gathering and assessing data to ensure its quality. This is followed by sections describing the steps of *Plant Model Set-up* and *Calibration and Validation* to produce a model with acceptable predictive capability. Finally this chapter concludes with *Simulation and Results Interpretation*.

Chapter 5.1

Project definition

5.1.1 INTRODUCTION

In setting up any project, an important first step is to define its purpose and goals. A good project definition sets expectations from the start, opens up clear communication lines with all the people involved in the modelling project – the "stakeholders" – and reduces significantly the risk of schedule and budget "slippage". Conversely, without a clear understanding of what a project is supposed to achieve, significant problems can ensue as the project proceeds. Modelling projects are particularly prone to having their scope expanded (commonly termed "scope creep"); without a full appreciation of the extra time and effort required to do the extra work. This is due to the fact that process models can be used for many different purposes with an almost infinite number of scenarios. Therefore, a good project definition is required to provide reasonable project constraints.

5.1.2 PROCEDURE

Figure 5.1.1 shows the elements of the *Project Definition* step in the Unified Protocol. The elements include: *Problem Statement*, *Objectives*, *Requirements* and *Client Agreement*. The protocol starts with the Problem Statement and proceeds in a stepwise manner to Client Agreement where all stakeholders check to make sure they are in agreement before proceeding to the next major step.

5.1.2.1 Problem statement

The starting point for any modelling project is the Problem Statement. This statement defines what it is the model will be used to solve. It should define the problem clearly, explicitly, and in plain language so that all the stakeholders understand what is needed. The Problem Statement should be focused and unambiguous. Examples of well-defined and not well-defined problem statements are given in Table 5.1.1. In some cases – such as the second example – it may be difficult to define the problem in exact terms but an attempt should still be made to make the intended use of the model as clear as possible.

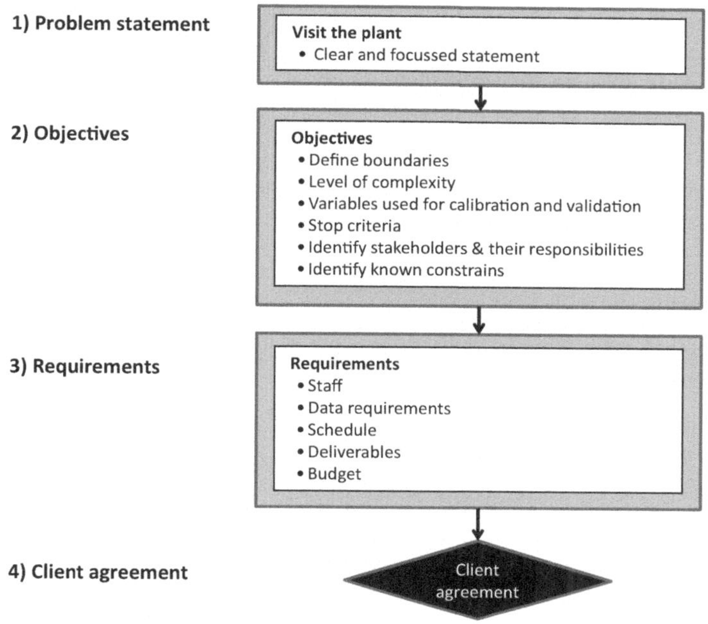

Figure 5.1.1 Flow diagram for Unified Protocol step 1: Project Definition.

Table 5.1.1 Examples for well-defined and weak problem statements.

Model use	Example of well-defined problem statement	Example of weak problem statement
To determine air requirements	Define the peak, average and minimum air flows required for the treatment system under the given design loading conditions	Size the aeration system using a model
Assess plant capacity for removing nitrogen	Determine the maximum flow that can be treated under design load conditions to meet required nitrogen removal and effluent limits	Use a model to assess the plant capacity

5.1.2.2 Objectives

Based on the Problem Statement, the Objectives of the project are established by considering the following model aspects:

(1) The model boundaries (e.g. whole plant or activated sludge stage only)
(2) The proposed level of complexity (steady-state versus dynamic, multiple trains versus single train etc.), noting that increased complexity increases costs
(3) The variables that will be used for calibration and validation and what accuracy will be expected for each output variable (i.e. establishing a stop criteria)

(4) The identity of the stakeholders and their responsibilities
(5) The identification of known constraints, including: time (by what date does the project need to be completed); money; the experience of available modellers; the capabilities of available simulators

The Objectives of the modelling project are often established using interviews and meetings involving all stakeholders (e.g. consultants and clients, researchers and supervisors). One method to engage all the stakeholders is to organise a workshop that encourages input on the objectives.

5.1.2.3 Requirements

Once the Objectives have been defined, the Requirements to meet those objectives can be established and will include the following:

(1) *Tasks:* individual tasks should be identified
(2) *Data requirements*: data gathering, extra sampling, specialised testing - time and costs must be compiled
(3) *Staff*: the type and required experience of staff should be identified
(4) *Schedule*: the time required to accomplish the tasks should be established and compared to the time available, (i.e. is the modelling effort part of a broader project which may have subsequent tasks dependent on the modelling outputs?)
(5) *Deliverables*: what reports, workshops and training are required? Will the model be passed on to the client? What documentation is required?
(6) *Budget*: the costs for the project must be established including labour, data acquisition, software, training and testing

The Problem Statement, Objectives and Requirements should all be clearly recorded in a *Project Definition Document* that defines the overall scope of the project.

5.1.2.4 Client agreement

This step is an iterative process that requires stakeholders to balance objectives and the requirements to achieve those objectives with the constraints of schedule, budget and staffing. In an ideal situation, the objectives can be met with the allotted budget, schedule and staffing requirements. However, it is often necessary to make adjustments to the original objectives because of these constraints. It may be acceptable to refocus the objectives and/or remove non-critical objectives. In some cases, the iterative discussion about the objectives and requirements may bring to light different objectives and requirements more suited to making the project a success.

5.1.3 DELIVERABLES

At the end of the discussions and negotiations, a final *Project Definition Document* should be produced. This document is the main deliverable from this step and acts as the reference agreement for the project scope. Stakeholders and their responsibilities are identified and budget and schedule constraints are agreed upon. If necessary, this document can be altered during the course of the project, but stakeholder agreement is required to do this i.e. this document should be a “living” or “dynamic” document. The project definition document should include:

(1) *Problem Statement*: A problem statement that clearly, explicitly, and in plain language defines what the model will be used to solve

(2) *Objectives*: The objectives of the modelling project have to include a description of:

- the boundaries for the model
- the level of complexity
- the list of variables that will be used for calibration and validation and what accuracy will be expected for each output variable to establish the stop criteria
- the identification of stakeholders and their responsibilities
- the identification of known constraints, including time, money, staff experience and software capabilities

(3) *Requirements*: Tasks, staff, data, budget and schedule required to meet the objectives

(4) *Client agreement*: The project definition document should be signed by responsible stakeholders to indicate their agreement with the Objectives and Requirements for the project.

FURTHER READING

HarmoniQua, Harmonising Quality Assurance in model based catchment and river basin management. Project website: http://harmoniqua.wau.nl/.

WEF MOP 31. (2009). An introduction to process modeling for designers. Manual of practice No. 31. *Water Environment Federation*, Alexandria, VA, USA.

WEF MOP 8. (2010). Design of municipal wastewater treatment plants. Manual of practice No. 8. *Water Environment Federation*, Alexandria, VA, USA.

WERF Project 04-CTS-5. (2009). Integrated methods for wastewater treatment plant upgrading and optimization. *Water Environment Research Foundation*, Alexandria, VA, USA.

Chapter 5.2

Data collection and reconciliation

5.2.1 INTRODUCTION

The results of the first GMP task group questionnaire (Hauduc *et al.* 2009) showed that data collection and reconciliation is recognised as one of the steps requiring the most effort (>1/3 of the overall effort of a modelling project, Chapter 6 *Application Matrix*). This conclusion places emphasis on the care that should be taken in providing data sets of high quality for use in a process model. Careful data reconciliation saves time in the subsequent steps of a modelling project.

The desired quantity and quality of the data is defined in step 1 of the Unified Protocol (Section 5.1). The input data quality will be shown to have a particular impact on the accuracy targeted in the model calibration step (Section 5.4).

5.2.2 PROCEDURE

A procedure for collecting, analysing and reconciling data is presented in Figure 5.2.1. The procedure starts with *Understanding the plant*, which includes getting familiar with the plant layout and process scheme based on available documents, plant visits and communication with the WWTP personnel.

The second phase *Collection of existing data* includes the description of data sources, data types and what data is required in relation to the modelling objectives.

The third phase is *Data analysis and reconciliation* and includes procedures to detect, identify and isolate faults in the data sets with the goal of creating reconciled data sets for the project.

Based on the outcomes of phase 3, *Additional measurement campaigns* can be planned. As these campaigns represent a major cost consideration, client agreement should be included in this stage. The data from the additional measurement campaign have to be checked for quality and reconciled in a similar way as existing data. If there is no client request to the contrary, the intensive data set should be from a period with normal plant loading and operation. It is important to collect "information-rich" data according to the calibration or prediction needs. These might include stress tests, measurements from specific reactors, or high-resolution data of specific components.

Final stakeholder agreement has to be received on the reconciled data sets in terms of data quality and quantity.

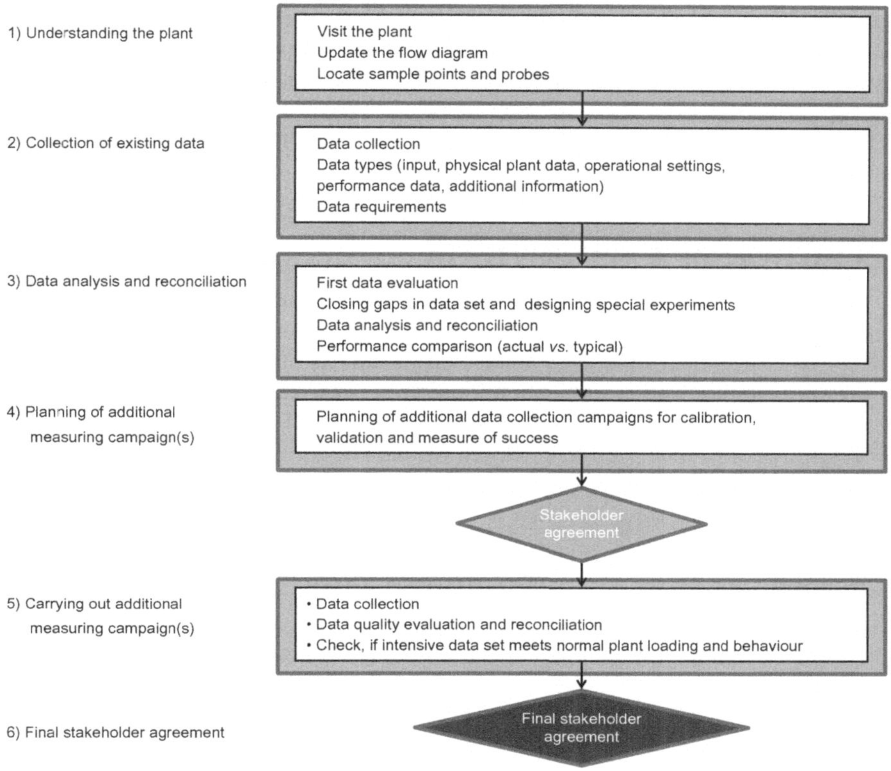

Figure 5.2.1 Data collection and reconciliation procedure.

5.2.3 UNDERSTANDING THE PLANT

The key starting point in analysing WWTP data is to correctly understand the plant *as operated*. This is best established by means of communication with the WWTP personnel and plant visits. In addition, information about the design values and design layout of the plant should be gathered in order to compare the current plant load and process scheme with the original design. Note that this is a continuous process as often the "as operated" status is unknown and has to be established during the set-up of the plant model.

Analysing the *process flow scheme* is the initial task. This task will help the modeller understand the treatment steps and the loading to each of those steps. Two important aspects are the configuration of the tanks (in series, parallel), different operational modes for different loading or temperature conditions (e.g. seasonal or temperature dependent) and knowledge about the configuration of internal recycle streams (extraction and discharge points). If applicable, information must be collected about *additional plant load* due to internal recycle streams (e.g. ammonia-rich digester supernatant) or plant external loads such as sludge from other WWTPs or for co-fermentation.

The *P&I-Diagram* (piping and instrumentation diagram) gives a more detailed insight into the plant layout. In addition, it shows the location of the process instrumentation and its utilisation for process control. It is advisable to compare the design with the actual status of the plant, for example installation locations of probes might have been changed since the design phase.

Most large WWTPs operate automatic samplers in order to document plant performance based on composite samples. The analysis frequency and sampling locations are dependent on country/state specific directives, utility guidelines and often on the plant capacity. These data are an essential base for understanding the plant behaviour and assessing its performance. In order to use the data correctly the exact installation location, the sampling schedule and further settings (time, volume, or flow-weighted sampling) of the automatic samplers need to be known.

5.2.4 COLLECTION OF EXISTING DATA

This phase is necessary in order to obtain the data that are required to carry out the subsequent modelling steps. In Section 5.2.4.3 *General data requirements*, an overview is given of what data are generally needed (i.e. data needed for most simulation studies, irrespective of the specific objectives). Specific data requirements for the examples of the Application Matrix are discussed in Chapter 6.

Data collection typically starts by looking at existing data from various sources. After an initial basic data analysis, gaps in the available data need to be closed by additional measurements or specific experiments.

This section discusses *Data Types*, typical *Data Sources*, and *Data Requirements* for simulation purposes (e.g. for influent characterisation).

5.2.4.1 Data types

Data required for simulation studies can be classified as outlined in Figure 5.2.2. This classification helps the modeller understand the different model inputs and outputs, and helps identify which data are essential, useful, or good to have.

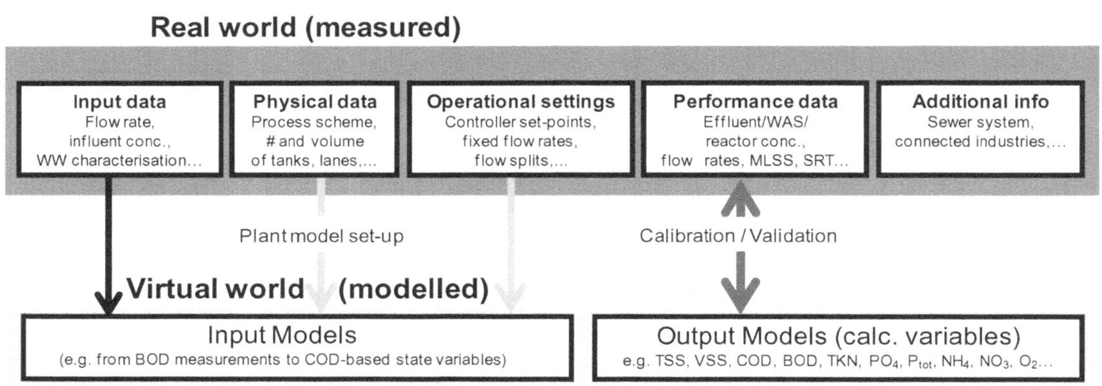

Figure 5.2.2 Data types required for simulation studies.

Input data

Input data is all the data related to the mass loads entering the modelled system. These normally consist of concentrations (flow-weighted) in the influent, and sidestreams, and additional inputs like chemical doses or external carbon sources. This data also includes the flow basis for each input (e.g. as daily average, diurnal, seasonal, or event-based flow) and reactor temperatures.

Physical plant data

Physical plant data includes all data describing physical attributes of the WWTP, such as:

- Tank volumes, depths and layout (lanes, cascades, etc.)
- Tank configuration, connections and hydraulic behaviour (plug-flow vs. completely mixed)
- Location of input points
- Aerators and mixers (location, specifications for blowers, diffusers, pipes, valves and other actuators, control scheme)
- Pumps (capacity, control scheme)
- Available sensors and control loops (see P&I Diagram)
- Type of sludge thickening, treatment and dewatering (continuous/discontinuous operation, discharge points and operation of return streams)

Operational settings

Operational settings describe *how the plant is operated*, (i.e. which operational measures and set-points are applied to deal with the plant load and its variation). Key data here include:

- Flow splits
- Set-points for pumped flows and pumping schedules (e.g. for return and waste sludge flows, internal recycle streams, chemical dosage)
- Parameters for aeration control (set-points and other control parameters, additional constraints such as forced on/off times, etc.)
- Parameters for other control strategies

Performance data

Performance data describe *how the WWTP responds* when specific loading conditions are applied and the plant is operated in a particular way. Effluent and wasted sludge concentrations are essential performance data; however important information can be gathered from reactor concentrations (e.g. ammonia profiles in a fully nitrifying plant provide better means for calibration than a low effluent ammonia concentration).

The following performance data are often helpful for WWTP process evaluations:

- Sludge disposal (sludge mass leaving the WWTP; often available from truck weighing)
- Energy consumption (preferably for specific processes; e.g. aeration, mixing, etc.)
- Gas production in anaerobic digesters (if applicable)

Additional information

Additional information is used here to describe further details about the plant under study. This data can help with understanding typical and "atypical" conditions for the plant and/or unusual behaviour.

Additional data might include information about process performance or chemical consumption in other process units of the plant (e.g. lime addition during sludge dewatering), industrial dischargers, seasonal loadings (e.g. tourism, wineries, etc.), or the sewer system (slope, average retention times, storage capacities, etc.). It is essential to have information on maintenance and service measures (changes of WWTP operation scheme for a specific period due to maintenance measures, such as taking tanks/lines out of operation) and it is often the case, that seemingly insignificant comments by plant personnel turn out to be quite important (i.e. we notice this happens during high flows, the colour of the material in that channel is always different on the right and left sides …).

5.2.4.2 Data sources

The main source of WWTP data is the *routine data* collected by the plant (e.g. influent and effluent composite samples, mixed liquor grab samples, manual meter readings, etc.) and data from the plant's Supervisory Control And Data Acquisition (SCADA) system (e.g. flow data and other on-line instrumentation). The available time resolution depends on the specific functionalities of the SCADA system. The content of the routine plant data is often defined by country (or state) specific legislation.

The plant SCADA system may collect further data not necessarily documented in the routine data protocol, e.g. pumping schemes of influent pumping stations or DO concentration time series at specific locations of the aeration tank.

In some countries, some plants are required to have periodic external audits (often from or ordered by legislative bodies). Those data are usually collected over a very short period (typically 1–2 days), but the analysis is often quite extensive (e.g. concentrations in parallel with plant laboratory analysis, detailed sludge analysis, metals, etc.).

Additional information can be gained from documentation of the sewer network and plant design and plant upgrade reports.

WWTP log books should be consulted to identify special events like equipment maintenance or failure.

In addition to the available written and electronic documentation, one of the most valuable sources of information is communication with the WWTP personnel.

5.2.4.3 General data requirements

This section describes general data requirements, which are typically necessary in all simulation studies and mostly independent of the specific goals of the study. Table 5.2.1 shows an overview of the general data requirements for a modelling project. Please note that the sampling frequency cannot be generalised and is directly linked to the objectives. Specific data requirements for the examples of the Application Matrix are discussed in Chapter 6.

Table 5.2.1 General data requirements.

Data type	Requirements	Use/remark
Input data	*Influent and other input flows:* Q_{INF}	For influent streams (raw WW, settled WW, sidestream, depending on the model boundaries)
	Influent organics and suspended solids: $COD_{tot,INF}$ TSS_{INF}, VSS_{INF}	Sludge production $COD_{tot,INF}/BOD_{5,INF}$ required if only $BOD_{5,INF}$ routinely measured
	Influent nutrients:	
	TKN_{INF}, $NH_x\text{-}N_{INF}$	N removal
	$P_{tot,INF}$, $PO_4\text{-}P_{INF}$	P removal, mass balancing
	Influent COD, N and P fractions	Wastewater characterisation
	Alk_{INF}	Critical for nitrification
Physical data	Tank volumes, depths and layouts Flow connections and hydraulic behaviour Equipment (aerators, mixers, pumps) P&I-Diagram	

(*Continued*)

Table 5.2.1 General data requirements (*Continued*).

Data type	Requirements	Use/remark
	Main characteristics of the sludge treatment train	
Operational settings	DO control strategy and set-points Pumping set-points / flow splits Other control strategies	
Performance data	*Effluent flow:* Q_{EFF}	Flow balance
	Effluent organics: $COD_{tot,EFF}$, $BOD_{5,EFF}$ TSS_{EFF}	Calibration of organics removal
	Effluent nutrients: TKN_{EFF}, $NH_x\text{-}N_{EFF}$, $NO_x\text{-}N_{EFF}$ $P_{tot,EFF}$, $PO_4\text{-}P_{EFF}$ Alk_{EFF}	Calibration of nutrient removal Mass balancing Critical for nitrification
	Mixed liquor MLSS, MLVSS $P_{tot,ML}$ DO (in-tank concentrations) Temperature	Sludge production Mass balancing DO control, aerobic sludge age
	WAS Flow: Q_{WAS} Solids: $MLSS_{WAS}$ $P_{tot,WAS}$	Sludge production Mass balancing

Wastewater characterisation

Wastewater characterisation (or fractionation) is the step during which influent measurements are converted into model state variables. Activated sludge models, for the most part, are based on COD balances, but the model equations do not include the measured COD *per se*. Rather, the models need to fraction the measured COD into model–specific components that are not explicitly measurable. The splitting of the COD into these component parts is termed characterisation (or fractionation).

Average values of these fractions are generally used to perform the simulations, although in reality they may vary over the course of a day, week, or with weather conditions. Special care with these fractions is needed when intermittent industrial loads are discharged or with seasonal load variations. These atypical conditions may require specific investigations in order to properly characterise the influent loads.

Table 5.2.2 provides a list of methods that can be used to help characterise a wastewater. Experience has shown that the proposed methods may lead to different fractions (Gillot & Choubert, 2010) and this may explain why initial values obtained through measurements have to be modified in some cases during the calibration step (Section 5.4.2.4). Although no clear standard measurement procedure has emerged, the authors recommend that a flocculation step be used to separate soluble, colloidal, and particulate fractions (Melcer *et al.* 2003).

N and P species such as NH_x-N, NO_x-N and PO_4-P and totals such as P_{tot}, N_{tot}, TKN should be obtained through standard analytical techniques. Organic fractions are typically calculated by difference (Appendix A: Input models).

Table 5.2.2 Experimental methods for COD fractionation.

Target fraction/parameter	Method	Principle	Material	Test duration	Result analysis	Reference
Total biodegradable COD ($COD_{tot,B}$)	Long-term BOD-test	BOD monitoring	BOD-meter	8 to 10 days	1st order rate constant determination	Roeleveld & van Loosdrecht (2002)
	Aerobic batch test High S/X ratio	Respirometry: OUR monitoring after addition of WW to activated sludge	Respirometer	Few hours	Area under the OUR curve	Sperandio *et al.* (2001)
Readily biodegradable COD (S_B)	Physico-chemical method	Pore size = 0.1 μm	Filter	Few hours (including chemical analysis)	$S_B = COD_{f0.1} - S_U$ Requires the determination of S_U	Roeleveld & van Loosdrecht (2002)
	Physico-chemical method	Flocculation (Zn+ filtration) Pore size = 0.45 μm	Filter Stirred vessel	Few hours (including chemical analysis)	$S_B = COD_{ff0.45} - S_U$ Requires the determination of S_U	Mamais *et al.* (1993)
	Flow-through aerobic test	Respirometry: OUR monitoring in a continuous reactor	Continuous reactor	3 SRT (pilot) Few hours (WW addition)	Area under the OUR curve	Ekama *et al.* (1986)
	Aerobic batch test Intermediate F/M ratio	Respirometry: OUR monitoring after addition of WW to activated sludge	Respirometer	Few hours	Area under the OUR curve	Ekama *et al.* (1986)
	Aerobic batch test Low S/X ratio	Respirometry: OUR monitoring in an aerated WW sample	Respirometer	10–20 hours	Modelling of OUR-curves Requires the separate	Wentzel *et al.* (1995); Sperandio *et al* (2001)

					determination of $COD_{TOT,B}$	
	Anoxic batch test	Anoxic respirometry Nitrate-N monitoring	Stirred batch reactor	4–6 hours	Simple calculations based on slopes	Naidoo *et al.* (1998)
Total unbiodegradable COD ($COD_{tot,U}$)	Long-term aerobic batch tests	COD monitoring in aerated reactors	2 aerated reactors (raw and filtered WW samples)	20–40 days	Residual microbial products determined using the reactor initially filled with filtered WW	Lesouef *et al.* (1992); Orhon *et al.* (1997); Stricker *et al.* (2003)
Unbiodegradable soluble COD (S_U)	Physico-chemical method	Effluent filtration	Filters	Few hours (including chemical analysis)	$S_U = COD_{EFF,f0.45}$ $S_U = 0.9\ COD_{EFF,f0.45}$ (low loaded system) $S_U = 0.9\ COD_{EFF,f0.45} - 1.5\ BOD_{5,EFF}$ (high loaded system)	Melcer *et al.* (2003); Roeleveld & van Loosdrecht (2002)
Unbiodegradable particulate COD (X_U)	Model calibration	SBR operation or full scale data + calibration of an AS model	SBR pilot plant	Minimum 3 SRTs	Calibration of the pilot plant model	Melcer *et al.* (2003)
Heterotrophic biomass (X_{OHO})	OUR-based respirometry	OUR monitoring of an aerated WW sample	Respirometer	10– 20 hours	Modelling the OUR curve	Kappeler & Gujer (1992); Wentzel *et al.* (1995)

This section has provided a brief overview on what to measure and how to collect the basis for estimating wastewater characteristics; the reader is referred to Section 5.3 and Appendix A for the characterisation procedures itself.

General remarks

Data frequency is a critical parameter in any modelling exercise. Average values may be sufficient for a steady-state modelling project. However, the data should be carefully examined to ensure that the plant has indeed been operating at an approximate steady-state during the entire sampling period. If the plant is not stable, then using the average values can be misleading. A simple way to do a rough check is to plot the MLSS or to create cumulative clarifier mass balance plots (in *vs.* out).

If dynamic processes are to be investigated, daily values or typical diurnal patterns might be required. In most cases the time modelling objectives will determine the required influent data resolution (*Application Matrix* Chapter 6).

As an example, Figure 5.2.3 presents the diurnal flow pattern obtained at the inlet of a municipal wastewater treatment plant. Hourly flow rates show a similar pattern, however the flow peak occurs later during weekends than weekdays and the weekend average is slightly lower. Figure 5.2.4 shows that COD, N and P patterns are sometimes shifted and not always identical even in the same wastewater. If greater data resolution is required to achieve the project objectives (i.e. peak hourly loads), it might be important to consider the diurnal variations in the influent concentrations. Data resolution is case specific and can only be determined by considering the project objectives.

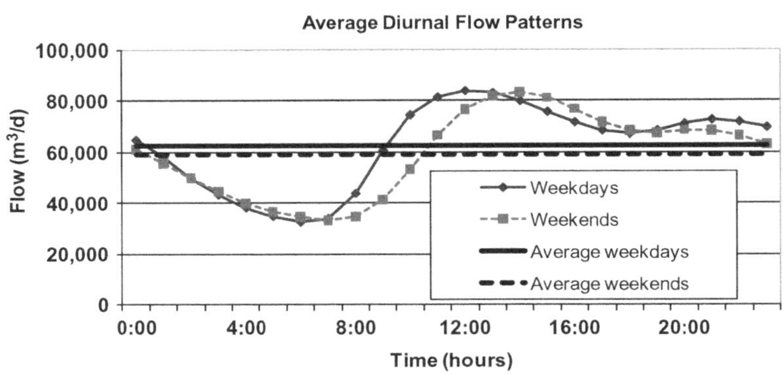

Figure 5.2.3 Example of diurnal flow pattern (hourly data) based on one year of data.

Some general remarks:

- If only BOD_5 data are available, additional measurements should be carried out to establish a correlation between COD and BOD_5 in the influent, activated sludge, and effluent as all activated sludge models discussed in this report are based on COD mass balances.
- Special care should be given to the flow and concentration data of the waste sludge withdrawal. It is a measure of the sludge production and the proper simulation of the sludge retention time (SRT) is essential to achieve accurate modelling results.
- Measuring the phosphorus concentration in the waste sludge (or MLSS) is very valuable for mass balancing. If not included in routine data, a few measurements should be performed, even if the plant is not required to remove phosphorus.

- The DO profile through the aerated tanks should be measured. As a minimum requirement, data from the DO probes should be collected and if possible a few additional measurements should be carried out to establish a DO profile.

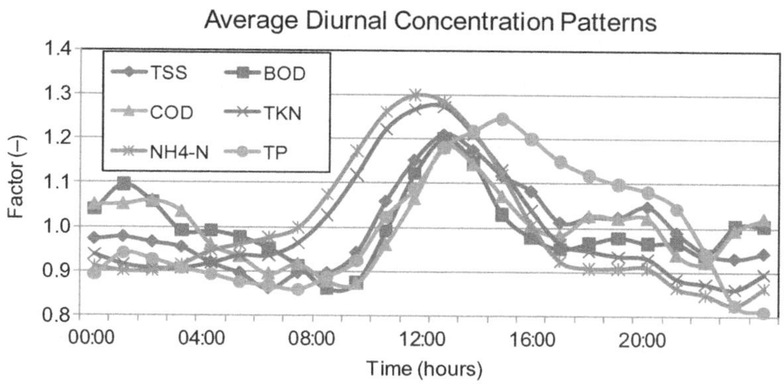

Figure 5.2.4 Example of typical diurnal concentration pattern (hourly data) based on four diurnal measurement campaigns.

5.2.5 DATA ANALYSIS AND RECONCILIATION

All collected data should be checked for consistency and quality and reconciled before proceeding to the next step of the Unified Protocol. The following section explains some fundamentals of data quality control and then proposes a step-wise procedure for data reconciliation.

5.2.5.1 Fundamentals in data quality control

The dilemma in data quality control is that the "true" value is never known and has to be determined by a reference method. However, in a closed technical system like a wastewater treatment plant, input-output relations and detailed process understanding can provide additional information on the accuracy of the measurements. Figure 5.2.5 shows the difference between systematic errors, random errors and outliers when following a symptomatic classification scheme.

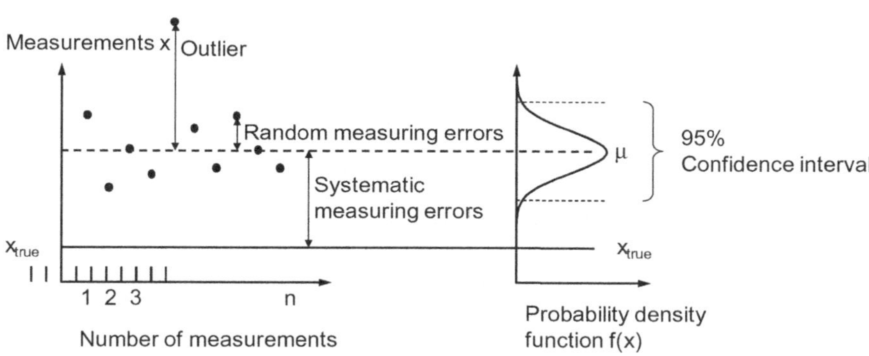

Figure 5.2.5 Definition of systematic errors (trueness), random errors (precision), and outliers (Rieger *et al.* 2005).

Systematic errors can be characterised as offsets (or shifts), signal drifts (change of signal over time), or calibration curve errors (linear or non-linear) (Thomann *et al.* 2002). Random errors can be split into random errors related to the instrument and the measuring principle (which are typically not reducible) and random errors related for example, to environmental conditions, sensor installation, measuring frequency, or signal transmission. For example, an increased signal noise due to sensor clogging would be seen as increased random error following a symptomatic classification (Villez, 2012). However, as this error is often reversible by cleaning, it cannot be considered random according to a root-cause classification. Outliers are gross errors and are typically removed based on statistical analysis.

It should be noted that data quality is a relative term and clear criteria should be defined to determine if a dataset is acceptable or not.

5.2.5.2 GMP data reconciliation procedure

Figure 5.2.6 shows the proposed procedure for data reconciliation including phases for error detection, isolation and identification (for definitions, Isermann & Ballé, 1997 and the Glossary).

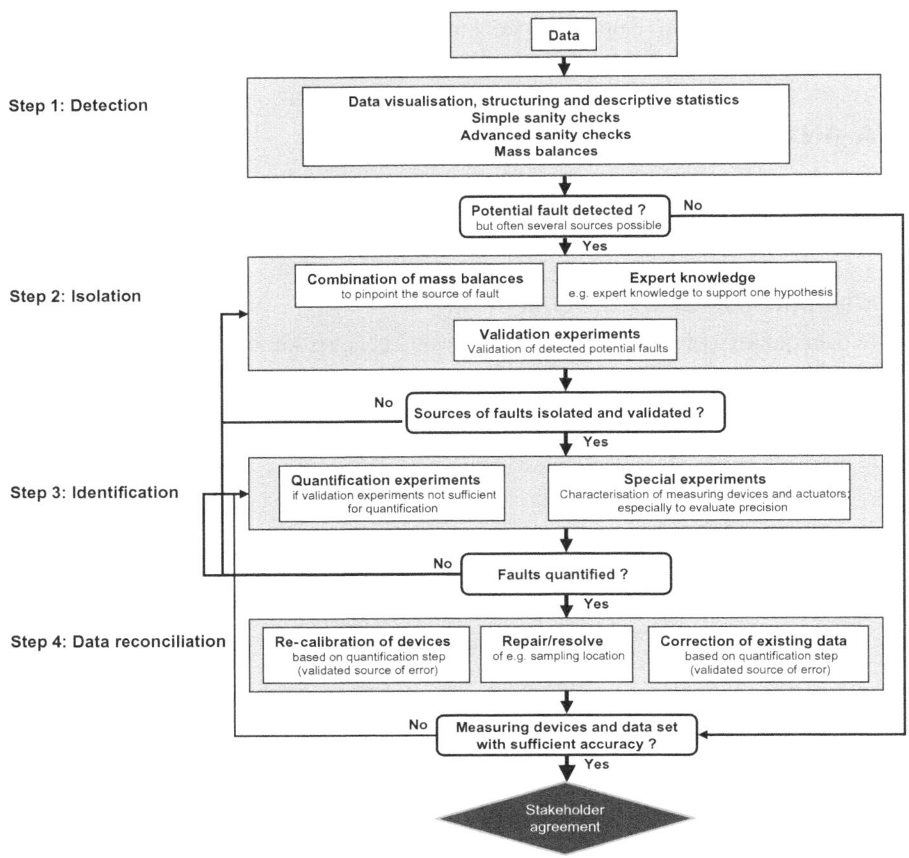

Figure 5.2.6 Data reconciliation procedure to detect, isolate, and identify faults, and finally reconcile data set (adapted from Rieger *et al.* 2010).

5.2.5.3 *Step 1: Fault Detection*

The first phase of the data reconciliation procedure is to detect potential faults. Figure 5.2.7 points out the main aspects of the proposed fault detection procedure.

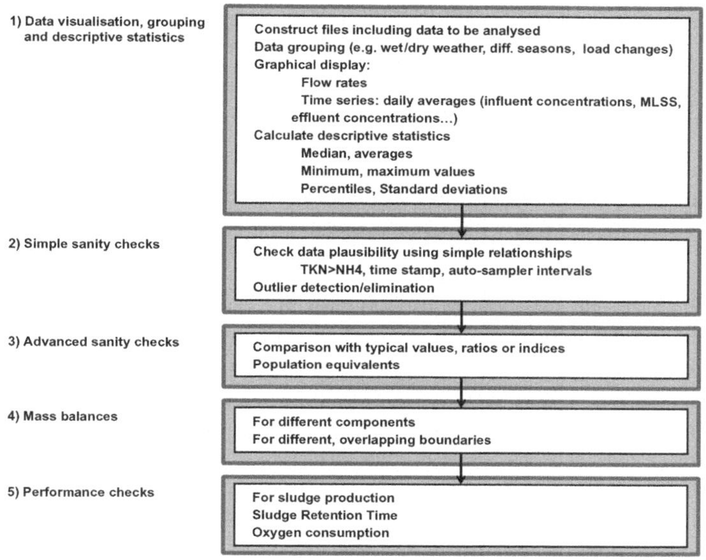

Figure 5.2.7 Overview of fault detection procedure.

Data visualisation, grouping and descriptive statistics

Construct files with data to be analysed. Trivial, yet sometimes serious faults occur when data points have incorrect time stamps. Such faults can usually be detected only through communication with the WWTP personnel. The following check list highlights some issues which should be considered before a detailed data analysis is initiated:

- The sampling period of automatic samplers (e.g. 08:00 am – 08:00 am, next day) and the averaging periods applied in the WWTP reports should be consistent.
- The proper assignment of lab results to the time of sampling (e.g. BOD_5-analysis results become available only five days after sampling).
- A proper understanding of the total hydraulic retention time (influent/effluent automatic sampler connected to which flow meter).
- Response time of sensors or analysers including sampling and filtration (according to ISO 15839, 2003).

Often some data are missing or the measurement intervals are inconsistent (e.g. flow and concentration or lab and online sensor measurements). Depending on the objectives, data might have to be interpolated.

Frequently, composite sampling data has to be translated into model inputs with a specific time stamp. The need for this and the method to accomplish this often depend on the simulator used. Three ways to overcome this issue are as follows:

- The same measured load is entered for the whole given interval
- The time stamp is set in the middle of the interval and the new value is used from the middle of the current interval to the middle of the next interval

- The data is interpolated between the measured intervals and loading values (pseudo-dynamic approach)

All three ways dampen the dynamics of the influent and introduce an error to the model. However, in most cases the resulting error is insignificant.

Note:

To keep track of all data to be analysed, it is suggested that supplemental information be stored with the data including:

- Units of measurement associated with all data
- Data source (collection, measurement, calculation, estimation ...)
- Time stamp when sampled (or interval for composite samples)

Data visualisation. Data visualisation can be used to gain an overview of the general trends and variability of the data (e.g. the nature of the distribution). This is best done by plotting the time series (preferably over longer periods to see seasonal changes). The plot provides an overview of plant load variability and process stability. Variables of interest include flow rates, influent load dynamics (e.g. COD, N_{tot}, P_{tot}), effluent concentrations (e.g. NH_x-N, NO_x-N, N_{tot}, PO_4-P, P_{tot}), reactor concentrations (e.g. PO_4-P in anaerobic zones or phases), temperature and MLSS in the aeration tanks, and SRT (considering solids in wastage and perhaps effluent streams).

Plotting also provides a simple yet powerful tool to identify shifts or outliers in concentrations or flows. Plotting measurements obtained through different means (e.g. laboratory measurements *vs*. online sensors, plant laboratory *vs*. external audits) can easily be used to identify problematic data. Data from parallel lanes may exist and when plotted, these data too may identify issues.

Figure 5.2.8 shows time series data for raw and primary effluent flows at a full-scale WWTP (solid lines). The plot clearly shows an offset and helped identify an averaging error in the operational data storage system that was corrected in July 2006.

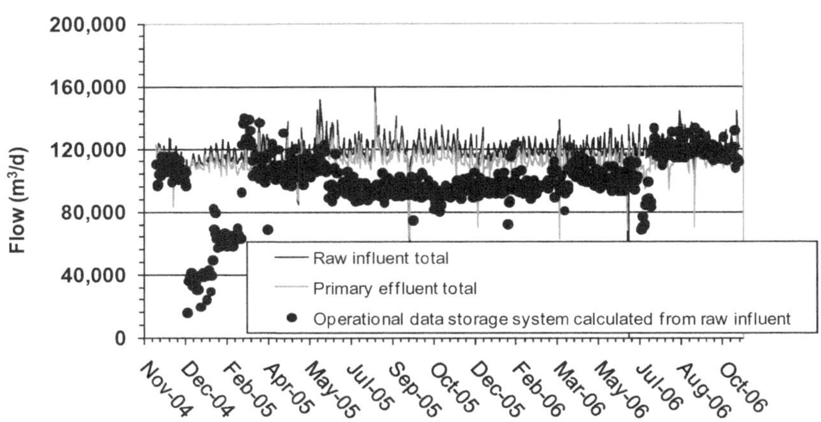

Figure 5.2.8 Averaging error in data base used for reporting compared with original flow measurements in SCADA system. Problem discovered during data evaluation for modelling project and fixed July 2006 (Third *et al.* 2007).

To check online sensors, good practice is to i) plot laboratory measurements *vs*. online data and calculate linear regression parameters and 95%-prediction intervals (Figure 5.2.9) and ii) plot residuals over time (Figure 5.2.10). The laboratory data should be on the X-axis as a standard regression assigns all errors to the Y-axis. A detailed description of one evaluation procedure can be found in Rieger *et al.* (2005).

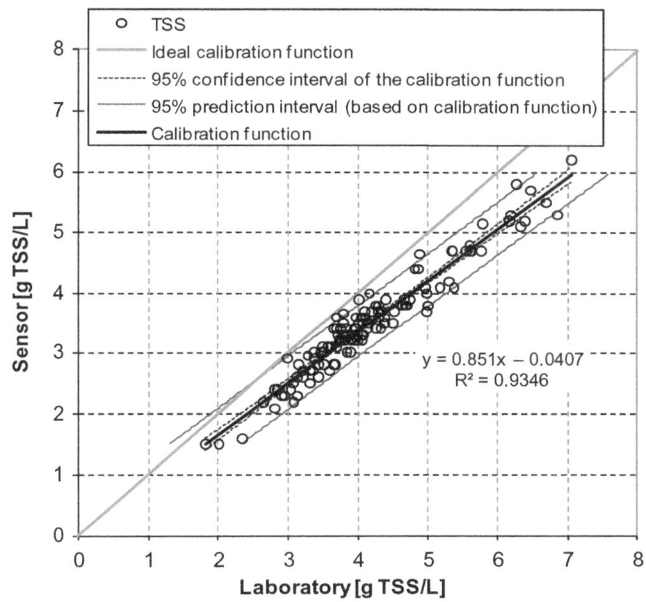

Figure 5.2.9 Laboratory measurements *vs.* sensor data for a TSS probe showing problems with calibration.

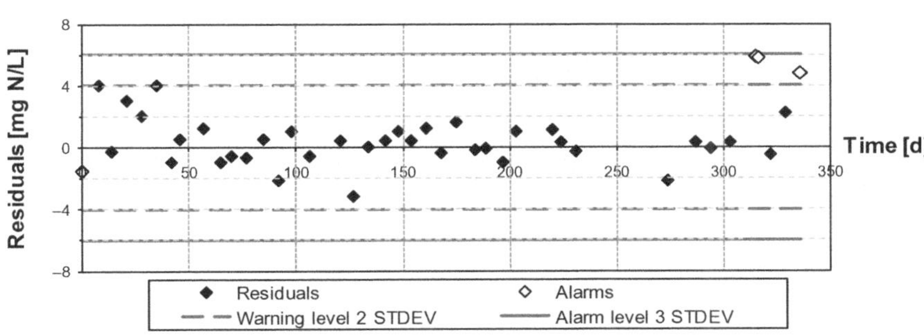

Figure 5.2.10 Control chart: Residuals over time for an ammonia ion-sensitive electrode probe including warning and alarm levels.

Data grouping. Data may be grouped into data sets representing specific loading or seasonal conditions. By doing this, the variability in the data within the different sets might be decreased. After grouping, *Descriptive statistics* can be performed on different sets of "typical" conditions.

An example that illustrates the distinction between dry and wet weather flows for a plant connected to a combined sewer system is shown in Figure 5.2.11. The dry weather average and wet weather average are shown as two separate peaks.

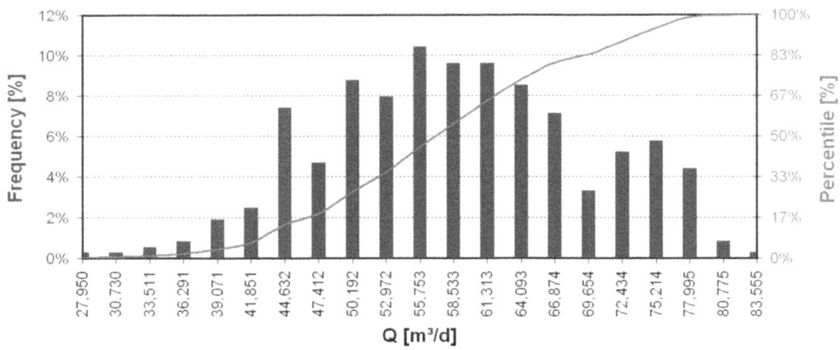

Figure 5.2.11 Influent flow distribution of a municipal WWTP connected to a combined sewer.

Descriptive statistics. *Descriptive statistics* are calculated in order to get more detailed information about the typical (average) plant load and its variability relative to the design load (Figure 5.2.12). Typical statistical measures for normally distributed data are:

- Minima/maxima
- Mean/median/mode
- Standard deviation, coefficient of variation, percentiles
- For skewed distributions, the *Coefficient of skewness* should be calculated.

Often exceptional loading or operational conditions can be identified by looking at performance indicators and comparing those to single measurements. These derived or calculated data can be plotted to get an instant overview of the nature of the data.

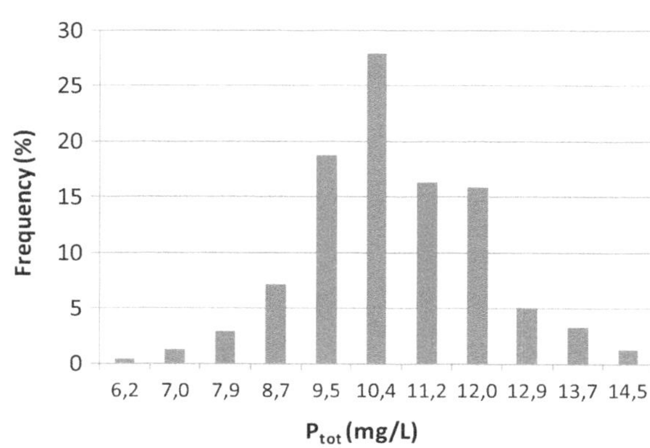

Figure 5.2.12 Histogram showing the influent total phosphorus (P_{tot}) normal distribution for a municipal wastewater.

Simple sanity checks

This step includes basic plausibility checks and the detection of potential outliers.

Plausibility checks. These checks usually start with verifying simple relationships for each sample, such as:

- $N_{tot} \equiv TKN + NO_3\text{-}N + NO_2\text{-}N$
- $TKN > NH_x\text{-}N$
- $P_{tot} > PO_4\text{-}P$
- $COD_{tot} > COD_{fil} > COD_{sol}$
- $COD_{tot} > BOD_5$
- $TSS > VSS$
- $MLSS_{RAS} > MLSS_{AST}$ (where AST = activates sludge tank)

If the data fails one of the simple tests, data quality investigations can be carried out to determine the source of the error. If the investigations do not reveal a cause for the inconsistency, then all data for that sample might have to be removed from the data set.

Potential outlier detection. Data outside typical ranges should be double-checked for plausibility and correctness and separated (or eliminated) from the data set if:

- Their correctness cannot be confirmed (e.g. double checking the data source)
- Causes for outliers are not evident (corroborating information about e.g. special load events within a specific period is not available)
- The data appear to be correct and plausible, but still are so far outside the "range of typical conditions" that they have a disproportionate influence on averages
- The data represent an unusual plant condition, (i.e. if the plant experiences a reduced capacity due to a maintenance period and this condition should not be analysed)

Using mass loads to detect outliers is a better approach than using concentrations because concentrations can be affected by flows (i.e. dilution due to rain events).

Advanced sanity checks

Comparison with typical ratios. Calculating ratios and comparing the average values of the investigated data set to typical ranges found in municipal wastewater can help to identify general characteristics of a plant. An example of this type of time series (before outlier elimination) is shown in Figure 5.2.13.

Mass load ratios are more universally applicable than concentrations, because as stated previously they are less dependent on rain, but mass-based ratios are also less dependent of regional factors such as drinking water consumption or sewer infiltrations. Reference values for municipal wastewater are given in Table 5.2.3.

For the case where this comparison reveals a wastewater characteristic that is clearly outside the range of municipal wastewater, a significant industrial wastewater contribution may be the cause (or the data is erroneous). Such a scenario requires special attention as most activated sludge models have been developed for municipal wastewater treatment. Aspects of applying activated sludge models to industrial wastewater are described in Chapter 8.

Mass balancing

Mass balances are powerful tools to detect inconsistencies within WWTP data sets and can help identify systematic errors (Barker & Dold, 1995; Nowak *et al.* 1999; Thomann, 2008). They allow the

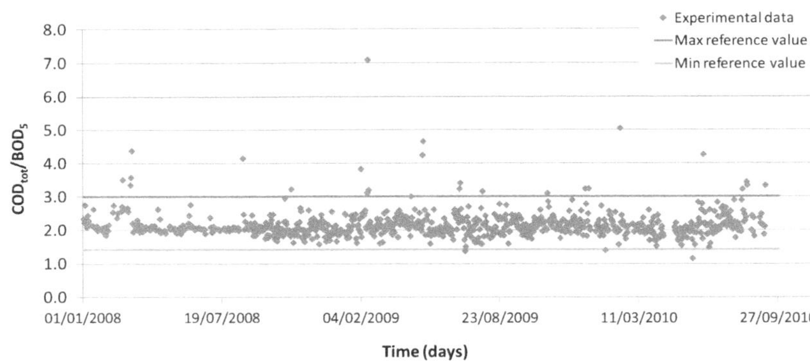

Figure 5.2.13 COD_{tot}/BOD_5 ratio in municipal wastewater (France). On a side-note: The external lab was changed in August 2008.

Table 5.2.3 Synthesis of typical ratios for municipal WWTPs from GMP questionnaire.

	Ratio	Unit	n[1]	Mean	*Std%* [2]	Median	Min	max
Raw influent	N_{tot}/COD_{tot}	g N/g COD	12	0.095	*17%*	0.091	0.050	0.150
	$N\text{-}NH_x/TKN$	g N/g N	13	0.684	*8%*	0.670	0.500	0.900
	P_{tot}/COD_{tot}	g P/g COD	12	0.016	*22%*	0.016	0.007	0.025
	$P\text{-}PO_4/P_{tot}$	g P/g P	12	0.603	*16%*	0.600	0.390	0.800
	COD_{tot}/BOD_5	g COD/g BOD	12	2.060	*11%*	2.050	1.410	3.000
	COD_{fil}/COD_{tot}	g COD/g COD	13	0.343	*29%*	0.350	0.120	0.750
	TSS/COD_{tot}	g TSS/g COD	12	0.503	*18%*	0.500	0.350	0.700
	COD_{part}/VSS	g COD/g VSS	11	1.690	*12%*	1.600	1.300	3.000
	VSS/TSS	g SS/g SS	12	0.740	*20%*	0.800	0.300	0.900
	BOD_5/BOD_∞	g BOD/g BOD	7	0.655	*7%*	0.650	0.580	0.740
	Alkalinity	Mol_{eq}/L	11	5.173	*35%*	5.000	1.500	9.000
Primary effluent	N_{tot}/COD_{tot}	g N/g COD	9	0.134	*35%*	0.120	0.050	0.360
	$N\text{-}NH_x/TKN$	g N/g N	11	0.755	*4%*	0.750	0.430	0.900
	P_{tot}/COD_{tot}	g P/g COD	9	0.023	*25%*	0.023	0.010	0.060
	$P\text{-}PO_4/P_{tot}$	g P/g P	10	0.741	*12%*	0.750	0.500	0.900
	COD_{tot}/BOD_5	g COD/g BOD	9	1.874	*31%*	1.900	0.500	3.000
	COD_{fil}/COD_{tot}	g COD/g COD	10	0.449	*31%*	0.495	0.150	0.750
	TSS/COD_{tot}	g TSS/g COD	9	0.380	*21%*	0.400	0.180	0.560
	COD_{part}/VSS	g COD/g VSS	9	1.718	*14%*	1.700	1.400	3.500
	VSS/TSS	g SS/g SS	9	0.794	*7%*	0.800	0.700	0.909
	BOD_5/BOD_∞	g BOD/gBOD	6	0.644	*10%*	0.656	0.533	0.760
	Alkalinity	Mol_{eq}/L	9	5.711	*40%*	6.000	1.500	9.000
Activated sludge	COD_{tot}/VSS	g COD/g SS	9	1.434	*7%*	1.420	1.266	1.600
	N_{tot}/COD_{tot}	g N/g COD	7	0.073	*35%*	0.060	0.045	0.116
	P_{tot}/COD_{tot}	g P/g COD	7	0.020	*64%*	0.015	0.010	0.044
	VSS/TSS	g SS/g SS	10	0.739	*8%*	0.750	0.650	0.900

[1] number of answers; [2] standard deviation in %

Source: Hauduc, 2010.

identification and confirmation of the mass flows into and out of a specific system. Based on the principle of mass conservation, a mass balance has the following general form:

$$\text{Input} + \text{Reaction} = \text{Output} + \text{Accumulation} \qquad (5.2.1)$$

Mass balances can be based on a number of different process variables (e.g. Q, COD, N, P, TSS). An advanced way of using mass balances for fault identification is to set up *parallel mass balances* (mass balance of the same system utilising different process variables) or *overlapping mass balances* (mass balances with different system boundaries with one common measuring point). Please refer to Appendix I for more information and examples.

Performance checks

To further assess the data sets, the actual plant performance can be compared with a typical performance for this type of plant. Estimating performance criteria, such as sludge production, sludge retention time (SRT) or oxygen demand, requires the mass flows and the mass of sludge in the reactors to be validated beforehand. Therefore these criteria should be calculated on the basis of reconciled data sets and not on raw data.

5.2.5.4 Step 2: Fault Isolation

In the previous steps, faults were detected without knowing the error location or the root cause of the error. This section describes how to isolate (locate) a fault. This task is often based on experience (e.g. the probability of erroneous wastage data is high), but can be done with additional experiments sometimes.

A more structured way is to use overlapping mass balances (Appendix I) to isolate the fault or at least to formulate hypotheses on fault locations to trigger validation experiments. Overlapping mass balances integrate specific measuring points in more than one mass balance. If one mass balance can be closed but another overlapping balance cannot, the error is probably in another variable (not the overlapping one). If both are not closing, the probability is high that the error is in the variable (flow or concentration), which overlaps.

5.2.5.5 Step 3: Fault Identification

After isolating where the source of the fault is located, the root cause of the fault should be identified and finally quantified. Specific measurements may be required in order to determine the precision of the measuring devices.

Sources of errors can be split into flow measurements, sampling and analytical methods (Figure 5.2.14). Typical error sources are described below. An extensive list of sources of measuring errors is provided in Appendix G.

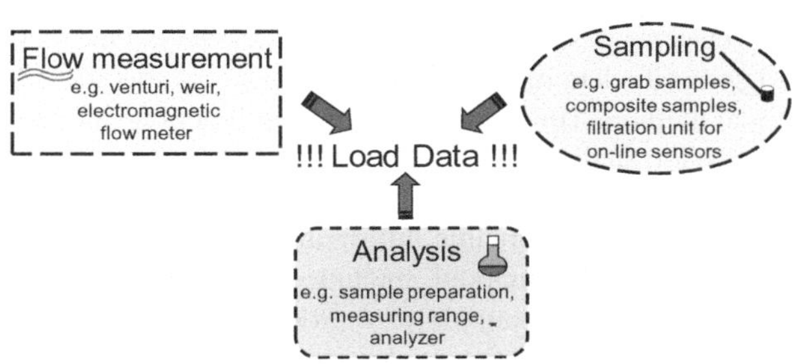

Figure 5.2.14 Influence on accuracy of load data (Rieger *et al.* 2010).

- *Flow measurements* can be a significant source of error. It is therefore advisable to check flow meters, for example by measuring the time required to fill up a known volume. Table 1 in Appendix G provides a list of typical sources of errors in flow measurements.
- *Sampling* (automatic or manual) is another common error source. A representative sampling point with homogenous mixing is important – especially if sampling for sludge or other wastes with high solids concentrations. See Table 2 in Appendix G for a list of potential sampling problems.
- *Analytical methods* (laboratory or on-line measurements) can be the source of errors due to a number of different issues. Sample preparation and processing (e.g. storage, homogenisation, sub-sampling, incomplete digestion, etc.) are common error sources. Table 3 in Appendix G lists potential analytical issues.
- *Online sensors* are increasingly used at WWTPs and provide the high measuring frequency appreciated by modellers. However, they introduce additional demands in terms of maintenance and quality control. It is therefore suggested that the quality of online sensor data be regularly checked against reference measurements and the use of control charts. With respect to modelling, special attention should be given to the response time of these devices including sampling devices and filtration units. A list of potential errors in relation with online devices is given in Table 4 in Appendix G.

5.2.5.6 Step 4: Data reconciliation

It has to be emphasised that all activated sludge models are based on mass balances and will result in a closed balance (with few exceptions). It is therefore essential to reconcile the available WWTP data set so that it also provides a closed mass balance. Feeding an activated sludge model inconsistent data will not give meaningful simulation results. It is good practice for an experienced process engineer to allocate the error even if the exact source of the error is not known. Failing to do so will force the model to assign the error and this may or may not be appropriately done.

The following flow diagram (Figure 5.2.15) suggests a structured procedure to reconcile data sets such that the derived mass balances close.

Depending on the available information, it may not be possible to link an identified fault to a specific location. In this case, the error has to be equally distributed between potential locations. However, often expert knowledge (e.g. experience from other plants or knowledge about plant-specific measurement problems) can be used to locate the fault based on probability, experience and/or intuition.

As soon as evidence on the fault location is available, the fault can be isolated (assigned to a specific location/probe). The last step is to identify the root cause of the fault. This might lead to a change in laboratory procedures or a new calibration curve for measuring equipment. Although this is important for ensuring that future measurements are correct, it does not help reconcile existing data sets.

To reconcile existing data sets a regression analysis may be carried out. One method would be to generate a linear regression model to correlate data of high accuracy (e.g. from specific control experiments) to the data measured by the historical method. The resulting regression function can then be used to correct all of the historic data. After the regression analysis and recalculation of mass balances with the new data set, it still has to be decided whether the data are reconcilable or if they should be discarded. Sometimes the data set cannot be reconciled but will still provide valuable information (e.g. data below the detection limit is not reconcilable). Questionable data can be marked, included and ignored, or simply discarded. This regression approach is a relatively simple example of a procedure that can be used, but there are many more advanced methods also available and the reader is referred to the *Further Reading* section for more information.

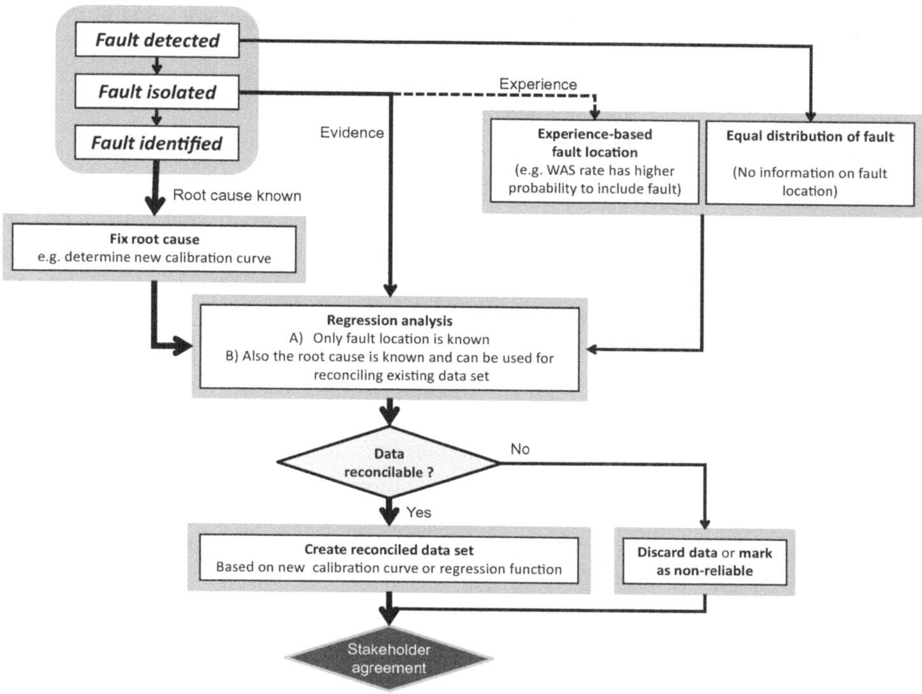

Figure 5.2.15 Structured data reconciliation procedure.

The final step is to receive stakeholder approval for the data set used in the simulation project.

Closing data gaps

Data reconciliation includes closing data gaps (by means of substitute values), which may be necessary if a specific period needs to be modelled and only an incomplete data set is available. If essential data are missing, literature values (e.g. for typical population equivalents or ratios) are often a possibility. Another option would be to utilise additional data sources (e.g. analysis reports of the disposed sludge). Data from another period may be used provided the operating conditions (temperature, influent load, $MLSS_{AST}$, operational settings, effluent data, etc.) during the two periods are similar. It needs to be emphasised that this has to be done with great care in order to avoid the introduction of dissimilar data.

In the case of limited data availability (e.g. design study of a not yet existing plant), methods such as population equivalent loadings can be used as a first guess. If a high data frequency is required, 'influent generators' can be used to generate typical patterns for influent flows and loads. These 'influent generators' provide patterns depending on parameters such as plant size or structure of the connected sewer network (Gernaey *et al.* 2006; Langergraber *et al.* 2008). Data sets provided by 'influent generators' represent average conditions of plants of similar size and can therefore give a coarse estimation of the load characteristics of the investigated plant. Tools like this are not a substitute for specific monitoring campaigns when detailed questions related to the plant loading dynamics are the core goal of the simulation study, but these tools are often valuable for less detailed studies.

It should be clear that additional measurements may be necessary to close the gaps.

5.2.6 ADDITIONAL MEASUREMENT CAMPAIGNS

After the detailed assessment of an existing WWTP data set it might turn out that critical data are missing (i.e. reference values are not suitable, higher data resolution is required, etc). In such cases an additional measurement campaign is required.

No additional monitoring should be carried out before the following aspects have been discussed and agreed upon with the client:

- Which process variables shall be monitored at which frequency?
- Which sampling locations are suitable?
- Which sampling and analytical methods are applied?
- Who provides the required sampling and analysis equipment?
- Who is responsible for sampling, analysis and reporting?
- In case sampling and analysis is carried out by different parties: Who is responsible for adequate sample labelling, storage and transport?
- How is the additional monitoring campaign documented? How is the data stored and exchanged?
- Who is responsible for data quality assurance of the additional analysis?

If on-line probes or analysers are installed, the following issues need to be considered:

- Can the probes be installed at locations which are i) suitable for providing meaningful measurements and ii) accessible for probe maintenance
- Who operates and maintains the installed probes?
- How are the data collected and processed?
- Which reference analyses are required to ensure data quality of the continuous measurements – and who is responsible to carry them out and document it?
- How long is the anticipated installation period of the additional probes/analysers?

After the above points have been clarified and agreed upon, the additional monitoring campaign can be carried out noting that special care should be taken to ensure that the sampling period is representative of the conditions under study. Table 5.2.4 gives an overview of additional data that are often required and how the data would be used.

Typical Pitfall:

Because of time pressures, intensive measurement campaigns are often started before routine data are properly analysed and reconciled. In these cases, existing errors may not yet be identified and could have significant impacts on the model results and consequently on decisions made.

5.2.6.1 Client agreement

As additional data requirements represent a large effort and potentially an important cost to the overall project, client agreement is required before carrying out the intensive measurement campaign(s).

Table 5.2.4 Additional data requirement *vs.* objectives.

Item in question	Process variable	Objective	Recommendations
Mixing behaviour Plant model	Tracer test	Set-up of hydraulic model (number of tanks in series)	Pre-simulate the system to design tracer experiment
Sludge production SRT	$P_{tot,WAS}$	Closing the P_{tot} mass balance	2 measurements/week Min. 5 values
Influent load dynamics	COD_{INF}, $N_{tot,INF}$, NH_x–N_{INF}, $P_{tot,INF}$, TSS_{INF}	Plant performance under dynamic load	2h-composite samples (influent, effluent, and selected reactors) min. 2x 24h-campaign Alternatively: Continuous monitoring, min. 1 month
Effluent concentration peaks	NH_x–N_{EFF}, NO_x–N_{EFF}, PO_4–P_{EFF}	Meeting effluent peak limits	2h-composite samples (influent, effluent, and selected reactors) min. 2x 24h-campaign Alternatively: Continuous monitoring, min. 1 month
Aeration control strategy	DO_{AST}, NH_x–N_{AST}, NO_x–N_{AST}	Optimising aeration control	DO, NO_x-N: in-situ probes NH_x-N: in-situ probe or analyser Min. 2 weeks including min. 1 storm flow event
MLSS dynamics in the activated sludge tank	$MLSS_{AST}$	Plant behaviour under storm flow conditions WAS withdrawal strategy	Turbidity probe, min. 2 weeks including min. 1 storm flow event

5.2.7 FINAL CLIENT AGREEMENT

The final task in this step is to get stakeholder approval for the deliverables listed below.

5.2.8 DELIVERABLES

- Reconciled data set(s) including:
  - Representative flows and loads for all major influent and effluent streams
  - Up-to-date flow diagram of the WWTP including tank volumes as operated
  - Aerated and mixed fractions of the tanks (space or time)
  - Flow and concentration of internal recycle streams
  - P&I-Diagram of the WWTP as operated
  - Sampling location and scheme of influent and effluent samplers
- Data outliers justified, separated or eliminated
- List of isolated and identified systematic faults in data sets

FURTHER READING

Nowak O., Franz A., Svardal K., Müller V. and Kühn V. (1999). Parameter estimation for activated sludge models with the help of mass balances. *Water Science and Technology,* **39**(4), 113–120.

Barker P. S. and Dold P. L. (1995). COD and nitrogen mass balances in activated sludge systems. *Water Research*, **29** (2), 633–643.

Melcer H., Dold P. L., Jones R. M., Bye C. M., Takacs I., Stensel H. D., Wilson A. W., Sun P. and Bury S. (2003). Methods for wastewater characterization in activated sludge modeling. *Water Environment Research Foundation (WERF)*, Alexandria, VA, USA.

Montgomery D. C. (2005). Introduction to Statistical Quality Control, 5th edn; John Wiley & Sons, New York, USA.

Russel S. and Norvig P. (2010). Probabilistic reasoning. In Artificial Intelligence – A modern approach, 3rd edn, Chapter 14, Prentice Hall, USA.

Chapter 5.3

Plant model set-up

5.3.1 INTRODUCTION

This section of the STR deals with the set-up of a plant model. A first part introduces a six-phase procedure for setting up a plant model and a second part discusses available sub-models and some criteria on model selection. Examples of model simplifications and suggestions on how to translate typical plant configurations into model flow schemes are given. Detailed descriptions of the main sub-models are provided in Appendix A.

5.3.2 PROCEDURE

Figure 5.3.1 shows the step-wise procedure to set up a model, starting from the plant layout agreed upon in Step 1 of the Unified Protocol. The first task towards setting up a plant model is the careful appraisal of the system, including the definition of the model boundaries, sub-systems or operational units, and their interactions (such as mass and energy flows or control signals). The objectives of the study determine which units of the plant need to be modelled and in how much detail. For example, it is the objectives that define if a point clarifier is sufficient or a layered model must be used. These same objectives dictate whether sidestream treatment is included and whether or not sensors, control handles and controllers have to be represented. The procedure then proceeds with the establishment of connections to data bases and files, and the setting up of output graphs and tables. Finally a number of initial test runs are carried out to check the model. The result is a functional model which provides outputs sensible enough to fulfil the objectives of the study.

Depending on the simulator package used, the procedure can vary and may include additional phases like model building (i.e. creating an executable code). For more specific information the reader is referred to the simulator manuals.

Detailed descriptions of every phase are provided in the next sections.

5.3.2.1 Plant layout

Based on earlier steps, the boundaries of the plant model and process units to be used are established. For every process unit the level of sub-model detail has to be defined. This includes any decisions on model simplifications.

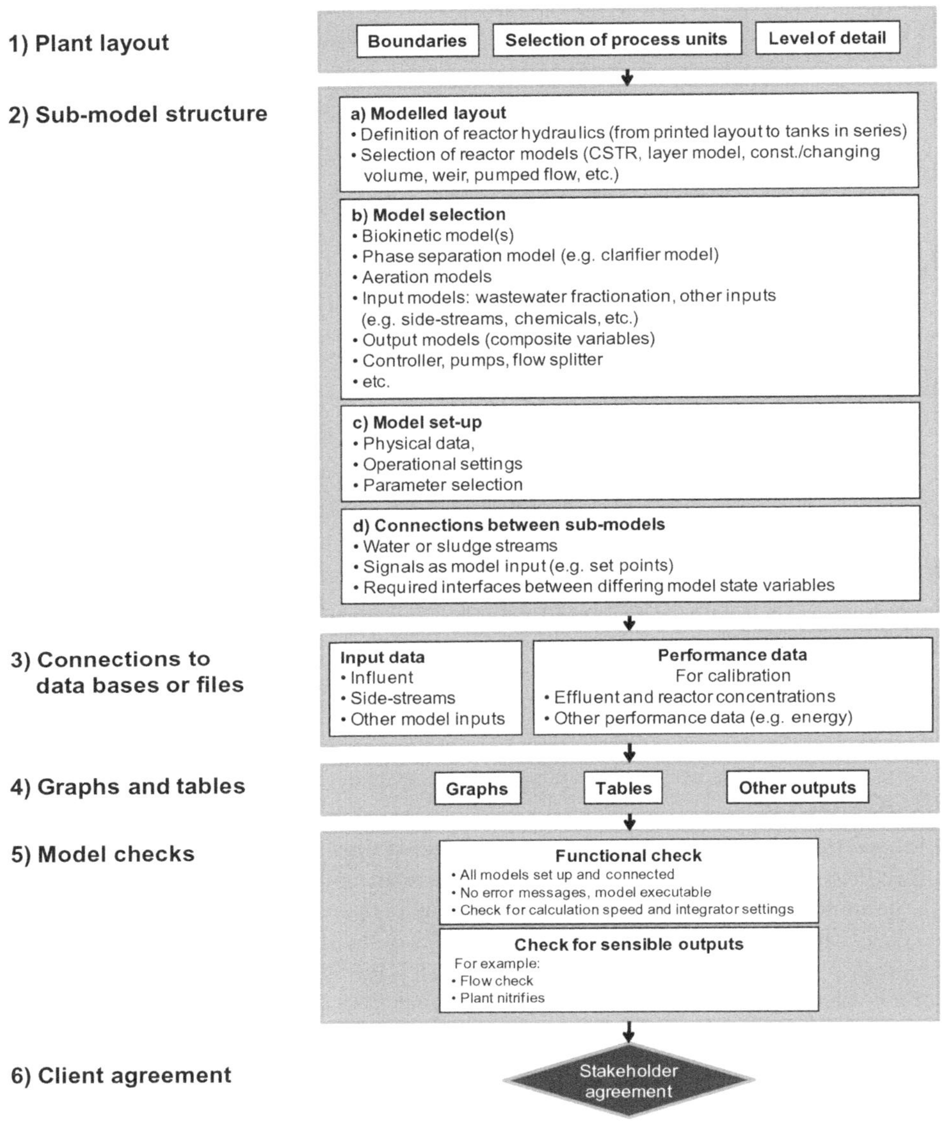

Figure 5.3.1 Step-wise procedure to set up a plant model.

The required level of detail is a major decision and should involve a senior modeller. Examples are given in Table 5.3.1. Incorporating too many details will prolong the time for model set-up and execution. Unnecessary details will require more data and will increase the effort required for calibration/validation. Too few details could introduce significant uncertainties in the model outputs due to an inadequate model structure. The appropriate balance depends on the objectives but it should be kept in mind that certain operational problems can be hidden in a model simplification (e.g. an unequal flow split cannot be detected if only one lane is modelled).

Table 5.3.1 Typical model simplifications and potential pitfalls.

Objective / Plant operation	Typical simplifications	Potential pitfalls
Several parallel equally operated lanes	Model whole plant as one lane with total volume (preferred)	Unequal flow distribution causes error in predictions
	Model only one lane	Incorrect flows to one lane or lane not representative (check all involved volumes and flows: e.g. influent, RAS, WAS, etc.)
Several parallel equally operated clarifiers (independent sludge systems)	Model all clarifiers as one with combined surface	A one sludge system is modelled but in reality settling characteristics may differ between clarifiers
	Model only one clarifier	The selected clarifier is not representative
Continuous dosage of plant recycles with approx. constant concentrations	Include plant recycles in influent	Impact of plant recycles cannot be evaluated
The SRT is known (e.g. from P mass balance) and the settling behaviour is of no interest	Use an ideal clarifier model (complete phase separation, no or fixed percentage of solids in effluent) and set the calculated SRT	Particle capture is not modelled. Modelled sludge production is slightly higher
Plant is operated for COD and N removal but not for Bio-P	Use a CN model (e.g. ASM1 or ASM3)	If anaerobic zones or phases occur, the model is not able to predict the real behaviour (because consumption of S_B associated with bio-P in these zones will not be modelled). P nutrient limited cases are not detected
Only activated sludge system is of interest	Model input to AS system, skip primaries	Impact of primaries not modelled (e.g. wet weather situation) Return fluxes (e.g. from sludge treatment) have to be adequately taken into account

Decisions on the level of detail are arrived at through an iterative process. A loop in the procedure allows the modeller to go back to the plant model set-up if model inaccuracies are detected during the calibration step.

5.3.2.2 Sub-model structure

(a) *Modelled layout*: Most simulator packages provide a graphical user-interface where each sub-model or a combination of models is represented by an icon. These icons are typically connected together in a specific "design" window by using drag and drop functionality. This task involves the translation of the existing process flow scheme (as operated) and the mixing behaviour (i.e. completely mixed to plug-flow) into a model concept. When doing this, the modeller must decide which reactor model to use (e.g. *completely stirred tank reactor* (CSTR), constant/changing volume, weir, pumped flow, etc.). The state-of-the-art is to use a "tanks in series" approach, which combines several CSTRs in series to mimic the mixing behaviour in the real facility. The more CSTRs used, the more plug-flow behaviour is modelled. Note that the tanks in series model impacts the simulated effluent concentrations as a result of the concentration profile and is thus an important consideration in the layout design.

(b) *Model selection*: For each modelled process unit, one or more models have to be selected. Section 5.3.4 *Sub-Model Selection* discusses typical sub-models and model selection criteria. A list of sub-models with explanations is provided in Appendix A.

(c) *Model set-up:* Physical data (e.g. volumes), operational settings (e.g. flows and set-points) and other parameters (e.g. model parameters, initial values for state variables, etc.) need to be provided for the selected models.

(d) *Connections between sub-models*: All sub-models have to be connected to allow for information flow. The connections represent the mass (liquid and solids) flow through the plant or a signal vector, providing the necessary input to the sub-models. The components in those connections are characterised by the sub-models chosen for the layout and are calculated for all applicable state variables at every time step simulated. If models with a different set of state variables are used, an interface is required for model connection (e.g. biokinetic state variables into clarifier state variables). This is frequently automated and does not necessarily require user input. A signal vector can consist of one or more variables (e.g. air flow into reactor) or constants (e.g. controller set-point).

5.3.2.3 Connections to databases

Once the plant model structure is set up in the simulator, input data has to be provided. For steady-state runs, the measured (or estimated) average concentrations and flows are required and for dynamic simulations, time series data are linked to the selected input model. Some simulators allow links to external data bases and others require the data be entered into an internal form. Either way, the influent model calculates values for all state variables, which then become the input to the biokinetic model.

5.3.2.4 Graphs and tables

Outputs (for example, graphs or tables) have to be set up in the software or exported for external viewing to enable the results to be evaluated. Calibration typically requires graphs or tables that include both simulated and measured values. Some simulator packages provide statistical evaluations for the quality of fit. The presentation of model output should be designed to give valuable feedback on the plant performance for different scenarios. Consequently, graphs or tables that are information-rich or provide information at essential locations for the evaluated processes or compounds should be selected (e.g. if the effluent ammonia concentration is always close to zero, a look at the reactor ammonia concentration profile could provide a better insight into the plant performance than a graph of the effluent ammonia).

5.3.2.5 Model checks

A functional test of the model consists of some initial runs to make sure that the plant model is properly implemented in the software environment (no error messages), contains all significant process units and provides sensible – though uncalibrated – outputs (i.e. outputs in normal ranges, for example the plant nitrifies if that is expected based on process knowledge).

The calculation speed should also be checked to verify that it is adequate for the project requirements. Simulation speed is an important factor to consider when assessing the complexity of the modelled system and the expected number of runs. For large projects, it might be advantageous to optimise integrator settings to minimise the total simulation time.

5.3.2.6 Stakeholder agreement

It is typical to get the client to agree on the model implementation (including all steps explained above) and adequacy of the model for meeting the objectives.

5.3.3 DELIVERABLES

The deliverables from the plant model set-up are:

- A functional (uncalibrated) plant model
- Working model connections to process databases
- Well-designed graphs and tables
- Description of all sub-models
- Justification for model selection, simplifications and assumptions
- Proposed modifications to the project definition document for calibration and validation

5.3.4 SUB-MODEL SELECTION

A plant model consists of several sub-models. This section discusses the most common sub-models and gives some guidance on how to select the appropriate model for the chosen objectives. Table 5.3.2 lists common sub models; for a more detailed description please refer to Appendix A.

Table 5.3.2 List of common sub-models (see Appendix A for details).

Sub-model type	Sub models
Hydraulic and transport models	Reactor models (e.g. CSTRs with fixed or variable volumes) Flow scheme: reactor configuration and combination (e.g. tanks in series) Return Activated Sludge (RAS) and Internal Recycle (IR) flows Waste Activated Sludge (WAS) flow Flow splitter: Ratio, fraction, flow, bypass, flow router, flow pacing
Clarifier models	Point clarifier models Ideal clarifier models Layered clarifier models (to mimic one-dimensional settling) More-dimensional CFD clarifier models (typically not available in WWTP simulators) Reactive clarifier models (coupled with biokinetic model)
Input models	Influent model: to convert measurements (typically COD, N, P) to model components (state variables). Operational settings and other inputs Constants e.g. controller set-points Energy inputs for surface aerators or other process units
Output models	Combined (or composite) variables (e.g. COD_{tot}, BOD, TSS, etc.) Energy or cost models
Biokinetic models	e.g. ASM1/2d/3, etc. Temperature dependency (Arrhenius equation)
Aeration models	Model to translate air flow into k_La Oxygen transfer model Aeration equipment models (diffuser, pipes, blower system)

(*Continued*)

Table 5.3.2 List of common sub-models (see Appendix A for details) (*Continued*).

Sub-model type	Sub models
Phosphorus precipitation models	Iron or aluminium salt dosage Calcium and magnesium phosphate precipitation
Other sub-models (not described in detail)	Controller models, sensor models, actuator models pH Gas transfer Sub-models for whole plant modelling (e.g. model for primaries, anaerobic digester, sludge handling, etc.) Models for enthalpy (temperature), operational costs, energy, carbon footprint, conversion of green house gas emissions into carbon units, etc.

5.3.4.1 Flow scheme

The mixing behaviour of the plant impacts the treatment efficiency. The lack of a calibrated mixing model can severely limit prediction accuracy. Examples for the translation of a real plant into a model configuration are shown in Table 5.3.3. In most simulators, the hydraulic behaviour is modelled as a series of *Completely Stirred Tank Reactors* (CSTR). The number and combination of tanks has to be calibrated based on experience, measured performance data, empirical equations, tracer experiments, or CFD (computational fluid dynamics).

Table 5.3.3 Example layouts for real systems and translation into model flow schemes.

Plant scheme	Real system	Model flow scheme
MLE	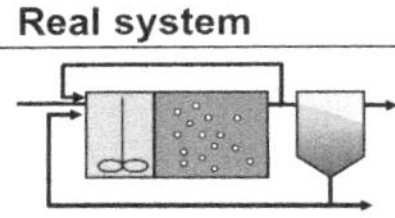	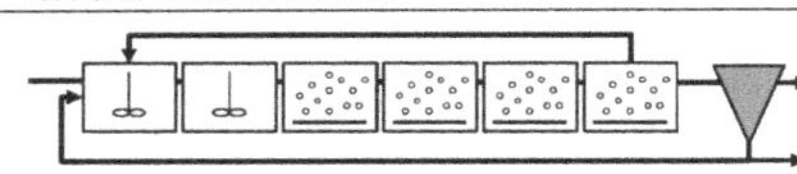
A_2O	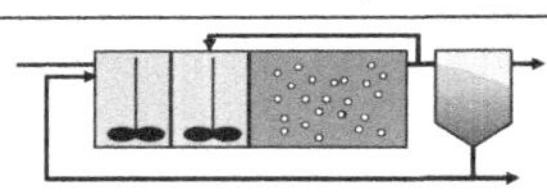	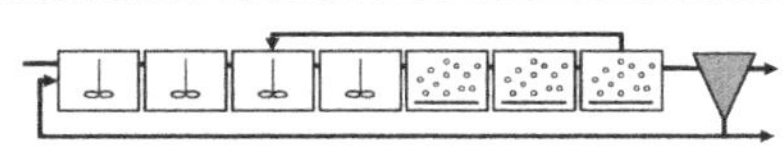
Oxidation ditch	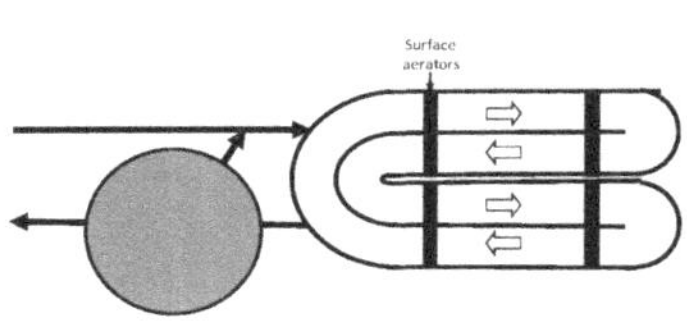	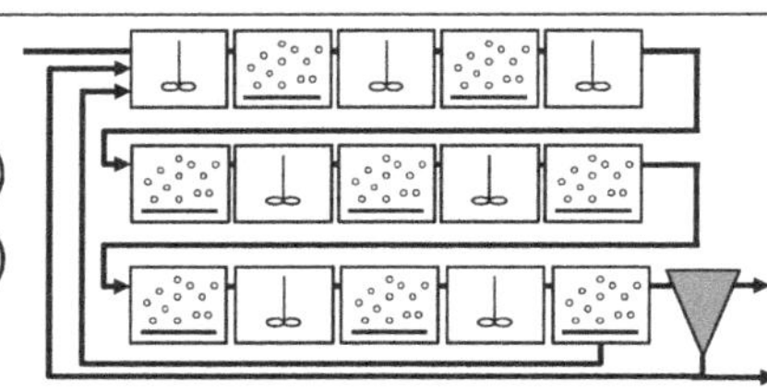
Alternating influent	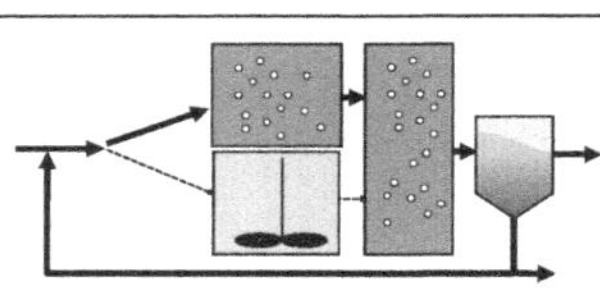	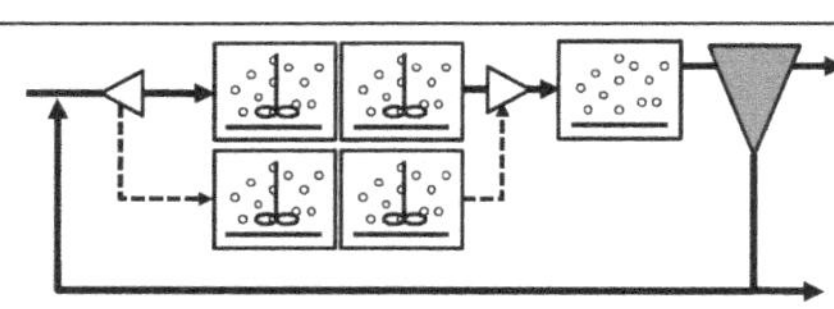

It is not uncommon for plants to have tracer test results for the basins, but if tracer test results are not available, several equations exist to estimate the required number of tanks in series. Fujie *et al.* (1983)

presents an approach, which gives satisfactory results for most standard cases (Makinia & Wells, 2005). A spreadsheet with different estimation formulae is provided on the GMP IWA WaterWiki website.

5.3.4.2 Selection of clarifier models

Clarifier models

Different clarifier models are available in commercial simulators and it is important to choose a model that is best suited for the modelling objectives. A graphical representation of the various models is provided in Table 5.3.4. Simple clarifier models (e.g. ideal clarifier) are applicable if the hydraulics and the suspended solids load are relatively constant. If rain events or other significant dynamic perturbations lead to considerable sludge shifts between activated sludge tanks and clarifiers, more complex models have to be applied so that the settling behaviour can be taken into account.

Table 5.3.4 Different types of clarifier models typically used in AS modelling.

	Complete split of particles	Loss of particles
a) Point clarifier models	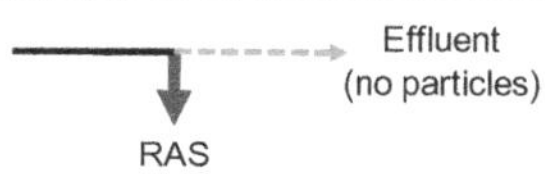	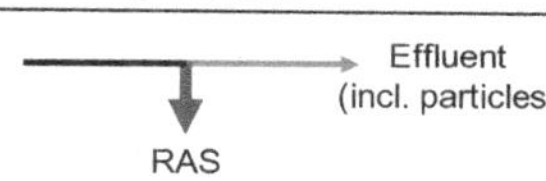
b) Ideal clarifier models (with volume)		
Volume for sludge storage	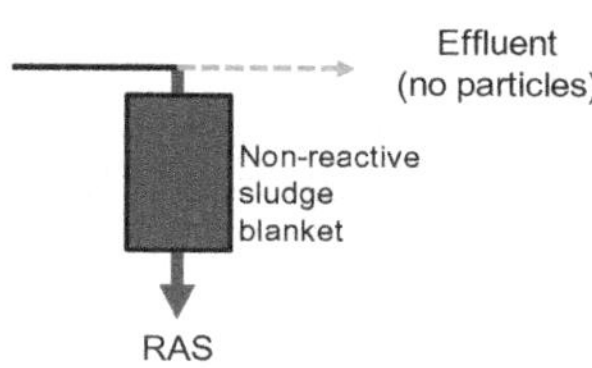	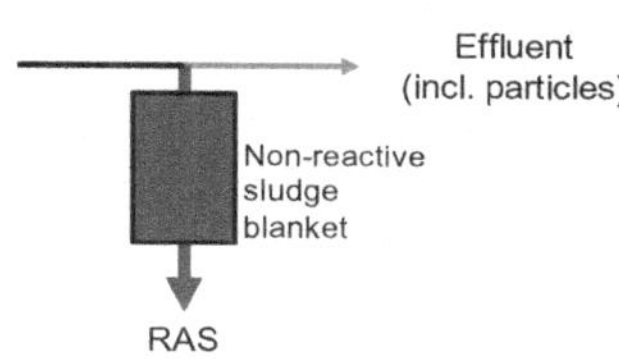
Volume for sludge storage and clear water	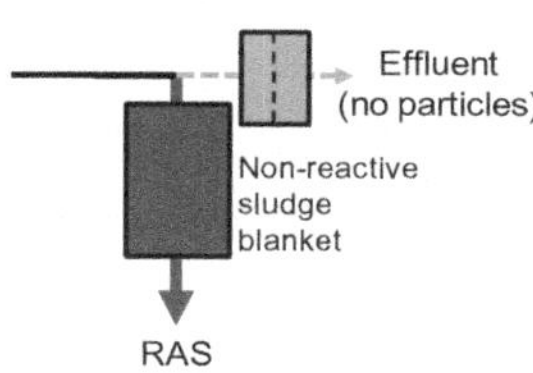	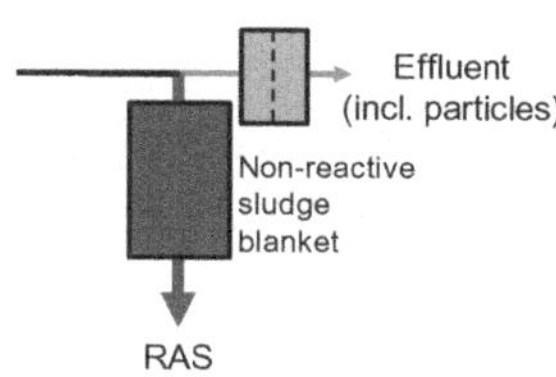
c) Layered clarifier models (1-D or flux models)		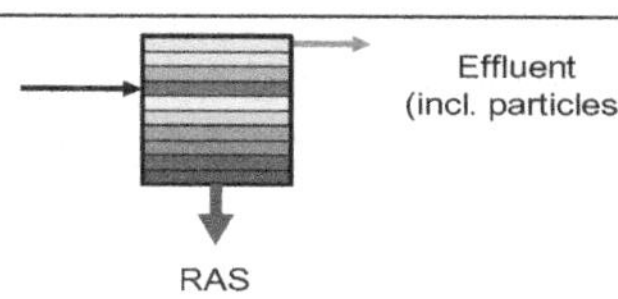
d) more-dimensional CFD models (typically not available in WWTP simulators)		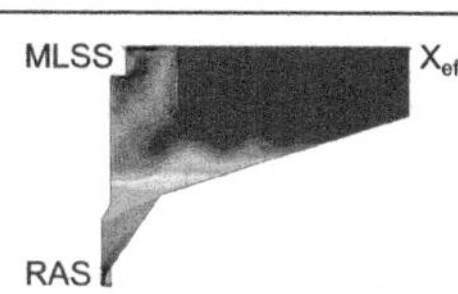
e) Reactive clarifier models	Applicable to models with volume as described above	

In general, ideal clarifier models are sufficient for a variety of typical objectives. More complex layer models, even if calibrated, may not predict effluent solids correctly. However, these models are capable of predicting rising sludge blankets which can indicate a potential failure and/or wash-out of solids. In some models the number of horizontal layers used in the model has a significant impact on the clarifier performance. It should also be noted that detailed data to calibrate a layered model is not typically available and has to be obtained by additional measuring campaigns. CFD models provide a great deal of detail with respect to settling behaviour, but these models are mainly used in research at the moment and not often applied during the set-up of a standard WWTP model.

Typical pitfall:

Endogenous denitrification in the sludge blanket can significantly contribute to the nitrogen removal of the plant (e.g. Siegrist *et al.* 1995). This is especially true when considering SBRs, as up to 30% of the denitrification can occur during the settling phase.

Reactive clarifier models

In addition to the hydraulic behaviour of the clarifier, it is important to recognise that the solids remain biologically active during the settling process. The sludge blanket is a reactive zone in which biological reactions such as endogenous denitrification or secondary phosphorus release are occurring. Coupling clarifier models with a biokinetic model gives the model the ability to predict these biological conversions, but it increases the computational load so the need for this additional capability should be assessed carefully.

5.3.4.3 Biokinetic models

An overview of seven published models is given in this section (Table 5.3.5). It should be noted that many more models and countless model extensions are available. Some of these are proprietary and some are available in specific simulators. These variations are not discussed here, but the selection below should not be interpreted as an endorsement by the authors of any model(s) in particular.

More information on the model representation (in the form of the Gujer Matrix) is given in Appendix B. Proofed Gujer Matrices in spreadsheet format are provided on the GMP WaterWiki website for the following models and even more information on a model evaluation study can be found in Appendix E as published in Hauduc *et al.* (2010):

- ASM1 (Henze *et al.* 1987; re-published as Henze *et al.* 2000a)
- Barker and Dold model (Barker & Dold, 1997)
- ASM2d (Henze *et al.* 1999; re-published as Henze *et al.* 2000c)
- ASM3 (Gujer *et al.* 1999; new version published as Gujer *et al.* 2000)
- ASM3 + EAWAG Bio-P module (Rieger *et al.* 2001)
- ASM2d + TUD (Meijer, 2004)
- UCTPHO+ (Hu *et al.* 2007)

Table 5.3.5 Overview of discussed biokinetic models.

Model characteristics	ASM1	ASM3	ASM2d	Barker & Dold	ASM3 + P	TUD	UCTPHO +
Year of publication	1986	1998	1995	1997	2001	2004	2007
Model type	**C/N**	**C/N**	**C/N/P**	**C/N/P**	**C/N/P**	**C/N/P**	**C/N/P**
Nitrification	✓	✓	✓	✓	✓	✓	✓
Denitrification	✓	✓	✓	✓	✓	✓	✓
Fermentation	–	–	✓	✓	–	✓	✓
Bio-P	–	–	✓	✓	✓	✓	✓
pH limitation [1]	Alk.[1)] warning	Alk. limit.	Alk. limit.	–	Alk. limit.	Alk. limit.	–
P precipitation [2] with Fe, Al	–	–	✓	–	–	–	–
# of processes	**8**	**12**	**21**	**36**	**23**	**22**	**35**
# of state variables	**13**	**13**	**19**	**19**	**17**	**18**	**16**
# of interacting processes/variables [3]	***31***	***72***	***136***	***153***	***148***	***154***	***169***
Total # of parameters	**26**	**46**	**74**	**81**	**83**	**98**	**66**
# Stoichiometric parameters							
Hydrolysis	–	1	1	2	1	1	–
OHO	1	4	1	5	4	2	3
ANO	1	1	1	2	1	2	2
PAO	–	–	3	8	5	12	7
Biomass general [4]	1	1	1	–	1	–	–
# Kinetic parameters							
Hydrolysis	3	2	6	4	2	6	–
OHO	6	13	12	9	13	12	13
ANO	5	6	6	5	7	6	4
PAO	–	–	18	11	21	26	14
Biomass general [4]	–	–	–	1	–	–	1
# of composition parameters	2	8	13	16	15	16	12
# of temperature adjustment coeff.	7	10	12	18	13	15	10

[1] Alkalinity is used to describe the pH limitation on biology
[2] P precipitation (with Fe and/or Al) is in many implementations added to the original model matrix.
[3] Number of non-empty cells in stoichiometry
[4] Parameters used for all biomass fractions
Source: adapted from Hauduc, 2010.

Selection of biokinetic model

When it comes to the question of which biokinetic model to select, the most quoted statement is "keep it as simple as possible to answer the question". This can be straightforward if the modelling objective is simple, but in practice the modeller typically has to deal with more complex systems and multiple questions. A "robust" engineering model requires that other criteria be considered and these other criteria may play a more important role in the decision than just selecting the model with the fewest processes and parameters. Some criteria are listed below:

- Processes: All processes that significantly affect the target variables under study have to be described. In addition, the model should be able to predict effluent concentrations for typical conditions. For instance, a WWTP is designed for C and N removal, but because of a load reduction, anaerobic conditions exist. A pure C/N model will not be able to predict the resulting bio-P processes. The challenge is to identify potential process conditions and to select a model, which covers these critical situations.
- Experience:
  - Consulting engineers typically do not have the time to learn about every model. Experience with a particular model behaviour is essential for "good modelling practice".
  - Are appropriate default parameters available for the model of choice? What other model information is available and what support is provided?
- Ease of use: The model has to follow rigorous scientific criteria but should be straightforward to understand and communicate.
- Availability in simulators
- Processing time

5.3.4.4 *Input models*

Input models convert measurements into model state variables or other variables (e.g. energy consumption) or constants (e.g. set-points) required by the model.

Influent model

Wastewater characterisation is used to fractionate influent measurements (typically for COD, N and P) into model state variables (Figure 5.3.2). The influent fractionation of the incoming wastewater depends on the biokinetic model being used, differs from plant to plant and has a significant impact on the model output. The choice of model used to fractionate the influent is a critical step in the modelling process.

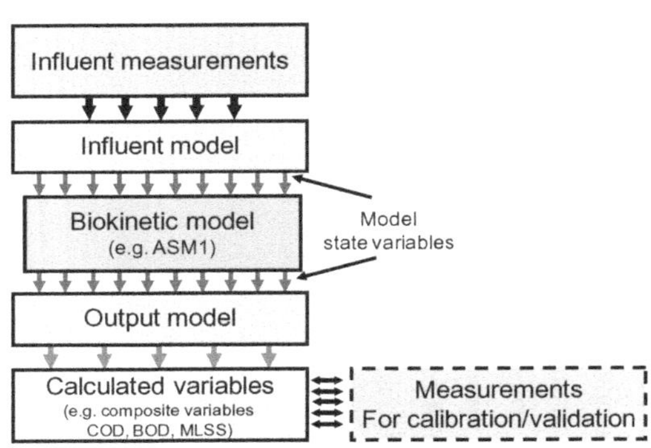

Figure 5.3.2 Translation of measurements into model state variables and back.

The concepts of influent fractionation for COD, nitrogen and phosphorus are discussed in Appendix A. (Table 5.2.2) provide a list with references of typical influent characterisation methods, which are applicable mostly for municipal raw influent and primary effluent. For influents containing a significant amount of industrial discharges some information is given in Chapter 8 *Use of Activated Sludge Models for Industrial Wastewater*.

FURTHER READING

Dochain D. and Vanrolleghem P. A. (2001). Dynamical Modelling and Estimation in Wastewater Treatment Processes. IWA Publishing, London, UK. ISBN 1-900222-50-7.

Gujer W. (2008). Systems Analysis for Water Technology. Springer, Berlin, Germany, ISBN: 978-3-540-77277-4.

Henze M., van Loosdrecht M. C. M., Ekama G. A. and Brdjanovic D. (2008). Biological Wastewater Treatment: Principles, Modelling and Design. IWA Publishing, London, UK, ISBN: 9781843391883.

Olsson G., Nielsen M. K., Yuan Z., Lynggaard-Jensen A. and Steyer J. P. (2005). Instrumentation, Control and Automation in Wastewater Systems. IWA Scientific and Technical Report No. 15, IWA Publishing, London, UK.

USEPA (U.S. Environmental Protection Agency, Office of Research and Development). (1989). Design Manual: Fine Pore Aeration Systems. EPA/625/1–89/023. U.S. E.P.A., Cincinnati, OH, USA.

WEF. (2006). Clarifier Design: WEF Manual of Practice No. FD-8. Water Environment Federation, Alexandria, VA, USA.

Chapter 5.4

Calibration and validation

5.4.1 INTRODUCTION

Mathematical models are simplified representations of a complex system and therefore do not reflect all of the processes that are on-going. However, models should describe the key processes of the real system. Simplifications in model structure, limited understanding of the real physical, chemical and biological processes, and limited knowledge of the influent and other variations may require that model parameters be adjusted to fit specific situations.

Model calibration can be described as an iterative adjustment of the model parameters until simulation results match an observed set of data. Calibration in this sense does not include any specific additional measurements or modifications of the plant model. If issues are detected with respect to data quality, availability or plant model set-up, the modeller must return to steps 2 (*Data collection and reconciliation*) and 3 (*Plant Model Set-up*). If the model is "calibrated" based on erroneous data, the predictive power of the model is compromised. Therefore going back to previous steps is preferred over modifying (e.g. biokinetic) parameters.

The quality of the simulation results is assessed for specific state or combined variables – the *target variables* – that have to be within a pre-defined error range of the measured values to stop the *calibration* procedure. The error ranges are defined beforehand during the Project Definition step but often need refinement after evaluating the data quality.

Validation is then performed using a number of tests also defined during the Project Definition step. An acceptable validation should ensure the use of the plant model with the level of confidence required to meet the modelling objectives.

Plant model parameters can be categorised as physical, operational, stoichiometric, or kinetic. In this report a plant model parameter is considered as:

(1) *original*, when its value is fixed to the one published in the original model publication;
(2) *default*, when it refers to a consensus value that can be used as a starting point in a calibration process. A plant model parameter set can be qualified as *default* if it has been validated on different wastewater treatment plants and has a clear explanation of how each proposed value was obtained. Preferably, experimental evidence is provided.

(3) *measured* when experiments were performed to determine its value, or
(4) *calibrated*, when its value has been changed during either a manual or an automatic calibration procedure.

The Application Matrix (Chapter 6) illustrates that the level of calibration and the required effort are directly linked to the objectives of the modelling project. In some cases (e.g. design, comparison of alternatives, etc.) a model might be able to meet the objectives using default values, but this is not always the case.

In addition to the aim of minimising the errors between measured data and model predictions, the calibration/validation step helps to establish the environmental conditions within which the model is robust and able to give reliable results, and can therefore serve as a basis for a decision (the domain of validity, according to Melcer *et al.* 2003).

5.4.2 PROCEDURE

The procedure proposed to calibrate and validate ASM models reflects actual practice and is based on the usage of these well-known models. Advanced methods developed by system analysis experts were intentionally excluded, as consensus on their use has not been reached. For information on such methods, the reader is referred to the "Further reading" section.

In short, the procedure consists of 5 main steps as outlined in Figure 5.4.1. It begins with the refinement of the stop criteria previously defined in the *Project Definition* step. The model is initially run with default or measured parameter values. Simulated data are then compared to measured data, and parameter values are adjusted if required. The obtained parameter set is validated during validation tests. Model prediction quality, uncertainties, and model limitations should be explained to the stakeholders and agreement obtained from them at the end of the procedure.

5.4.2.1 Model prediction quality

In manual or automated calibration procedures, the deviations between the model output and observations are quantified using performance criteria. The criteria are used either as quantitative values to stop the calibration process (in this case "objectives functions") or in the validation tests to assess the prediction quality of the model. Table 5.4.1 presents some functions that have been used in the field to characterise the model prediction quality.

Although no consensus on the standard methods has yet been achieved, progress is being made (Hauduc *et al.* 2011). A detailed description of the sources of uncertainty in wastewater treatment modelling can be found in Appendix H. The next phase of development for good modelling practice will be explicit uncertainty evaluations. Work towards this goal has already been initiated by the International Water Association (IWA)/Water Environment Federation (WEF) Task Group on Design and Operations Uncertainty (Belia *et al.* 2009; DOUT, 2011).

5.4.2.2 Refinement of the stop criteria and validation tests

The first task in the calibration step is to refine the criteria which dictate when the adjustment of parameters should stop. The stop criteria are adjusted on the basis of data quality and availability (e.g. number of samples and frequency). Typical stop criteria are provided in Table 6.5 in Chapter 6 *Application Matrix.*

Stop criteria may combine quantitative (minimum) values for the objective functions, but also other criteria such as:

- The maximum number of model runs;

- The minimum value of the change in the objective functions;
- The maximum value of the change in the parameters.

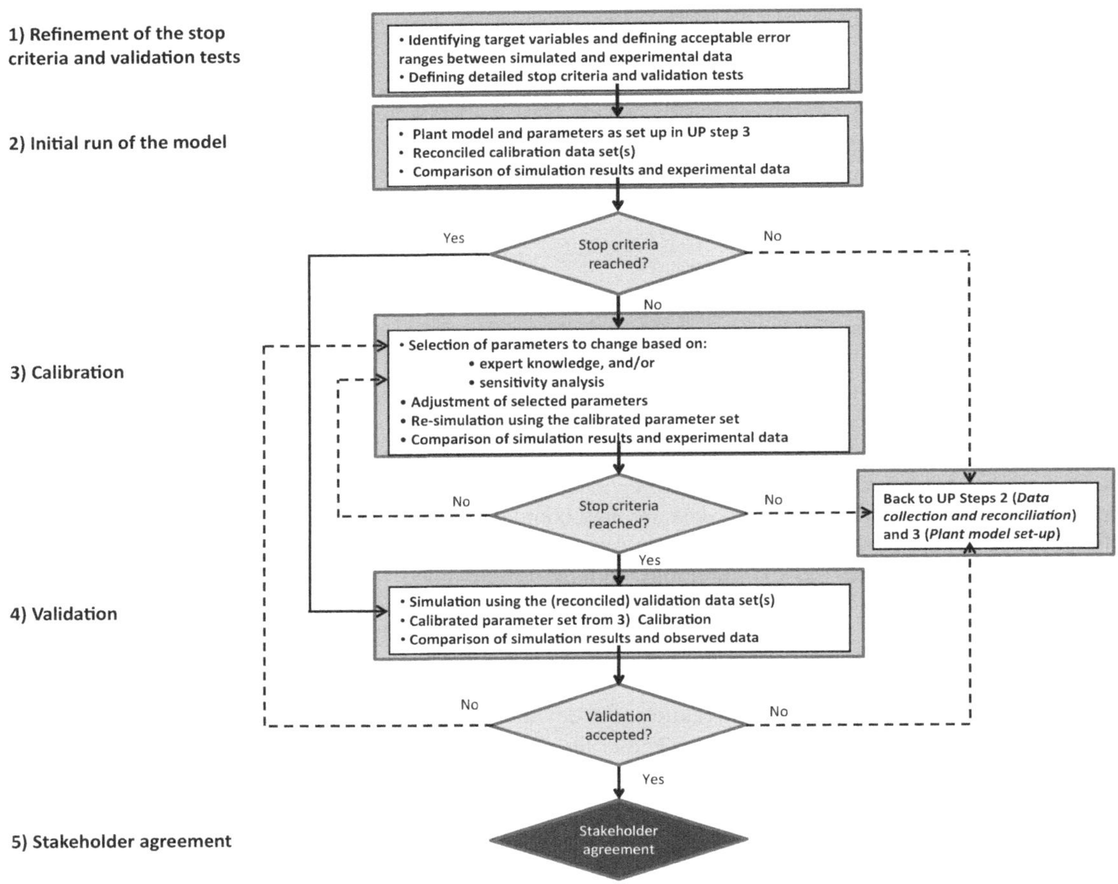

Figure 5.4.1 Procedure to calibrate and validate a plant model.

Validation tests are also refined at this stage. These tests may be based on engineering/expert knowledge, or dedicated runs can be performed using specific data sets (validation data sets). If the model passes the tests, then it is assumed that it can be used to meet the assigned modelling objectives. If not, re-calibration may be required or the model can be used with the understanding that its scope is limited.

5.4.2.3 Initial run of the model

Reconciled data were previously organised into two or more data sets: the calibration data set(s) and the validation data set(s). The initial run of the model was performed during the model set-up using the calibration data set, and this run typically provides the first outputs to be compared with the measured performance data. This run is the start of an iterative procedure to change parameters, settings and sometimes the model structure to achieve the calibration.

Table 5.4.1 Examples of criteria/objective functions used to quantify model prediction quality.

Index	Definition/formula	Significance
Residuals	$r_i = P_i - O_i$	Residuals (r_i, difference between observed O_i and predicted P_i values respectively at time step i) should be the lowest possible and no additional information should remain in the residuals (i.e. they should be random). The individual residuals can be graphically observed, and the maximum values can be compared.
Mean of residuals	$m_i = \frac{1}{n}\sum_{i=1}^{n} r_i$	The mean of residuals (m_i) allows highlighting the existence of systematic bias, that is a systematic over-predictive or under-predictive model.
Mean absolute error	$MAE = \frac{1}{n}\sum_{i=1}^{n} \lvert r_i \rvert$	The mean absolute error is an indicator of variability. This criterion should be the lowest possible.
Root mean squared error	$RMSE = \sqrt{\frac{1}{n}\sum_{i=1}^{n} (r_i)^2}$	As MAE, this criterion avoids error compensation and indicates the average magnitude of the errors. RMSE is the maximum likelihood estimator of the variance of a model prediction if the distribution is normal. However, it emphasises high errors. This criterion should be the lowest possible.
Janus coefficient	$J^2 = \frac{\frac{1}{n}\sum_{i=1}^{n} (P_{validation,i} - O_i)^2}{\frac{1}{n}\sum_{i=1}^{n} (P_{calibration,i} - O_i)^2}$	The Janus coefficient indicates the change in model accuracy between the calibration step and the validation step. A Janus coefficient of 1 means that the model has the same predictive performance in both steps. A high coefficient could indicate a change in model structure or that the model has been over-fitted and lost its robustness. However, this coefficient does not indicate a good predictive performance per se.

5.4.2.4 Calibration

When the target values are outside the specified range (i.e. outside the acceptable error range), double-checking the data and model set-up must first be undertaken. If no additional measurements or plant model changes are performed to confirm the results, selected parameters are modified manually or with the help of a numeric optimisation algorithm.

As schematised in Figure 5.4.2, the following general *calibration procedure* is proposed:

(1) Hydrodynamics: the *number of tanks* in series represents the hydrodynamics of the plant and is the first input parameter to consider modifying. If no hydraulic measurements are available, a peak in the effluent (i.e. ammonium peak) can be calibrated by changing the number of tanks in series. The *flow distribution* should also be checked. Hydrodynamics have indeed a high impact on the diurnal variations of the dissolved concentrations.

(2) *Influent wastewater or recycle stream characteristics* can be modified, to fit the simulated sludge production. Adjusting recycle streams characteristics is in effect establishing the correct SRT for the system. If accurate results are not available, modifying the plant model set-up may be considered at this stage (e.g. a different input model might change the output).

(3) Parameters for the *clarifier* model are site-specific and typically need adjusting when a simple point or ideal clarifier model is not sufficient to meet the project objectives.
(4) *Aeration* sub-model parameters can be modified, especially when dissolved oxygen measurements are not adequate (badly positioned sensors) or are inaccurate.
(5) Modifications in *biokinetic parameters* should be the last parameters considered, and should be changed only when the change is supported by a plausible bioprocess reason.

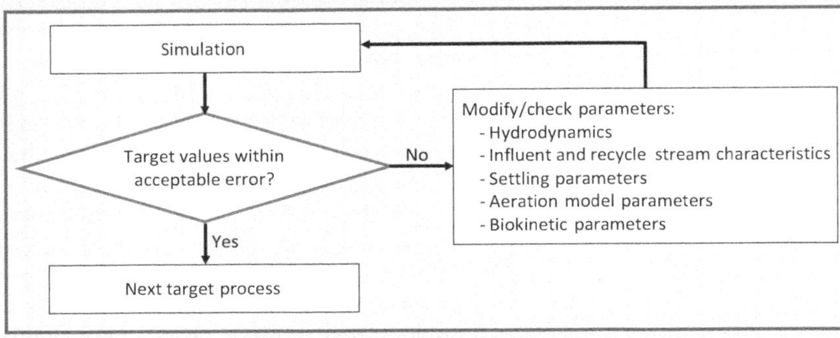

Figure 5.4.2 Iterative procedure to calibrate each target process.

Selection of calibration parameters

The choice of the parameters to be adjusted during calibration depends on different factors:

(1) Preferably, parameters should be measured (directly) or calculated from measurements (indirectly);
(2) Strong evidence should be obtained to support parameter changes from default or measured values. For example a large contribution from an industrial wastewater may necessitate parameter changes;
(3) The influence of a parameter on the simulation results has to be sufficiently significant to justify modifying the default values. This is generally verified with the help of a sensitivity analysis;
(4) Data quality has to be proven accurate enough to allow modifying parameter values on the basis of experimental results. Calibration is strongly linked to the *Data Collection and Reconciliation* step, and data may have to be reassessed when problems are encountered in calibrating the model.

In the case that the parameters of the biokinetic model have to be changed, those that have been measured should be set to the measured values (assuming that the measurement was achieved through an acceptable procedure). Other parameters may not have been measured but are known to be stable under varying conditions (e.g. the heterotrophic yield). Other parameters may be selected based on expert knowledge and an analysis of the model structure. Sensitivity analysis can be used to identify the parameters that have the strongest effect on the target variables. Only influential parameters should be modified and expert knowledge should be used to assure that the values of the parameters are still in a realistic range. Methods and examples to perform sensitivity analyses can be found in different references (e.g. Melcer *et al.* 2003; De Pauw & Vanrolleghem, 2006).

Procedure to calibrate a BNR plant

Figure 5.4.3 represents an iterative procedure that can be used to calibrate a biological nutrient removal (BNR) plant. Parameters subject to changes are proposed, although they may differ in specific applications. The procedure can also be used to calibrate less complex plants (e.g. nitrification-only) by omitting the steps for processes that are not relevant. Examples based on this procedure are developed in Chapter 7 (*Using the GMP Unified Protocol by Example*).

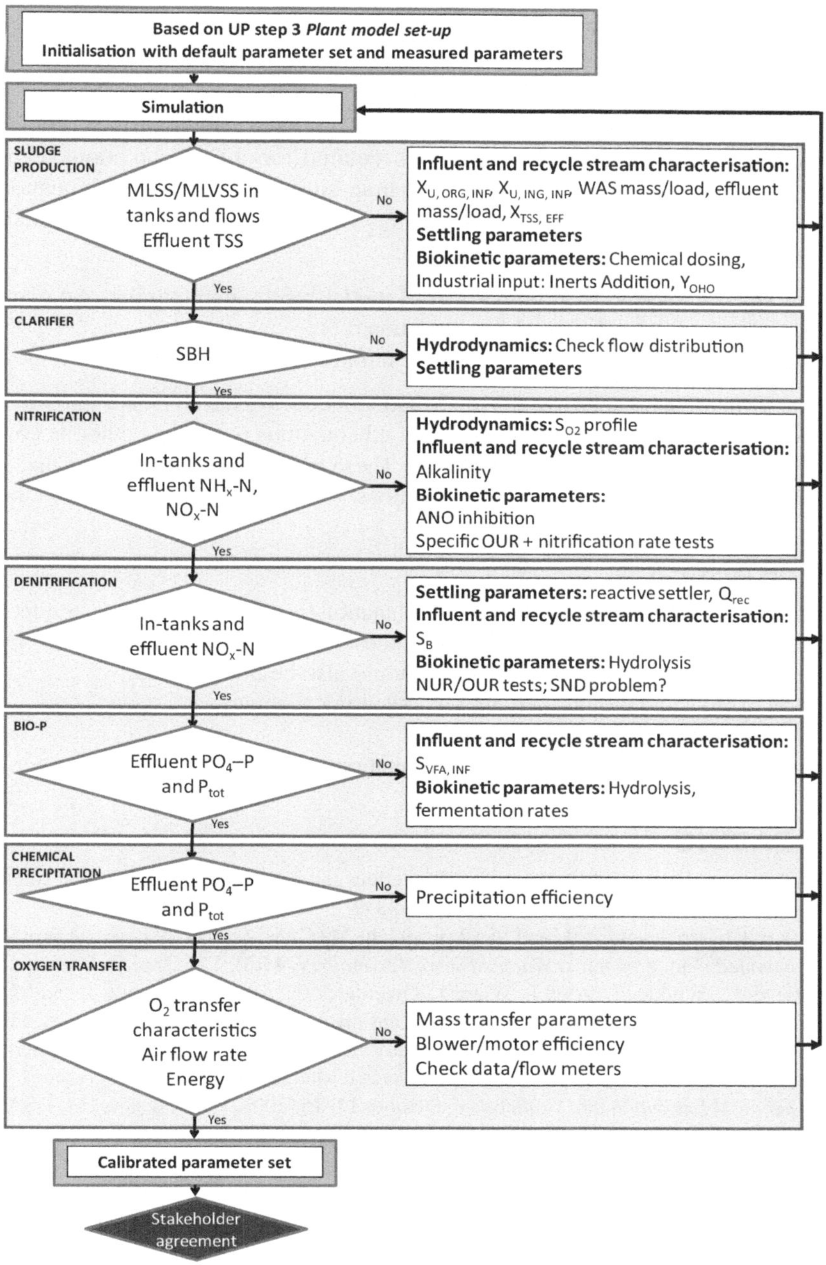

Figure 5.4.3 Proposed iterations to calibrate a biological nutrient removal (BNR) plant model.

The proposed approach aims at first calibrating the model to accurately predict the sludge production – and therefore the SRT of the system – then the parameters linked to nitrogen and phosphorus removal are adjusted. The oxygen transfer model parameters are the last ones to consider, though it might be that a calibrated model is required before the nitrification/denitrification calibration steps, in particular if no DO data are available or if they are not representative of all reactors.

5.4.2.5 Validation

Activated sludge models have been in use for years and general validation is provided by growing experience; however, specific validation tests are still required for critical conditions, depending on the objectives of the project (i.e. winter conditions, dynamic simulations…). These situations have to be specified and the selected validation data sets should resemble the conditions under which the model is to be used for predictions. Validation tests may include:

- Engineering checks, such as the comparison of model results with similar real plant data or with results from other methods (design charts, equations…);
- Validation runs performed with data sets for the defined specific/critical conditions.

In practice data are difficult and expensive to collect and to reconcile, which means that often the available data are dedicated to the calibration phase. In addition, all situations typically cannot be covered, (i.e. it is impossible to collect data for all possible conditions). These practical concerns mean that model validity is rarely proven and is only assessed for specific situations where validation is critical to the project.

5.4.3 DELIVERABLES

- A calibrated parameter set, including an explanation/justification when parameters have been changed from the default values or from the values assigned in the *Plant Model Set-up* step. The impact of those changes on modelling results should also be documented;
- A description of the validation tests (data sets, resulting accuracy);
- A measure of the model prediction accuracy;
- Known or potential uncertainties and model limitations.

FURTHER READING

Dochain D. and Vanrolleghem P. A. (2001). Dynamical Modelling and Estimation in Wastewater Treatment Processes. IWA Publishing, London, UK. ISBN 1-900222-50-7. p. 342.

Hulsbeek J. J. W., Kruit J., Roeleveld P. J. and van Loosdrecht M. C. M. (2002). A practical protocol for dynamic modelling of activated sludge systems. *Water Science Technology*, **45**(6), 127–136.

Langergraber G., Rieger L., Winkler S., Alex J., Wiese J., Owerdieck C., Ahnert M., Simon J. and Maurer M. (2004). A guideline for simulation studies of wastewater treatment plants. *Water Science Technology*, **50**(7), 131–138.

Vanrolleghem P. A., Insel G., Petersen B., Sin G., De Pauw D., Nopens I., Weijers S. and Gernaey K. (2003). A comprehensive model calibration procedure for activated sludge models. In: *Proceedings WEFTEC 2003, 76th Annual Technical Exhibition and Conference*. October 11–15, 2003, Los Angeles, CA, USA (on CD ROM).

Chapter 5.5

Simulation and result interpretation

5.5.1 INTRODUCTION

The final step in the Unified Protocol, *Simulation and Result Interpretation*, covers how the model can be used once it has been designed, assembled and calibrated. There is a strong link between this step and the first step *Project Definition*. In *Project Definition* the modeller and other stakeholders set out the scope and expectations of the model and in *Simulation and Result Interpretation* the model is used to fulfil the scope requirements, meet the majority of the expectations and provide justification if expectations cannot be met (e.g. due to unforeseen problems with the model or insufficient funding or time). All important aspects of the modelling, including the results and interpretation have to be documented in a report.

5.5.2 PROCEDURE

Presuming the previous steps have been carried out correctly, the model is ready, calibrated and can be used now with an expected level of confidence to answer questions laid out in the original objectives. This step typically consists of a series of steady-state or dynamic runs, termed "scenarios". These scenarios are performed by the modeller using the calibrated model and can be termed "scenario analysis". The number and scope of the scenarios are normally planned for in the *Project Definition* step, but calibration uncertainties and unexpected simulation results frequently necessitate additional scenarios. Preparing scenarios and performing the runs usually requires less effort than *Data Collection and Reconciliation*, however, the number of investigated scenarios can easily grow out of control if not properly defined upfront. A systematic procedure for carrying out this step is shown in Figure 5.5.1. The following sub-sections describe each of the procedure phases.

5.5.2.1 Define scenarios

Usually a significant number of simulations will be necessary to answer the questions and satisfy the objectives set in *Project Definition*. It is important to have a realistic and practical list of simulations defined before running and analysing scenarios. Running several conditions using several layouts, for example, can quickly multiply to a large number of scenarios that take a lot of time to run and analyse. The following is a checklist of items to take into consideration when planning a scenario analysis:

- *Steady-state vs. dynamic:* Is steady-state or dynamic simulation required to find the answer? For example, in design, it is customary to use models in steady-state mode to investigate sludge quantities, but calculating blower capacity for peak loading requires dynamic runs. Table 5.5.1 lists the type of model needed to describe the behaviour of a plant based on whether the influent and operation of the plant is constant, periodic or fully dynamic. Periodic behaviour refers to influent or operational inputs that follow a repeating but consistent pattern such as the operation of a sequencing batch reactor (SBR) or an intermittent aeration scheme.
- *Include all required inputs*: Are all the required inputs available? For instance if peak blower capacity is to be determined, is there reliable information for diurnal peak loading (flow and concentration peaks and their occurrence in time)?
- *Simulating control actions:* Is there a need to simulate control actions that will likely be undertaken manually by operators in response to changing conditions or through an automated control system? This aspect (both in steady-state and dynamic simulations) is frequently overlooked. Ignoring operator or controller intervention will provide unrealistic results. In a proper analysis of future loading conditions, even on a steady-state basis, it is misleading to simply increase the plant loading to the expected future values. The real plant, if maintained and operated correctly, will make corrective actions (e.g. increase wastage or recycles, install new blowers, etc.). Ignoring these in the simulation is equivalent to running the plant in the future with today's operating policy.
- *Recognise simulation time constraints:* How much time will it take to run each scenario and is the total simulation time required for all simulations realistic and within scope? In practice, most scenarios will have to be repeated with minor corrections or changes and this additional effort is frequently not planned for. The time it takes to run a scenario may vary from seconds to hours. It is possible to speed up numerical simulation by using relaxed solver settings (switching to a fixed steps solver or accepting higher error criteria), however this is not recommended as it may lead to inaccurate solutions (Appendix C). In extreme cases "fast" variables (those with the highest derivatives, such as dissolved oxygen, nitrite or hydrogen in the fermentation process) will show oscillations and the results (even for smooth looking variables) become meaningless. The use of faster computers with more processing power is an obvious consideration if there is a need to run complex models.
- *Select the right outputs:* Which outputs (e.g. state or calculated variables, process indicators at which time steps) are necessary to evaluate the simulation run? Due to practical constraints, none of the simulators will save all data generated during the simulation; therefore defining this list is of great importance. If a simulation is performed but the results are not saved or the wrong data are saved or with insufficient resolution, the model will have to be rerun, wasting time.

Steady-state scenarios

Steady-state simulations are usually quick to perform and provide useful results in a number of situations. These types of runs are based on average influent loads and fixed operational conditions and are normally used to predict "typical" or long term performance. There is little or no dynamic information gained using steady-state simulations. The main types of steady-state analysis are:

- *Yearly average performance* – annual averages
- *Monthly simulations* – typically maximum monthly loading conditions
- *Seasonal averages* – winter, spring and summer plant performance using monthly loading conditions.

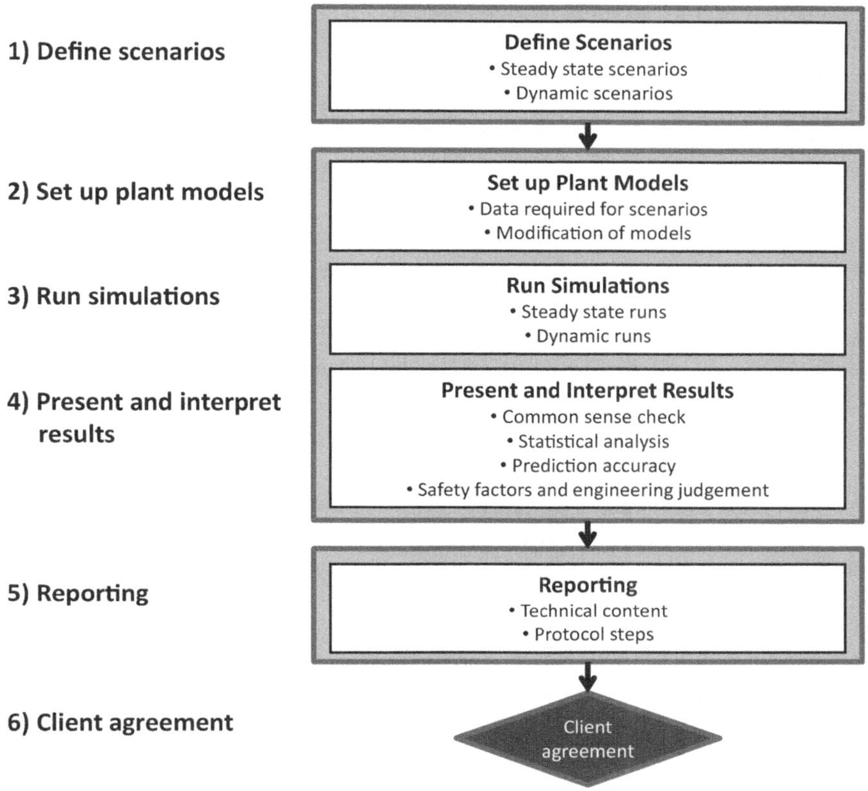

Figure 5.5.1 Unified Protocol flow diagram for Step 5: Simulation and Result Interpretation.

Table 5.5.1 Consideration of model type required to describe the plant based on influent and operation.

		Operation		
		Constant	**Periodic**	**Dynamic**
Influent	**Constant**	Steady-state	Cyclic	Dynamic
	Periodic	Cyclic	Cyclic	Dynamic
	Dynamic	Dynamic	Dynamic	Dynamic

Two common uses of steady-state scenarios are:

- *Overall mass balancing*. Steady-state simulations are very efficient at identifying potential data sampling or analytical errors. The model always maintains mass balances, so steady-state simulation results enable errors from laboratory or monitored data to be identified.
- *Long term performance check*. Steady-state can be used to estimate plant performance under various loading conditions, for example under current or future design load conditions.

Steady-state simulations normally should NOT be used for sizing equipment that operates under dynamic conditions (such as blowers) or equipment that's design and reliable operation is dependent on peak and minimum flow or loads.

Time periods shorter than a month may not be represented well in steady-state. A plant's behaviour is a function of the operating conditions and the sludge retention time. Even if the operating conditions of a plant are held constant, it could take upwards of a month for a typical plant to reach equilibrium after a change, so this is why a month of data is recommended.

Dynamic scenarios

Dynamic simulations consider the changing conditions that plants are normally subjected to, such as variation in influent flow or loading, temperature and operational conditions. The "raw" unprocessed simulation results consist of detailed information on the dynamic behaviour of the plant. Five typical uses for dynamic simulations are summarised briefly below. The true power of structured dynamic models lies in dynamic simulation, as these calculations cannot be repeated easily on the back of the envelope or even in spreadsheets.

- A common dynamic simulation is the *diurnal run* which contains one day (24 hours) of "typical dry weather" influent and operational data. The 24 hour period can be repeated as many times as necessary. Diurnal runs have to be used to describe cyclic processes. For example, if there is significant variation (a high peaking factor) in the influent loading throughout the day, or if the plant is inherently non-steady-state (i.e. SBRs, or plants containing intermittently aerated zones), a steady-state run is not feasible or may not provide useful results so dynamic runs are required.

  The results of dynamic simulations can be further processed to calculate averages, simple or more advanced statistics (min/max, distribution, etc.) and these can be tabulated against different running conditions to provide an easier overview for reporting. Calculating diurnal peak and minima conditions in a dynamic simulation are frequently used to determine equipment limits such as blower requirements.
- *Long term dynamic simulations* are frequently performed to investigate weekly, monthly or even seasonal effects. Typical tasks for this type of simulation involve:
  - The investigation of operational differences due to weekday and weekend loading conditions and operational control changes (e.g. no sludge wasting at weekends, or sludge liquors returned during certain shifts).
  - A monthly dynamic run can be compared to the steady-state monthly average conditions to investigate the impact of plant dynamics on the results.
  - The investigation of seasonal variation of the process. The length of the simulation in this case could be several months, a year, or longer.
- A "*birthday cake*" analysis is a useful dynamic analysis for design purposes. The simulation is run over several weeks and includes the design loadings that are typically used to size equipment, including annual average, maximum monthly average and peak day events. The flow input to the model resembles a birthday cake (Figure 5.5.2).
- Dynamic simulation provides a unique opportunity for investigating and planning for optimal management of *short term* events. These can include:
  - Wet weather periods or specific storms, for example bypass planning or preparation for, and proper timing of, step-feed operation
  - Taking equipment or reactor volume offline for maintenance or construction activities
  - Planning for equipment (e.g. pump, blower) failure

- *Controller design* and *control parameter tuning* are special cases of dynamic simulation. Both the plant model and relevant parts of the plant control system need to be implemented in the dynamic model. This may also be accompanied by more details such as sensor models, noise filtering and actuator response curves.

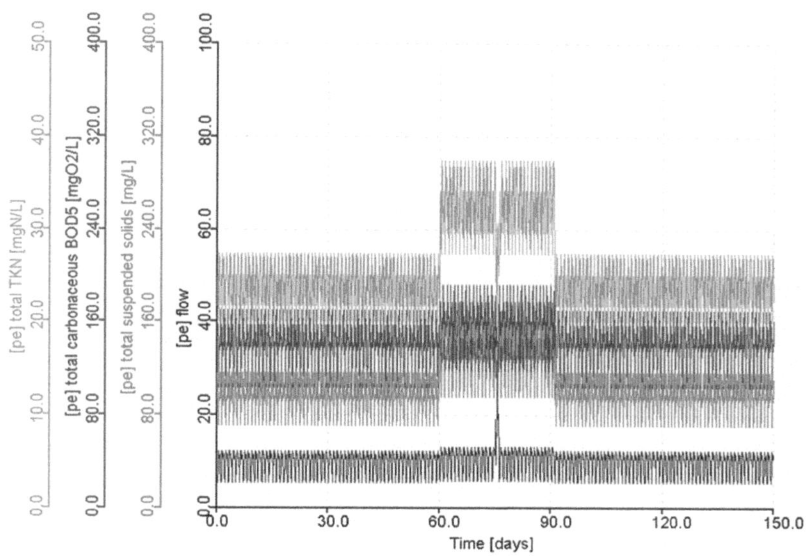

Figure 5.5.2 Example input data for a birthday cake analysis.

Dynamic simulations add another dimension (time) to the analysis and therefore require more details for their setup. Dynamic runs are slower to implement and take longer to run. In accordance with the principle of "keep it as simple as possible to answer a specific question", dynamic simulations should be used only when a process question cannot be answered by steady-state simulations, or where a dynamic simulation may give a more detailed understanding of the process. For dynamic process schemes (e.g. SBRs) or controlled processes, dynamic simulation is mandatory. Results for highly dynamic processes like bio-P can differ when simulated to steady-state or dynamically.

5.5.2.2 Set up plant models for scenarios

In preparing to run scenarios there are two things to consider: 1) preparation of the data required to run the simulations; and, 2) if necessary, modifying the plant model.

Data requirements

The data for steady-state scenarios are similar to the data used for calibration. Influent fractions are usually based on calibration data and may need to be adjusted if influent loads from industries and other significant changes are the subject of the scenario. Future loading conditions add uncertainty to the model simulations and this must be considered when analysing and interpreting the results.

Dynamic simulations require attention to different aspects such as the shapes of the diurnal flow and concentration profiles. These are site-specific and depend on several factors including the type of waste collected in the sewer system (domestic, commercial, industrial); the type of sewer (combined or separate, gravity or pumped), length and complexity of the sewer system; the size and location of the treatment facility; and the influence of plant recycles on the influent.

Diurnal flow data is usually available for existing facilities however diurnal concentration data is rarely available. Special sampling must be carried out to develop typical patterns which can then be superimposed on daily average loads. If diurnal variations are not available, it is sometimes possible to estimate a profile using an influent generator that takes the plant size into account (Langergraber *et al.* 2008; Gernaey *et al.* 2006). A spreadsheet implementation of the Langergraber *et al.* (2008) influent generator can be found on the GMP WaterWiki website: http://www.iwawaterwiki.org/xwiki/bin/view/Articles/GuidelinesforUsingActivatedSludgeModels.

Modifying the model for scenarios

Many scenarios require the plant model itself to be modified. These include:

(1) *Plant expansion* – how is the plant expected to perform if new tanks or new trains are added to the plant? This may require additional tank volumes or additional units. Hydraulics may change and the new tanks may have different depths and different aeration equipment which must be taken into consideration.

(2) *Plant upgrade* – how is the plant expected to perform if the plant is modified to meet more stringent effluent limits? This is a very common use of process models and particularly useful for comparing different process options. This may require significant modifications to the calibrated model including the addition of new process units or the construction of a completely new layout. In these extreme cases, the influent characteristics may be the only facet of the original, calibrated model carried forward to the new layout.

(3) *Plant optimisation* – how will operational changes or improved control affect the plant performance? The model can be used to investigate the impact of operational and/or control parameters on the plant performance. This may require the addition of more sophisticated controller models.

(4) *Maintenance and construction impacts* – how will the plant perform if tanks or equipment are taken out of service? This will require adjustment to the size of reactors and/or the maximum range for equipment may need to be reduced.

(5) *Extreme event assessment* – how will the plant respond to wet weather or other extreme events (such as a toxic spill)? These scenarios may require different influent data files. They may require modifications to operational or control responses to match the responses used in the plant. This type of scenario may also require modification of the model hydrodynamics.

5.5.2.3 Run simulations

Performing simulations has become significantly easier in recent years; however, modern user-friendly software does not alleviate the need to understand the basic functionality and the modelling concepts behind them.

Activated sludge models, from the mathematical point of view, are expressed as a set of ordinary differential equations (ODEs) that describe various rates of change. The concentrations of components (the state variables) cannot be directly calculated by solving this equation system. The equations determine the change in state variables over time (the derivatives) and these depend on the concentrations themselves, as well as on model parameters. Because the output of each equation is a state variable change, the initial values for the state variables can have a profound effect on the model output making the determination of those initial values a critical part of the scenario runs for both steady-state and dynamic simulations.

A brief introduction in the numerical engines is given in Appendix C.

Steady-state run

There are fundamentally two ways of getting steady-state results out of a model:

(1) *Running a model dynamically into steady-state*: An *initial guess* for the concentration of each state variable is established and a dynamic simulation started from that point but with *fixed inputs* (i.e. not varying with time). With each model step, the state variables are changed by the simulator according to the values obtained by the equations, "adding" the calculated change to the previously known (or estimated) concentration until this concentration becomes steady and the derivatives approach zero. The initial guess has a large effect on the trend and absolute values of the change in concentrations.

(2) *Using a steady-state solver*: The model is iteratively run with a set of inputs and operational variables (no changes in time), until the derivatives (the change) all become zero. This is a unique solution called a *steady-state*, and (ideally) it does not depend on initial guesses.

Most simulators provide one or several steady-state solvers. The reader is referred to the simulator manuals for more information. A brief introduction in solvers is given in Appendix C.

In certain cases the model may have two or more valid steady-state solutions for a given set of inputs, (e.g. there may be a nitrifying and a non-nitrifying solution to the set of equations). Care and common engineering sense must be used to achieve the most appropriate solution. Software techniques can be used to drive the simulation towards one solution. Example techniques to do this include:

- Choosing the initial conditions in the right operating range
- Providing a small "numerical" seed of active microorganisms
- Turning off inhibitions at early stages in the convergence process

Dynamic run

An important precursor to carrying out a dynamic run is establishing initial conditions. The initial state variable concentrations are not known (there are too many of them, dozens or hundreds even in the simplest models) and estimates for each need to be provided. These estimates can be prepared in several different ways. The following stepwise approach is recommended:

(a) Take a typical value for each state variable at the start of the simulation ($t = 0$). Unless the model implementation (software) is prepared with specific "initial conditions" or seeding, this method is not feasible for practical reasons. There are simply too many state variables. However, if possible, this approach can be used as a first step in conjunction with other methods in order to establish initial conditions more rapidly.

(b) Run a steady-state simulation before the dynamic one and use the results as initial conditions. This solution assumes that the plant was operating in a typical way (e.g. dry weather conditions, without upset) for an extended period just before the simulated time period. In some instances this step may be sufficient to establish initial conditions.

(c) An improvement on the conditions established in b) is to start from the steady-state solution, and follow this with several days or weeks of diurnal (or other time-base such as weekly) variation to bring the initial concentrations from daily averages to the correct concentrations for the start time (typically midnight). All model runs should be started at stable conditions in which subsequent daily variation patterns are consistently repeated as it is very hard to accurately recreate initial concentrations after a major upset (e.g. a large storm). The simulation should be run for several sludge retention times (typically at least 3 SRTs). The model can be run using a diurnal or weekly pattern for several weeks until similar output patterns are observed (pseudo steady-state). This approach is processor intensive and can take a long time to run. Alternatively, this step can be carried out without running a steady-state simulation first (i.e. skipping step b). For the case of SBRs and other alternating processes or when real time control is modelled a steady-state solution cannot be established and so step b) has to be skipped.

5.5.2.4 Present and interpret results

An important consideration in analysing and interpreting results is presenting them in a way that makes the analysis as clear and as simple as possible. Table 5.5.2 lists common methods for presenting results for steady-state and dynamic simulations. Example outputs are provided in Chapter 7.

Table 5.5.2 Methods for presenting results from steady-state and dynamic simulations.

Steady-state simulations	Dynamic simulations
Tables	Graph of time-based outputs
Bar graphs, line graphs and pie charts	Animated versions of bar graphs, line graphs and pie charts
Process flow diagram	Table of output data and summary statistics

Common sense check of results

Models are tools used by process engineers to make decisions, but models are unconstrained so quality control and quality assurance is essential. The modeller must check that the model results are sensible before attempting to draw any conclusions. Many pitfalls can be avoided using the checks and guidelines suggested below:

(1) *Tunnel vision*: Always look at the whole model. Looking at only a few variables may cause the modeller to miss variables that make no sense and make the model as a whole invalid. For example, a modeller may look at the MLSS and sludge wasted but neglect to check the effluent TSS, giving an incorrect picture of the true sludge production.

(2) *Nonsense values*: Check that all values (MLSS, OUR, effluent TSS, etc.) are within reasonable ranges. For example, problems might be identified if the model predicts an OUR greater than 100 mg/L/hr in an aerated reactor. Typical aeration systems cannot supply that kind of oxygen so assuming a fixed DO of 2.0 mg/L in that reactor might be unrealistic.

(3) *Sanity check*: Compare model results with similar real plant performance. For example, if a model predicts that enhanced biological phosphorus removal will not occur, but the plant is actually achieving biological phosphorus removal consistently then it is likely that there is an incorrect assumption somewhere in the model or erroneous data are used.

(4) *Alternative methods*: Compare model results with results from other methods such as standard design curves or design equations. For example, sludge production from the model should be consistent with values in design guidelines (e.g. the German ATV-DVWK-A 131E, 2000 or WEF MOP 8, 2010).

Statistical analysis

Statistical analyses of two general types can be carried out:

(1) Descriptive statistics for dynamic simulations
(2) Comparison of scenarios

The presented descriptive statistics for dynamic simulation results is limited to basic descriptors such as averages, minima, maxima and standard deviation or variance for outputs of interest. Table 5.5.3 provides general uses for common descriptive statistics (more information can be found in text books such as Montgomery & Runger, 2010).

Table 5.5.3 Uses for descriptive statistics from dynamic simulations.

Statistic	Uses
Average	Long-term performance, economic evaluations
Maximum or Peak	Short-term performance, equipment sizing
Minimum	Equipment sizing (required turn-down)
Standard Deviation or Variance	Process variability, process instability
Percentile	Variability, confidence in meeting limits

Scenario comparisons can be carried out for steady-state or dynamic simulations. Results from steady-state simulations can easily be compared to each other, providing basic statistics such as the ranking or best and worst in terms of different performance indicators (e.g. effluent concentrations). The significance of the differences between the outputs from the various scenarios may be expressed as absolute values, as a percentage of a baseline scenario or as a percentage of the original calibrated model. Comparison of the outputs from dynamic simulations is slightly more complicated but may be carried out by comparing the descriptive statistics for each scenario or by a visual comparison of time-series graphs of the outputs.

Table 5.5.4 lists performance indicators and other outputs commonly reported and compared using statistical analyses.

Table 5.5.4 Common performance indicators and other outputs.

Output	Purpose
Model effluent concentrations or loads	Overall process performance for target or licence limits
Waste sludge quantity	Design of, or impact on, sludge handling systems
Blower air flow or aeration system power	Required capacity or expected performance for blower or other aeration system

(*Continued*)

Table 5.5.4 Common performance indicators and other outputs (*Continued*).

Output	Purpose
OUR or air demand profile	Diffuser layout and aeration system control considerations
Nutrient profiles	"Health check" of conditions through the plant to ensure good performance (e.g. P-release and uptake in bio-P)
MLSS	Impact on clarifier
Sludge retention time (SRT)	Indicator of overall sludge retention time (SRT) that can be compared to critical SRTs for individual biomass populations (e.g. ensuring sufficient SRT for ammonia oxidising bacteria)

Prediction accuracy

The accuracy of the model predictions depends on many factors, not least of which is the quality of the plant model set-up and the model calibration and validation. Accuracies also depend on the variables of interest and the purpose of the model. The expected and acceptable accuracy for the model are established in the *Project Definition* stage (Section 5.1) but might have been refined in the *Data Collection and Reconciliation* (Section 5.2) and *Calibration and Validation* (Section 5.4) steps.

Safety factors and engineering judgment

Traditional design methods include safety factors in order to deal with uncertainties. Models do not contain safety factors and so good engineering judgement and practical operational experience must be used when interpreting the results from a process model. Detailed mechanistic models can replace some of the safety factors by describing the actual mechanism (e.g. peak loads) mechanistically, however the model predictions should not be trusted blindly. Established design and operational criteria rarely contradict modelling results if the model has been properly set up and the correct design and operational criteria are used.

5.5.2.5 Reporting

Technical documentation of this step should contain details about the scenarios that have been run, the results and the evaluation. Beyond communicating the results of the modelling project, reports and documentation should also contain any data and assumptions that are necessary to replicate the project and assess its quality. These documents are part of the deliverables agreed upon during the *Project Definition* step. They may comprise specific reports and documentation generated during each step, but may also include other documents, such as slides from presentations. The content of the main reports is discussed in the following section, as are typical pitfalls commonly encountered when reporting on modelling tasks.

Reporting on the main steps of the Unified Protocol

The structure of the final report should follow the main steps of the *GMP Unified Protocol*. The report should include an executive summary, and the last version of the *Project definition document* as an appendix. The final report encapsulates the outputs from all modelling steps, usually in summary form with the detailed reports from each step in appendices. Other important information (data files, models, input files…) should be made available in electronic format.

Table 5.5.5 lists the suggested content that should be included in the final report in order to properly document the model. The table includes suggestions for other deliverables to supplement the main body of the report. These may be included within the report, added as appendices or referenced as separate deliverables.

Table 5.5.5 Reports and documentation at each step of the Unified Protocol.

Unified Protocol step	Report content	Other deliverables
Project definition	Objectives of the project Requirements: staff, data, schedule, deliverables, budget Project definition document	
Data collection and reconciliation	Plant description Data analysis Mass balances Data reconciliation Proposed modifications to the Project definition document for Plant model set-up	Process flow diagram, P&I diagram, Raw and processed data for the main (lumped) variables Mass balances Data reconciliation: assumptions, methods Validated data files
Plant model set-up	Hydraulics Sub-models description Model simplifications/assumptions Proposed modifications to the Project definition document for Calibration and validation	Configuration information for the process units Input files (txt or other standard format) Base model
Calibration and validation	List of parameter values and comparison with defaults Domain of validity Uncertainty of calibration results Model prediction accuracy Proposed modifications to the Project definition document for Simulation and result interpretation	Calibrated results vs. measured data for the considered (lumped) variables Idem for validation data Input files Calibrated and validated model(s)
Simulation and result interpretation	Description of the scenarios Uncertainty analysis Last version of the project definition document Final report – Executive summary	Input files Output files Graphs required to explain the results Final version of the model, plus variants from different scenarios

Presentation of the modelling results themselves can differ depending on the project objectives as well as the expected accuracy. For example, relative differences may be more adequate than absolute values when comparing several design alternatives.

In addition, commercial simulators can provide customisable reporting features that may be useful to report relevant data (to define the process units, model/simulator set-up data and assumptions).

5.5.2.6 Client agreement

A final, but important modelling step is to obtain an agreement with the client that the objectives have been met. If the modeller has focussed on the objectives set out in the *Project Definition* and presented data in a clear manner, then this should be a straightforward process.

In some cases, particularly if the model is being used to make critical decisions it is good practice to have the model audited by a third party. For models used to predict future conditions, it is also good practice to re-evaluate model predictions with actual plant data as new information becomes available.

5.5.3 TYPICAL PITFALLS

The following is a list of common pitfalls that occur while carrying out simulations, interpreting results and reporting:

(1) *Incomplete or inadequate reporting*: The reproducibility of a modelling project is contingent upon providing complete reports containing enough documentation and the associated files. Failure to document this information makes it difficult to evaluate the quality of the model.
(2) *Unstated assumptions*: All assumptions should be stated clearly in the report.
(3) *No sanity checks*: Overconfidence in the modelling results can lead the modeller to overlook common sense and engineering judgement.
(4) *Incomplete or missing data files*: Validated data files and input files are not always provided. These files are particularly important if the model is to be re-used after the current project is completed.
(5) *Poor communication or inappropriate reporting style*: Technical results should be understandable and useable by the stakeholders. It is a good practice to provide detailed technical information supporting the main findings in appendices rather than in the main body of the final report. This will provide clarity in the main text.

5.5.4 DELIVERABLES

The main deliverable for this step is the *Final Report*. In addition, the following materials should be provided to supplement the *Final Report*:

- Input files
- Output files
- Summary data tables of key parameters including:
  - All parameters that were changed from defaults, and justification for changing them.
  - All parameters changed in scenarios, and the reason for changing them.
- Graphs required to explain the results
- Final version(s) of the model, plus variants from different scenarios

It is important that the content of the final report be checked against the requirements of the *Project Definition Document*. It is also important for all stakeholders to review and comment on the report to ensure that it meets expectations. It is recommended that modelling results be presented in a workshop setting or formal meeting prior to assembling a draft final report. The draft report should be reviewed and revised before the final submission.

FURTHER READING

WEF MOP 31 (2009). An Introduction to Process Modeling for Designers. Manual of Practice No. 31. Water Environment Federation, Alexandria, VA, USA.

WEF MOP 8 (2010). Design of Municipal Wastewater Treatment Plants. Manual of Practice No. 8. Water Environment Federation, Alexandria, VA, USA.

Chapter 6

The GMP Application Matrix

Short summary

This chapter describes a proposed tool – the *GMP Application Matrix* – meant to help modellers assess the level of effort required to carry out a modelling project depending on the particular modelling objective.

The Application Matrix includes 12 typical municipal modelling examples as well as 2 dealing with modelling industrial wastewater. Modellers can use these examples as a starting point for considering the level of effort required for their particular modelling project objectives.

6.1 INTRODUCTION

Carrying out all 5 steps of the Unified Protocol is not always required. For example, collecting existing data for a design project may be irrelevant for a greenfield site. The percentage of total project time allocated to the 5 steps in practice has been estimated by the results from a recent survey (Hauduc *et al.* 2009). The time allocation is shown graphically in Figure 6.1.

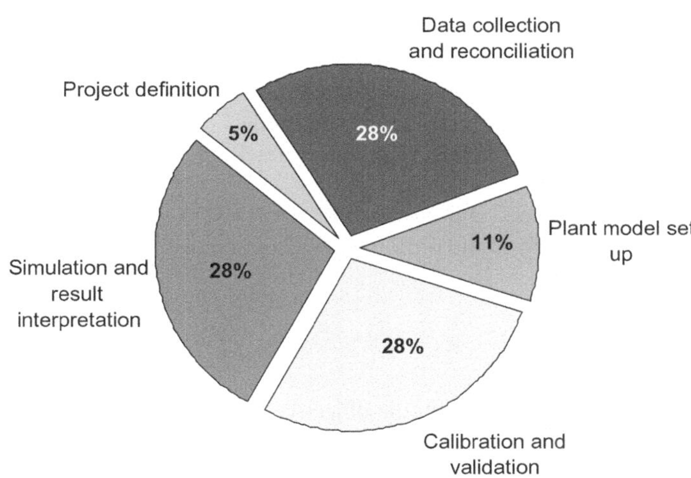

Figure 6.1 Time allocation for the 5 protocol steps (from Hauduc *et al.* 2009).

The required level of effort associated with these steps depends on the objectives of the modelling project, and the Application Matrix has been designed to help estimate that effort.

Process models can be used for many different purposes. In developing the Unified Protocol it was recognised that it would be difficult to formulate a rigorous procedure that would fit all possible modelling purposes. In order to provide guidance on how the level of effort required for each step of the protocol varies depending on the type of modelling project, the concept of the Application Matrix was developed.

The Application Matrix consists of a list of 14 typical modelling applications and assigns a subjective score for each of the Unified Protocol steps. This subjective score indicates the relative rigour with which a given step should be carried out for the particular application. A weighting is then applied to each step to account for the fact that each step, in general, requires a different level of effort. The 14 examples and the scoring system for the Application Matrix are described in more detail below. Throughout this chapter, the term "level of effort" is intended to be an indication of the relative amount of time required to carry out the modelling project and therefore can also be considered as an indication of the required costs.

The Application Matrix as an MS Excel spreadsheet is provided on the IWA WaterWiki website.

6.2 EXAMPLE APPLICATIONS

Fourteen (14) typical modelling objectives were selected for the Application Matrix. The objectives were selected as representative examples of the most common purposes for modelling; however, it is not an exhaustive list. For objectives not included here, the user can select an objective that is similar and adjust the scoring accordingly.

The plant envisaged as the subject for the applications is a typical medium-sized treatment facility for a small city (100,000 Population Equivalents) in a developed country that consists of a nitrifying activated sludge plant with fine bubble diffusers, and secondary settling tanks, and designed within local design standards. The service area of the facility is primarily domestic, with no major industrial dischargers but several commercial facilities (offices and restaurants). The sewer system is mostly for sanitary waste (separate sewer system), but it is assumed that there are a few connections that capture storm water and many that have leaks into the system. The influent characteristics are typical for a system of this type. The modelling project has a modest budget that includes up to 200 hours of labour, up to 15,000 EUR for special sampling and will be carried out by a modeller with sufficient expertise to do it using a commercial simulator which contains all the required sub-models (i.e. no specialised training, customised model development or third-party input required). The modelling project is required to be completed within 3 months.

The modelling objectives are grouped under 4 headings: (A) Design, (B) Operation, (C) Training and (D) Industrial. A brief description of each of the objectives is given below.

(A) Design examples

(1) Calculate sludge production

In this application, a model is constructed with the objective to estimate the quantity of sludge produced. Information from the model will be used to size equipment to handle the sludge. For this purpose monthly average quantities will need to be determined using the model. A steady-state model can be used for this objective.

(2) Design aeration system

This model is needed to design a new aeration system for the activated sludge process to replace an aging existing system. The model is used to determine the average and peak hourly air flows to size blowers. The model is also used to determine the distribution of air (i.e. the taper) needed through the aeration basins to select the correct number and layout for diffusers. A dynamic model is required for this objective.

(3) Develop a process configuration for nitrogen removal

In this application, a number (up to three) models are constructed to determine which configuration gives the best nitrogen removal under different design load conditions. It is assumed that the models are developed for an existing plant that currently provides full nitrification, but no denitrification, and so it is being upgraded to improve existing nitrogen removal. Nitrogen removal is assessed based on monthly average values with a target effluent limit of 10 mg/L total nitrogen. This assessment can be carried out using steady-state models.

(4) Develop a process configuration for phosphorus removal

In this application, a number (up to three) models are constructed to determine which configuration gives the best phosphorus removal under different design load conditions. It is assumed that the models are developed for an existing plant that is currently nitrifying only, and is being upgraded to provide biological phosphorus removal. Phosphorus removal is assessed based on monthly average values with a target effluent limit of 1 mg/L total phosphorus. The different configurations can be evaluated using steady-state models.

(5) Assess plant capacity for nitrogen removal

In this application, a model of an existing nitrogen removal plant is constructed i.e., the plant has already been upgraded from the base case to meet an effluent total nitrogen concentration of 10 mg/L. The model is used to determine the capacity of the plant to remove nitrogen under increasing load conditions and tightening effluent nitrogen requirements to 8 mg/L total nitrogen. Capacity is assessed based on monthly average values. The capacity is usually expressed as the maximum flow that can be treated by the plant to meet future flow and effluent requirements. A steady-state simulation is adequate for this application though a dynamic simulation would provide additional information and increase the confidence in the capacity rating.

(6) Design a treatment system to meet peak effluent nitrogen limits

In this application, a number of models (up to three) are constructed to determine which configuration gives the best nitrogen removal under different design load conditions. It is assumed that the models are developed for an existing plant that currently provides full nitrification, but no denitrification and is being upgraded to improve existing nitrogen removal. Nitrogen removal is assessed based on instantaneous values of 1 mg/L for ammonia (measured continuously using an online analyser) and monthly average values for total nitrogen of 10 mg/L, and so greater confidence is required in model predictions. This assessment must be carried out using a dynamic model.

(B) Operation examples

(7) Optimise aeration control

A model is developed for the existing nitrifying activated sludge plant with basic DO control. The model is used to adjust or modify the aeration control system to reduce energy costs whilst ensuring that existing effluent quality is maintained. A dynamic model must be developed for this application.

(8) Test effect of taking tanks out of service

In this application, a model is developed for a nutrient removal plant to test the effect of taking tanks out of service for maintenance activities. The model is used to test if the change will cause the plant to fail its effluent requirements (2 mg/L for ammonia, 10 mg/L for total nitrogen and 1 mg/L total phosphorus) all of which are calculated on a monthly average basis. A dynamic model will be used for this application.

(9) Use model to develop sludge wastage strategy

A model is developed for a biological phosphorus removal facility to investigate the impact of different sludge wastage strategies on the downstream solids handling system. Different waste frequencies, quantities and control strategies are tested. The investigation will look at the stability of the activated sludge treatment system, the quantities of sludge produced, the scheduling of wastage and dewatering, and the potential impact on sludge treatment processes. The plant effluent limits are 2 mg/L for ammonia and 1 mg/L for total phosphorus calculated on a monthly average basis. This particular example does not directly look at sludge treatment and return liquor impacts, which would require a whole plant model. In order to properly assess different wastage frequencies, a dynamic model is required.

(10) Develop a strategy to handle storm flows

A model of an existing plant is constructed and it is subjected to stormflow conditions. A number (up to three) different control strategies are tested to determine which strategy provides the plant with the best opportunity to cope with storm conditions. The focus of the investigation is two-fold: 1) what control strategy provides the best process robustness to mitigate the effects of storm conditions (e.g. peak effluent concentrations are reduced); and, 2) how quickly does the plant recover from the effects of the storm. This application requires dynamic simulations. The plant has an instantaneous effluent ammonia limit of 1 mg/L, and monthly average limits for total nitrogen and total phosphorus of 10 mg/L and 1 mg/L, respectively.

(C) Training examples

(11) Develop a general model for process understanding

A generic model of an activated sludge plant is constructed to increase process understanding of biological phosphorus and nitrogen removal. The model is loosely based on an existing plant that currently does nitrification but will be upgraded to nitrogen and phosphorus removal. The model will be used for a 3-hour training session and the level of effort includes time to produce presentation materials. A simple dynamic model is used for this.

(12) Develop a site specific model for operator training

A model of an existing activated sludge plant is constructed. The plant has recently been upgraded from a simple nitrifying plant to a full nitrogen and phosphorus removal facility with effluent limits of 10 mg/L total nitrogen and 1 mg/L total phosphorus based on monthly averages. The model is used to train operators, engineers and/or other personnel on the impact of making process changes to their specific plant. The model and presentation material are used to provide a total of 3 days of training including some interactive "hands-on" work and model demonstration runs. A dynamic model is used for this training.

(D) Industrial examples

(13) Develop a process configuration for nitrogen removal treating waste from a food production factory (soy sauce)

An existing factory consists of the full process of soy sauce production including soy bean processing, brewing and bottling. Current volumetric loading for the liquid waste is 500 m^3/d and expected to be increased up to 700 m^3/d within 5 years. The influent BOD_5 concentration is approximately 1200 mg/L and TKN is unknown. The existing treatment plant aeration tank total volume is 1000 m^3, which consists of 3 tanks in series. A post coagulation process is installed for phosphorus removal. Rising sludge is occasionally observed in the secondary settler. In this application a number (up to three) models are constructed to determine which configuration gives the best nitrogen removal under current and future design load conditions. It is assumed the models are developed for an existing plant that currently provides full nitrification, but no denitrification and is being upgraded to improve nitrogen removal. Based on a preliminary site audit, intermittent aeration is proposed as one option. Nitrogen removal is assessed based on daily grab sample values with a target effluent limit of 10 mg/L for the sum of nitrate and nitrite nitrogen. This assessment needs to be carried out using dynamic models.

(14) Assess acceptability of new influent at a petrochemical site

A petrochemical complex has a wastewater treatment plant with a reactor volume of 10,000 m^3, the average influent flow is 10,000 m^3/d and the average COD concentration 1000 mg/L. The wastewater collection system is composed of three major streams. A new plant installation is scheduled within 2 years which will increase the influent by 2000 m^3/d. An effluent sample can be obtained from a pilot plant of the new facility. In this application, a model for the existing industrial plant is constructed to evaluate the impact of 20% more influent flow and load. Treated effluent COD, sludge production, aeration capacity and inhibitory effects are the main concerns. This assessment can be evaluated using steady-state models.

6.3 MATRIX SCORING SYSTEM

For each example application the relative level of effort is scored using a scale of 0 to 5 for each of the model development steps of the Unified Protocol. The score is an indication of the relative level of effort required to carry out a particular step, with 5 meaning "*maximum effort*" and 0 meaning "*no effort*" for that particular step. Table 6.1 provides a general guide to this scoring system.

For the *Data Collection and Reconciliation* and *Calibration and Validation* steps the "Tiered Approach" suggested in the WERF Guidelines (Table 6.2; Melcer *et al.* 2003) were adopted.

It should be noted that this approach links data collection with calibration effort. If a better calibration is needed then more and better data are needed. This means that the effort required for *Data Collection and Reconciliation* is driven by the effort required in calibration.

Table 6.1 Scoring system for the application matrix.

Effort	Score	Notes
No effort	0	Step does not have to be carried out and can be skipped
Minimal effort	1	Step needs to be considered but does not require great effort. Can potentially be carried out by an individual with little or no consultation.
Below average effort	2	Requires some effort and 1 or 2 key points described in the Unified Protocol may require more astute attention.
Average effort	3	Medium level of effort for this step. Several tasks described in the Unified Protocol will have to be carried out.
Significant effort	4	More effort than average with almost all of the details of the Unified Protocol relevant, minus 1 or 2 less critical ones.
Maximum effort	5	This is a critical step and all of the details for it described in the Unified Protocol should be considered. Will require considerable consultation and collaboration with all involved.

Table 6.2 The "tiered approach" to calibrate models.

WERF Designation	Suggested Score for Application Matrix	Notes
Level 1: Defaults & assumptions only	0 – 1	Generally only applicable to greenfield sites or applications where no data is available
Level 2: Historical data only	2 – 3	Use is made of available historical data, no extra sampling is done
Level 3: On-site, full-scale testing	3 – 4	Additional sampling and testing is carried out to supplement historical data
Level 4: Direct parameter measurements	5	Specialised bench-scale testing is carried out to determine rates and/or waste characteristics

Source: Melcer *et al.* 2003.

Weightings

The different steps require different levels of effort even if individual steps are carried out to their fullest possible extent, and therefore weightings are applied to each of the raw scores to give an overall weighted "level of effort."

The Task Group sent out two questionnaires that included questions on the relative level of effort that practitioners spent on each of the modelling steps. The results are shown in Table 6.3. The Task Group agreed on its own weightings that also are shown in Table 6.3. The Task Group weightings are similar to those obtained from the two questionnaires except that the Task Group places a greater emphasis on the *Data Collection and Reconciliation* step. Through the discussions to produce the Unified Protocol it became evident to the Task Group that this step has a level of importance requiring more effort than some practitioners may appreciate. Significant effort during calibration might be better spent on data analysis.

Table 6.3 Weighting the different steps of the Unified Protocol.

Protocol step	Weighting		
	Questionnaire 1	Questionnaire 2	Task Group Selection
Project Definition	5	8	**10**
Data Collection and Reconciliation	28	28	**40**
Plant Model Set-Up	11	12	**10**
Calibration and Validation	28	28	**15**
Simulation & Results Interpretation	28	24	**25**

6.4 THE APPLICATION MATRIX

Table 6.4 shows the Application Matrix with the scores. Figure 6.2 is a graphical representation of the results. A number of observations can be made from the Application Matrix scores:

Table 6.4 Scores of the example applications selected in the application matrix.

		Relative Level of Effort Scoring for Each Protocol Step (0 - 5 per step)					
	Application	Project Definition	Data collection & Reconciliation	Model Set-Up	Calibration & Validation	Simulation & Results Interpretation	Total effort spent (Weighted)
	Weightings	10	40	10	15	25	
	DESIGN						
1	Calculate Sludge Production	1	3	2	3	1	44
2	Design Aeration System	2	3	3	2	3	55
3	Develop a Process Configuration for Nitrogen Removal	3	4	3	4	4	76
4	Develop a Process Configuration for Phosphorus Removal	4	4	3	4	4	78
5	Assess Plant Capacity for Nitrogen Removal	4	5	3	5	4	89
6	Design a Treatment System to Meet Peak Effluent Nitrogen Limits	5	5	5	5	5	100
	OPERATION						
7	Optimise Aeration Control	3	4	4	4	4	78
8	Test Effect of Taking Tanks Out of Service	2	3	3	3	3	58
9	Use Model to Develop Sludge Wastage Strategy	3	3	3	3	3	60
10	Develop a Strategy to Handle Storm Flows	5	4	4	4	4	82
	TRAINING						
11	Develop a General Model for Process Understanding	2	1	1	0	2	24
12	Develop a Site Specific Model for Operator Training	3	4	4	4	4	78
	INDUSTRIAL						
13	Develop Nitrogen Removal for Food Production Waste	3	4	3	4	4	76
14	Assess Acceptability of New Influent at a Petrochemical Site	4	5	4	4	3	83

- Simple modelling applications require less overall effort.
- More complex problems require more effort.
- Projects where inaccurate results have more serious consequences require more effort (e.g. meeting peak nitrogen limits or handling storm flows).
- Data collection and reconciliation is a significant effort for almost all applications (the generic model for training is the one exception). This observation is heavily influenced by the high weighting given to the Data Collection and Reconciliation step.

In the development of the Application Matrix, different weightings and scores were tested which gave different overall scores, however the relative order of effort for the different tasks remained roughly the same for example, designing a plant to meet a peak effluent nitrogen limit always had the highest level of effort and the generic model for training had the lowest level. This means that the matrix scores give a reasonable indication of the level of effort that should be expected for a given modelling project and the rigour with which the protocol steps should be undertaken, depending on the type of application. It is not possible to give specific estimates for labour and costs in this STR as these are project and location specific.

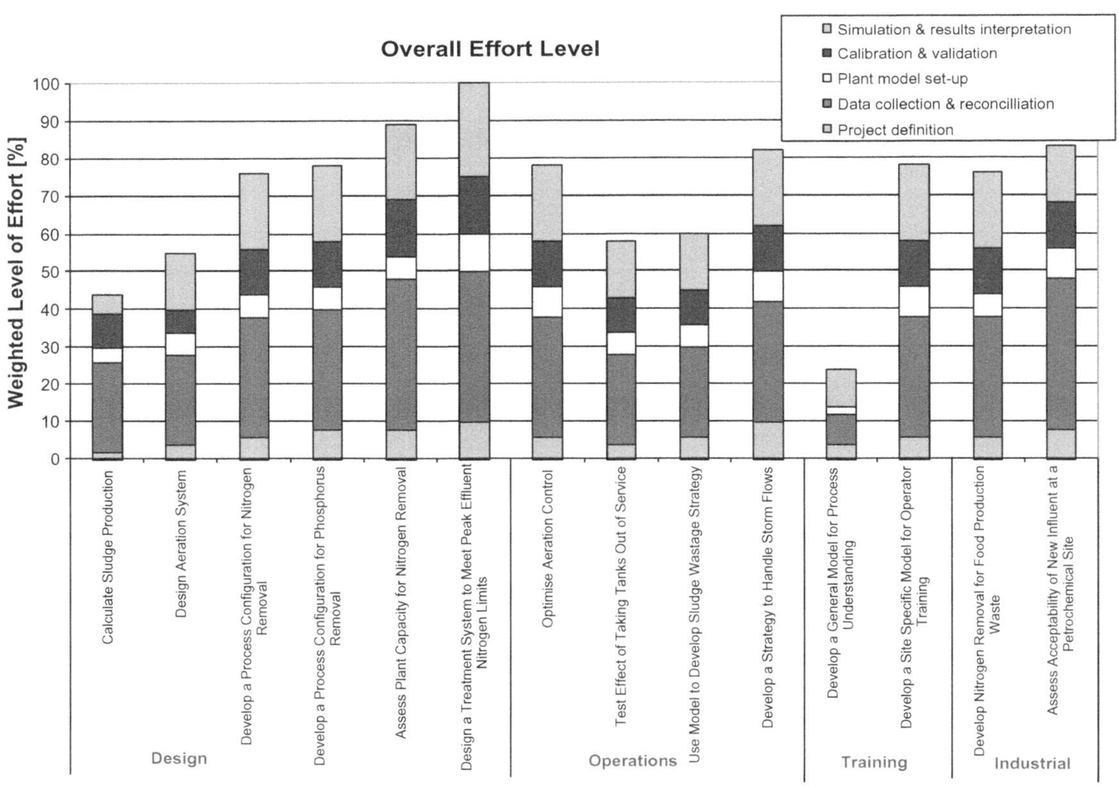

Figure 6.2 Overall effort level for selected examples.

6.5 OTHER IMPORTANT CONSIDERATIONS

It was not possible to include all factors and examples in the Application Matrix so a discussion of other factors that should be taken into consideration when developing estimates for project time and costs is included here.

- *Averaging time basis*: the shorter the averaging period for the effluent limits, the more attention to detail is required for example, limits based on daily or even hourly maximums require more focus than those based on monthly or annual averages.
- *Plant size*: larger plants usually have more complicated process problems to solve (i.e. multiple trains, sludge handling, stricter effluent limits) and these may require more time to investigate. Large plants might also benefit from smaller incremental improvements in operations so more accurate modelling for optimisation may be more cost effective at larger facilities. However, it should be noted that smaller plants generally have higher capital and operating costs per capita so good modelling can be beneficial in optimising design and reducing operating costs.
- *Plant complexity*: a more complex plant (e.g. different treatment trains with different configurations at the same facility) will require more effort.
- *Effluent limits*: tighter effluent limits require greater confidence in the results. Hence, greater effort is required, particularly with respect to ensuring data quality and carrying out the calibration. Consideration of the effluent limits also influences the need to carry out dynamic modelling versus relying on steady-state simulations.
- *Budget constraints*: the requirements of the modelling project and the scope of what can be done may be limited by the availability of funds.
- *Time constraints*: similar to budget constraints. There may be time constraints driving the project for example, the modelling project may be the first task in a fast-track design project, or the results of the modelling project may be required to determine the best course of action for a plant that is in danger of failing its effluent limits.
- *Personnel*: the level of detail and type of modelling that can be done may be limited by the lack of available personnel with the expertise to carry it out. Additional training which increases the cost and time required for the project may be required.
- *Client level of interest*: if a client has an appreciation for the benefits of modelling work, they can more-easily justify the budget, time and personnel required.

6.6 GUIDANCE BASED ON THE APPLICATION MATRIX

The following sub-sections provide some supplemental guidance and tables of information for certain aspects of the Unified Protocol using the Application Matrix examples.

6.6.1 Stop criteria for calibration

The stop criteria associated with the examples of the *Application Matrix* are presented in Table 6.5. The values included in this table were a result of expert opinions representing the collective expertise of the GMP core group. The IWA/WEF Task Group on Design and Operations Uncertainty (DOUT, 2011) is working specifically on uncertainties in wastewater treatment modelling and intends to develop a detailed, rigorous and quantitative approach to the stop criteria and this should refine the values further.

Table 6.5 Target variables and error ranges proposed for the application matrix.

	Example	Influent	Simulation	Averaging period	Target variable	Acceptable error range (±)
1	Calculate sludge production	Steady-state	Steady-state	Monthly average	MLSS MLVSS/MLSS WAS mass load Effluent TSS SRT	10% 5% 5% 5 mg/L 1 d or 15% for SRT < 5 days
2	Design aeration system	Dynamic	Dynamic	Daily average Hourly Peaks	Air flow rate DO (profile) OUR	10% 0.3 mg/L 10 mg/L/h
3	Develop a process configuration for nitrogen removal	Steady-state	Steady-state	Monthly or annual average	NH_x-N NO_x-N N_{tot}	1.0 mg/L
4	Develop a process configuration for phosphorus removal	Steady-state	Steady-state	Monthly or annual average	PO_4-P P_{tot}	0.5 mg/L
5	Assess plant capacity for nitrogen removal	Dynamic	Dynamic	Monthly or annual average	NH_x-N NO_x-N N_{tot}	1.0 mg/L
6	Design a treatment system to meet peak effluent nitrogen limits	Dynamic	Dynamic	Instantaneous values	NH_4-N NO_x-N N_{tot}	0.5 mg/L
7	Optimise aeration control	Dynamic	Dynamic	Hourly average	Air flow rate DO profile Effluent NH_x-N	10% 0.5 mg/L 0.5 mg/L
8	Test effect of taking tanks out of service	Dynamic	Dynamic	Monthly average	NH_x-N NO_x-N N_{tot} PO_4-P P_{tot}	1.0 mg/L 1.0 mg/L 1.0 mg/L 0.5 mg/L 0.5 mg/L
9	Use model to develop sludge wastage strategy	Dynamic	Dynamic	Weekly average Daily average	WAS load SRT NH_x-N PO_4-P	10% 1 d 1.0 mg/L 0.5 mg/L
10	Develop a strategy to handle storm flows	Dynamic	Dynamic	Hourly	MLSS Effluent TSS NH_x-N NO_x-N PO_4-P	10% 10 mg/L 1.0 mg/L 1.0 mg/L 0.5 mg/L

(Continued)

Table 6.5 Target variables and error ranges proposed for the application matrix (*Continued*).

	Example	Influent	Simulation	Averaging period	Target variable	Acceptable error range (±)
11	Develop a general model for process understanding	*No calibration*			None	Not applicable
12	Develop a site specific model for operator training	Dynamic	Dynamic	Monthly	MLSS WAS load Effluent TSS NH_x-N NO_x-N PO_4-P Air flow rate DO	10% 5% 5.0 mg/L 1.0 mg/L 1.0 mg/L 1.0 mg/L 10% 0.5 mg/L
13	Develop a process configuration for nitrogen removal treating waste from a food production factory (soy sauce)	Dynamic	Dynamic	Minutes	MLSS WAS load Effluent TSS NHx-N NOx-N Air flow rate DO	10% 5% 5.0 mg/L 1.0 mg/L 1.0 mg/L 10% 0.5 mg/L
14	Assess acceptability of new influent at a petrochemical site	Steady-state	Steady-state	Monthly, seasonal average	MLSS WAS load Effluent TSS NHx-N NOx-N Air flow rate DO Effluent COD_{sol}	10% 5% 5.0 mg/L 1.0 mg/L 1.0 mg/L 10% 0.5 mg/L 3.0 mg/L

6.6.2 Data requirements

The level of detail and types of data required for a modelling project depend on the objectives of the project. In general, the more detailed and accurate the results of the simulation study need to be, the more detailed the data need to be. A major impact on the data frequency is the time scale of the effluent limits. For instance a daily average effluent limit often requires a higher input measurement frequency than an annual limit. Some typical examples are:

- Diurnal, weekly, or seasonal variations of important process variables (e.g. DO, NH_x-N, NO_x-N, PO_4-P, MLSS, pH, temperature, COD) at various locations within the plant
- Measurements of influent characteristics and kinetics (e.g. nitrifier growth rate)
- Results from tracer tests (investigation of mixing conditions and hydraulic retention times (HRT)) for example when the reactors have an unusual layout and it is difficult to define the number of tanks in series required for the model

- Sludge blanket level in the clarifier tanks and TSS profile, when the sludge content in the clarifiers is important to understand the plant performance (e.g. endogenous denitrification)
- Energy consumption of specific plant units (e.g. aeration or mixing of bioreactors)

Table 6.6 lists examples of additional data requirements and frequency for the Application Matrix examples.

Table 6.6 Specific data requirements for the examples of the application matrix.

	Example	Data type	Acquisition frequency
1	Calculate sludge production	Input data	Economic analysis: annual averages Equipment sizing: monthly averages
2	Design Aeration System	Air flow rate	Hourly average, maximum, minimum
		Input data, performance data	Hourly average, maximum, minimum
3	Develop a Process Configuration for Nitrogen Removal	Input data, performance data	Monthly or annual average
4	Develop a Process Configuration for Phosphorus Removal	Input data, performance data	Monthly or annual average
5	Assess Plant Capacity for Nitrogen Removal	Input data, performance data	Hourly average
6	Design a Treatment System to Meet Peak Effluent Nitrogen Limits	Input data, performance data	Hourly average
7	Optimise aeration control	Air flow rate DO profile	Every 1 to 5 minutes
8	Test effect of taking tanks out of service	Input data, performance data	Hourly average
9	Use model to develop sludge wastage strategy	Input data, performance data	Hourly average, monthly average
10	Develop a strategy to handle storm flows	Input data, performance data	Hourly average, monthly average
11	Develop a general model for process understanding	Input data	Hourly average, 15-minute average (aeration)
12	Develop a site specific model for operator training	Input data, performance data	Hourly average, 5-minute average (aeration)
13	Develop a Process Configuration for Nitrogen Removal Treating Waste from a Food Production Factory (Soy Sauce)	Input data, performance data	1 minute average (for intermittent aeration modelling)
14	Assess acceptability of new influent at a petrochemical site	Input data	Monthly average

6.6.3 Selecting scenarios for analysis

Scenarios will be devised to address the requirements of the project definition and therefore they will be project specific. Table 6.7 provides a list of simulation types and purposes that might be run for the different examples in the Application Matrix and also gives suggestions for possible sensitivity analyses that could be run.

Table 6.7 Suggested simulations for different objectives from the Application Matrix.

		Simulation Type Steady State Simulations		Dynamic Simulations				
#	Objectives	Type	Purpose	Resolution	Type	Purpose	Notes	Sensitivity Analyses
	DESIGN							
1	Calculate Sludge Production	Annual	Economic Analysis					
		Monthly	Equipment Sizing					
2	Design Aeration System	Annual	Economics, blower and diffuser type	Hourly	Typical Day	Define normal operating range for blower	Alternatively a Birthday Cake Analysis or a full year of hourly data simulation can be run to fulfill all 3 purposes	
				Hourly	Maximum Day	Define peak air demand		
				Hourly	Minimum Day	Define minimum air demand		
3	Develop a Process Configuration for Nitrogen Removal	Monthly Seasonal	Performance					COD:N
4	Develop a Process Configuration for Phosphorus Removal	Monthly Seasonal	Performance					COD:P, rbCOD and vfa fractions
5	Assess Plant Capacity for Nitrogen Removal	Monthly Seasonal	Capacity					COD:N, SRT vs Temperature
6	Design a Treatment System to Meet Peak Effluent Nitrogen Limits	Monthly Seasonal	Short-list options	Hourly	Seasonal Weeks	Peak Effluent Nitrogen		COD:N
	OPERATION							
7	Optimise Aeration Control			Minutes	Week	Impact of control options		
8	Test Effect of Taking Tanks Out of Service			Hourly	Week	Impact of operational change		
9	Use Model to Develop Sludge Wastage Strategy			Hourly	Month	Impact of operational changes		
10	Develop a Strategy to Handle Storm Flows	Annual or Monthly	Establish initial condtions	Hourly	Week	Repsonse to storm flows		
	TRAINING							
11	Develop a General Model for Process Understanding			Hourly	Day or Week	Sludge Production		
				Minutes	Day	Aeration		
12	Develop a Site Specific Model for Operator Training			Hourly	Day or Week	Sludge Production		
				Minutes	Day	Aeration		
	INDUSTRIAL							
13	Develop a Process Configuration for Nitrogen Removal Treating Waste from a Food Production Factory (Soy Sauce)			Minutes	Week	Nitrogen removal		
14	Assess Acceptability of New Influent at a Petrochemical Site	Monthly Seasonal	Capacity					

Table 6.8 lists suggested inputs and outputs for the different objectives in the Application Matrix along with formats for presenting the output data.

Table 6.8 Suggested inputs and outputs for different objectives from the application matrix.

		Inputs & Outputs		Suggested Format			
#	Objectives	Parameters/Manipulated Variables	Main Outputs	Summary Table	Time Series Graph	Bar Chart	Cumulative Distribution
	DESIGN						
1	**Calculate Sludge Production**	Q_{WAS}	FWAS kg d	x x			
2	**Design Aeration System**		Qair % coverage airflow per diffuser	x	x x x		x x x
3	**Develop a Process Configuration for Nitrogen Removal**	V, Qrec, conf	Ntot	x		x	
4	**Develop a Process Configuration for Phosphorus Removal**	V, Qrec, conf	Ntot, Ptot	x		x	
5	**Assess Plant Capacity for Nitrogen Removal**	Q, qAir	Ntot	x		x	
6	**Design a Treatment System to Meet Peak Effluent Nitrogen Limits**	Qair, V, config	NHx, Nox		x		x
	OPERATION						
7	**Optimise Aeration Control**	Qair, DO set-points	kWh, Qair, NHx, NO3, NO2, DO		x	x	
8	**Test Effect of Taking Tanks Out of Service**	number of tanks	NHx, Ntot, Ptot		x		
9	**Use Model to Develop Sludge Wastage Strategy**	FWAS kg/d	MLSS		x		
10	**Develop a Strategy to Handle Storm Flows**	Q, Eq Basin V	NHx, Ntot, Ptot, TSS		x		
	TRAINING						
11	**Develop a General Model for Process Understanding**	Typical control variables	Typical process indicators	x	x	x	
12	**Develop a Site Specific Model for Operator Training**	Typical control variables	Typical process indicators	x	x	x	
	INDUSTRIAL						
13	**Develop a Process Configuration for Nitrogen Removal Treating Waste from a Food Production Factory (Soy Sauce)**	Conf, aerator timings	NHx, NO3, NO2, DO	x	x	x	
14	**Assess Acceptability of New Influent at a Petrochemical Site**	Q	COD,QWAS, Qair, DO	x		x	

6.6.4 Using the GMP Unified Protocol: benefits and averted risks

Table 6.9 lists some of the benefits and risks that might be averted if the GMP Unified Protocol is used for the Application Matrix examples.

Table 6.9 Benefits and risks averted by using the GMP Unified Protocol.

#	Objectives	Benefits of good process model developed using GMP	Averted risks (dangers if GMP not used)
1	Calculate Sludge Production	Accurate prediction of sludge production. Equipment for handling sludge treatment correctly sized	Sludge handling equipment oversized (excessive cost), or undersized (operational problems)
2	Design Aeration System	Aeration system correctly sized to account for system dynamics	Aeration system oversized - wasting energy, or undersized and unable to meet peak demands.
3	Develop a Process Configuration for Nitrogen Removal	Robust design for Nitrogen Removal	Sub-optimal design
4	Develop a Process Configuration for Phosphorus Removal	Robust design for Phosphorus Removal. Good understanding of process sensitivity to external factors.	Unstable or inadequate design
5	Assess Plant Capacity for Nitrogen Removal	Good assessment of plant capacity	Optimistic or pessimistic opinion of treatment capacity leading to bad decisions
6	Design a Treatment System to Meet Peak Effluent Nitrogen Limits	Robust design. Good quantification of risk of not meeting permit and frequency of such events	Low confidence in design
7	Optimise Aeration Control	Good suggestions for control optimisation	Inadequate control idea development
8	Test Effect of Taking Tanks Out of Service	Confidence in taking tanks out of service	Potential for wrong decision to take tanks out of service and potential permit violation
9	Use Model to Develop Sludge Wastage Strategy	Good idea development for sludge strategies	Inadequate assessment of strategies
10	Develop a Strategy to Handle Wet Weather Flows	Robust strategy for wet weather events	Inadequate wet weather strategy
11	Develop a General Model for Process Understanding	Clear and understandable training tool	Ambiguous and confusing example
12	Develop a Site Specific Model for Operator Training	Useful tool to investigate process dynamics and plant response	Confusion or poor appreciation of plant dynamics

(*Continued*)

Table 6.9 Benefits and risks averted by using the GMP Unified Protocol (*Continued*).

#	Objectives	Benefits of good process model developed using GMP	Averted risks (dangers if GMP not used)
13	Develop a Process Configuration for Nitrogen Removal Treating Waste from a Food Production Factory (Soy Sauce)	Robust design for industrial treatment	Inadequate plant design
14	Assess Acceptability of New Influent at a Petrochemical Site	Confidence in treatment capabilities of new treatment facility	Inadequate plant design

Chapter 7

Using the GMP Unified Protocol by example

Short summary

This chapter takes the steps of the GMP Unified Protocol and applies them to four examples of increasing complexity. The first three examples are taken from the Application Matrix discussing *Sludge production*, *Assessment of plant capacity for nitrogen removal*, and *Development of a site-specific model for operator training*. Following the three hypothetical examples, a real-life example of the application of the GMP Unified Protocol is presented.

7.1 INTRODUCTION

This chapter shows the use of the Unified Protocol through example by walking the reader through the different steps. This should highlight important aspects required in setting up and using a model based on the GMP Unified Protocol.

The first three examples are drawn from the Application Matrix examples presented in Chapter 6 and are listed below. The numbers in parentheses denote the example number in the Application Matrix:

- Calculate Sludge Production (#1)
- Assess Plant Capacity for Nitrogen Removal (#5)
- Develop a Site Specific Model for Operator Training (#12)

A real-life example of the application of the *GMP Unified Protocol* at the Beenyup WWTP is presented at the end of the chapter.

Included for each example is a suggested procedure to calibrate the model (UP Step 4). Each procedure is based on the suggested calibration procedure for a biological nutrient removal (BNR) plant model shown in Section 5.4, but tailored to match the specific example. The calibration described for example #1, *Calculate Sludge Production*, is the starting point for all the examples as this establishes the SRT which is a fundamental first step for all model calibrations.

7.2 CALCULATE SLUDGE PRODUCTION

UP Step 1: Project definition

The *Project Definition* step focuses on the development of a clear *Problem Statement* that defines objectives and identifies the requirements needed to meet those objectives.

In this case the objective is to estimate the quantity of sludge produced by the activated sludge plant with the intention of using this estimate to size sludge handling equipment (pumps, pipes and thickeners). The WAS should be determined with an accuracy of $\pm 5\%$.

The boundaries for the system are the influent to the activated sludge process, the effluent from the secondary clarifier and – most importantly for this example – the waste activated sludge. The waste activated sludge (WAS) is the main output of interest and can be expressed in one of two ways: (1) commonly, the WAS is expressed as a mass load of solids (kg/d) with assumptions being used to calculate the concentration of solids; or, (2) the model outputs both the volumetric quantity (m^3/d) of sludge and the sludge concentration (mg/L or % solids). If the downstream sludge processing now or in the future may include digestion, incineration or other solids reduction technologies, it may be helpful to provide more detailed characterisation of the WAS composition (e.g. biodegradable fraction and biomass fraction) based on the state variables predicted by the model, but irrespective of the downstream processing, the volatile solids content of the WAS should be provided.

Typically, the sizing of sludge handling equipment is based on weekly or monthly average quantities so a steady-state model using the maximum monthly average data is sufficient for this objective. If the model outputs are to be used for an evaluation of alternatives, a model of the overall annual average condition might be better or, multiple months from several seasons can be simulated to provide a detailed picture of seasonal variations.

An important consideration for sludge handling equipment is its operating schedule. The quantities predicted by a steady-state model assume a continuous flow of WAS, but if the equipment is operated for a few hours per day or only a few days per week, a dynamic model should be used to account for this.

Finally, when modelling any system with sludge handling, the load associated with return sludge liquors must be considered.

UP Step 2: Data collection and reconciliation

Data Collection and Reconciliation entails developing a comprehensive understanding of the plant and the preparation of a data set of sufficient quality and quantity to build a suitable model.

For the purposes of establishing sludge production estimates, particular attention should be paid to ensuring that appropriate data is developed to estimate WAS loads. An accurate estimate of the WAS mass flow is extremely important for *all* process models – not just for estimating sludge production – as it directly impacts the concentration of micro-organisms in the model. WAS flows are often poorly measured, sometimes incorrectly aggregated due to intermittent WAS flows and sometimes not measured at all. Sampling issues contribute to the problem as WAS samples are often grab samples as opposed to composites. It is often assumed that the WAS concentration is the same as the measured RAS and it is frequently assumed that samples taken from just one clarifier are sufficient to represent the output from all the clarifiers. Occasionally different methods are used to measure the solids concentrations so the modeller may have to be aware of that as well.

Section 5.2 contains detailed guidance on gathering and reconciling data, but two methods that are particularly useful for this application are (1) a phosphorus mass balance; and, (2) an overlapping mass balance. A phosphorus mass balance can be used to check WAS flows and concentrations because phosphorus is conserved within the system. If the influent phosphorus load does not balance with the effluent phosphorus load plus the WAS phosphorus load then this indicates a problem somewhere and more often than not with the WAS solids mass flow. Overlapping mass balances are also useful for checking WAS loads. For example it might be possible to mass balance the reported WAS thickening and sludge storage data with the flow of sludge tankers leaving the site (the latter having an economic consequence, is usually measured more diligently).

UP Step 3: Plant model set-up

In the third step of the *Unified Protocol*, the model plant layout is established, sub-models are selected, input data files are prepared and output graphs and tables are set up.

A simple kinetic model such as ASM1 is usually sufficient to estimate sludge production unless the plant has phosphorus removal in which case chemical addition or poly-phosphate accumulation will add a significant mass to the overall calculations. In this case a kinetic model that includes phosphorus should be selected. Sub-models for aeration are not normally required, except to specify the dissolved oxygen concentration in the reactors. A simple clarifier model can be used but it is important to include a clarifier model with volume in order to account for solids that are stored in the clarifier as this impacts the estimate of overall solids retention.

UP Step 4: Calibration and validation

The *Calibration and Validation* step begins with refining the stop criteria and proceeds with an iterative process of running the model and adjusting parameters until selected model outputs match plant data within an accepted error. Tests are then performed to validate the calibrated model.

Calibration of a model for sludge production requires that the mass distribution in the reactors and the clarifiers be considered. The calibration steps for sludge production and clarifier performance must both, therefore, be considered, as shown in Figure 7.1.

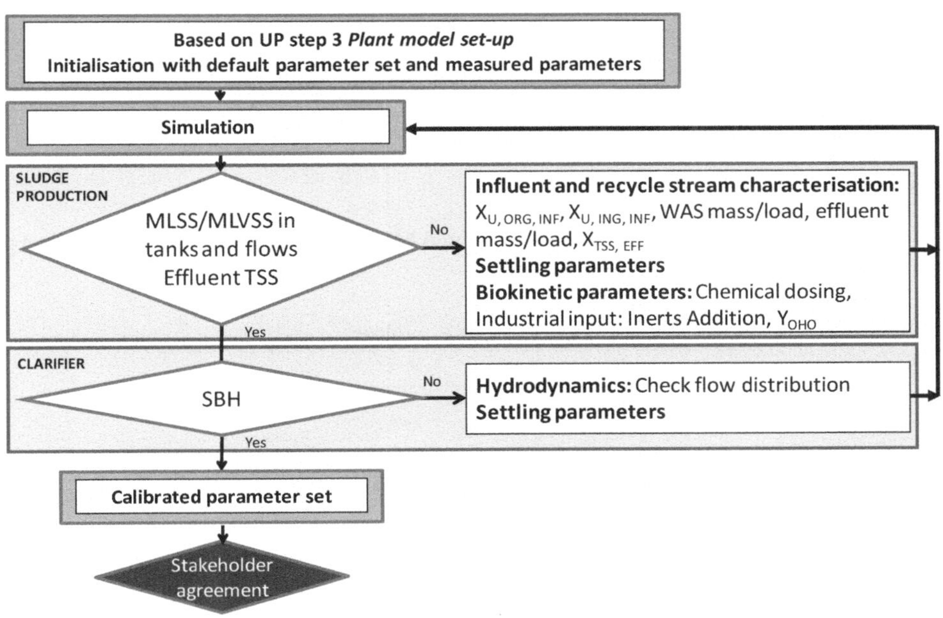

Figure 7.1 Calibration of a model for sludge production.

Sludge production

Although total suspended solids (TSS) is the variable of interest for this example, all biokinetic models discussed in this report are COD-based. Particulate COD is related to the prediction of volatile suspended solids (VSS) so the modeller will need to match both the VSS (in tanks and in the waste activated sludge (WAS)) and the TSS when calibrating the model.

Data used to match the sludge production include the WAS concentration and flow rate, and the effluent TSS concentration. These process variables are difficult to sample accurately, so specific care should be taken to double-check the accuracy of flows and TSS/VSS concentrations at this stage.

Calibration of the model will include adjustment of the influent particulate inert COD ($X_{U,\ INF}$), as this influences the VSS sludge production. Modifying the influent $X_{U,\ INF}$ involves changing the influent fractionation so after matching the solids, the influent COD fractionation should be checked to ensure that it is still consistent with the available influent data. TSS values are then fitted by modifying the inorganic suspended solids (ISS) in the influent if required. If chemicals are dosed for P-removal, precipitates should be accounted for to match sludge production. Figure 7.2 details the procedure to calibrate sludge production which should be used in conjunction with the iterative procedure presented in Section 5.4. The decision box on the left shows the variables that should be compared and the box on the right lists parameters that should be checked and adjusted to refine the model.

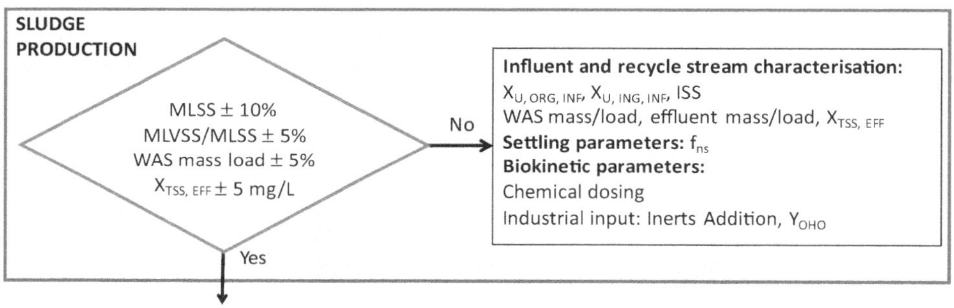

Figure 7.2 Calibration of sludge production.

Figure 7.3 shows an example of measured and simulated MLSS, using the ASM1 parameter set proposed by Choubert *et al.* (2009) for low loaded WWTPs.

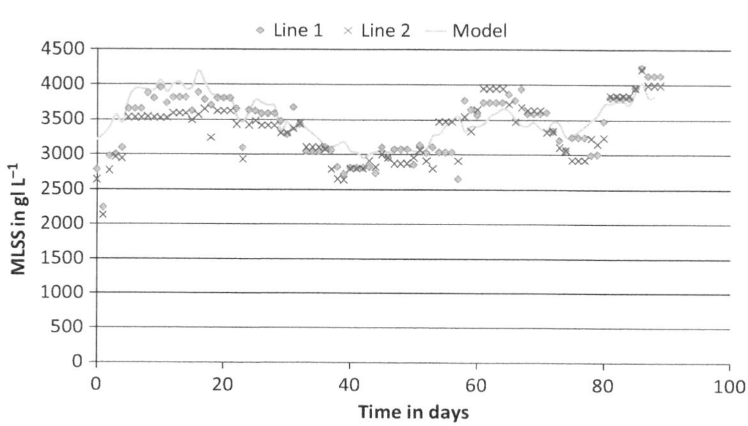

Figure 7.3 Example of the observed and simulated MLSS concentration in 2 aerated lines of a low loaded activated sludge plant (250,000 PE).

In this example, the simulated MLSS lies between $\pm 15\%$ of the observed values (7% on average) without modifying any model parameter when the first values are excluded. The initial overestimation of the model is attributed to the initial conditions).

An alternative approach to calibrate sludge production would consist of matching the observed and simulated sludge retention time (SRT), as presented in Figure 7.4. This approach requires the adjustment of the same parameters, but using SRT as the target variable.

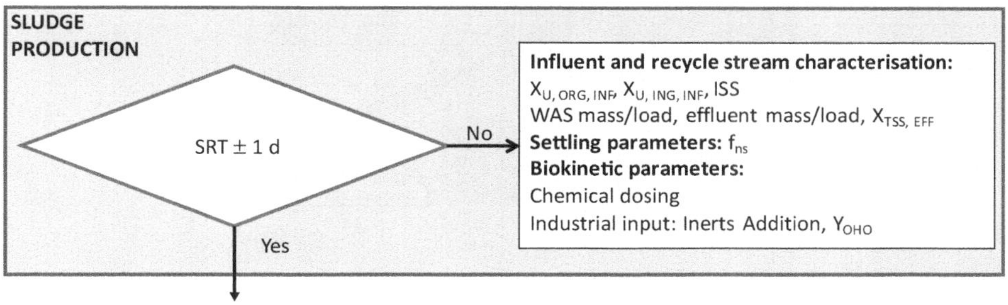

Figure 7.4 Alternative approach to calibrate sludge production on sludge retention time (SRT).

In activated sludge models that use TSS as a state variable (e.g. original ASM2d and ASM3), it is advised to calibrate the sludge production using COD as the target variable, and to modify the COD/TSS ratios to match the predicted X_{TSS} to the measured TSS, as shown in Figure 7.5. Using TSS directly as the target variable may lead to incorrect COD values, which is a typical pitfall.

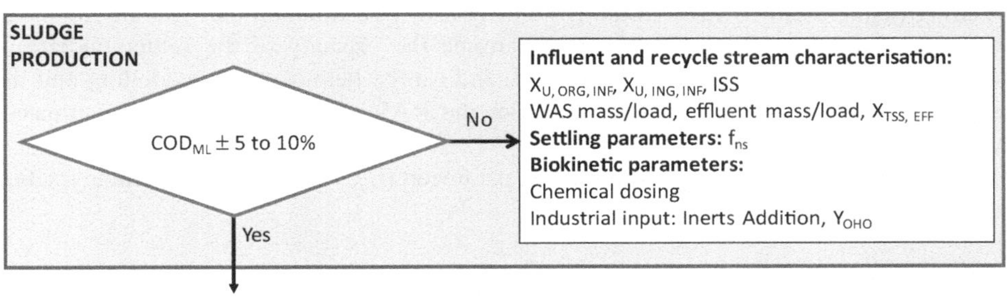

Figure 7.5 Recommended adjustment parameters for the sludge production for models with X_{TSS} as a state variable.

Clarifier

In a number of modelling projects the use of simple point or ideal clarifier models (phase separators) will be sufficient for this project objective. In these models effluent solids or removal efficiency is a direct model input. Layered flux models are usually required only under dynamic conditions to model settling and to better represent effluent and underflow concentration changes. Dynamic models also will predict sludge mass shifts when these are relevant to model the behaviour of the plant to set the correct SRT. A pie chart such as the example shown in Figure 7.6 can be used to show the relative distribution of sludge mass throughout the modelled system.

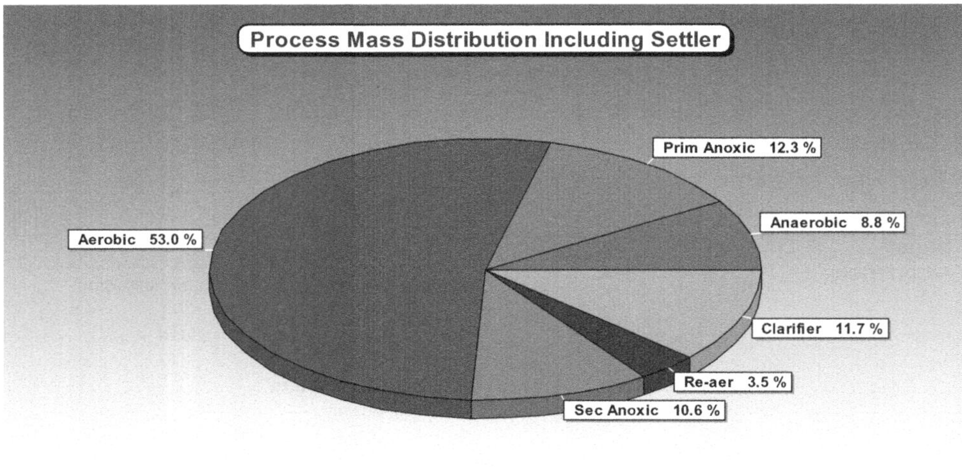

Figure 7.6 Pie chart showing mass distributions in different treatment zones.

For a dynamic model, calibrating a layered flux-based settler model involves matching the measured and simulated effluent solids and sludge blanket height (SBH) as shown in Figure 7.7. For example, in the double exponential (Takács *et al.* 1991) model the following steps may be used:

- Measuring (using a Stirred Zone Settling Velocity test) or estimating the Vesilind hindered settling parameters (v_0 and r_{hin}) can be used to determine the capacity of the settler underflow (RAS) concentration if the settler is critically loaded, and can be helpful when modelling and the sludge blanket movements. In an under-loaded settler the RAS concentration can be estimated from a simple mass balance.
- To match effluent solids the flocculent settling parameter (r_{floc}) and the non-settleable fraction (f_{ns}) or concentration can be adjusted.

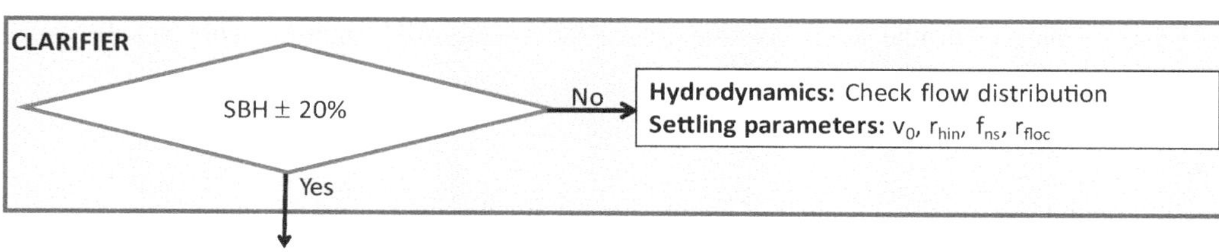

Figure 7.7 Calibration of settling parameters of the double-exponential model (Takács *et al.* 1991) with the sludge blanket height (SBH) as target variable.

An example of the measured and simulated sludge blanket height is presented in Figure 7.8.

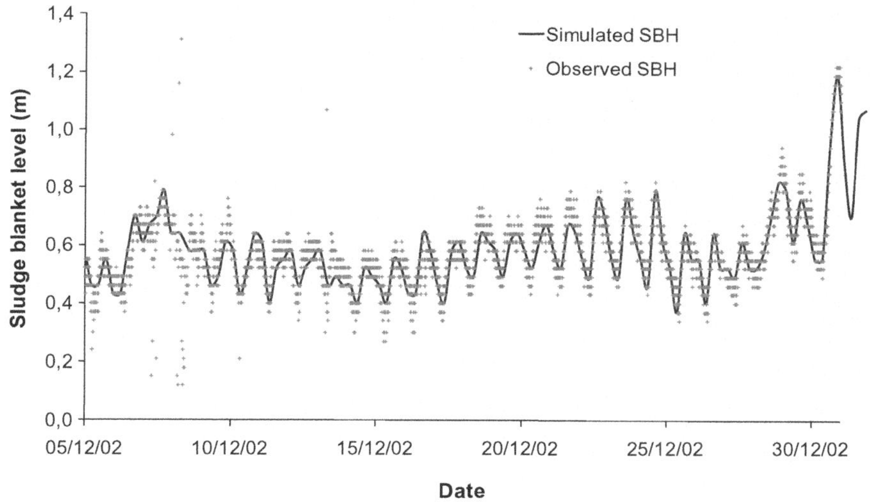

Figure 7.8 Example of the sludge blanket height (SBH) evolution (Marquot, 2006).

UP Step 5: Simulation and result interpretation

In the last step of the *Unified Protocol*, scenarios are defined, the model is adjusted as needed for those scenarios and simulations are run. The results from the model runs are then presented, interpreted, and recorded in a report along with other pertinent information from the other modelling steps.

The calibrated model can now be used to simulate other conditions and answer the modelling objectives for example, estimated sludge production at different planning horizons (say 5, 10 and 20 years) using different flows and loads. As noted in Section 7.2 step 1, if alternatives are being compared then several design conditions may be simulated to provide seasonal and overall annual sludge production estimates.

7.3 ASSESS PLANT CAPACITY FOR NITROGEN REMOVAL

UP Step 1: Project definition

In this application a model is developed for an existing nitrogen removal plant designed to meet an effluent total nitrogen concentration of 10 mg/L. A model is required to determine the capacity of the plant under increasing load conditions. Furthermore, the plant is required to meet a tighter effluent total nitrogen of 8 mg/L based on the monthly average concentration. The goal is to determine the "rated capacity" expressed as the maximum flow that can be treated by the plant to meet future flow and effluent requirements.

An important discussion that must take place in establishing the project definition for this modelling project is whether to use steady-state simulations or dynamic simulations to establish the capacity rating. It might be concluded that a steady-state simulation is adequate and could be based on historical data and some additional measurements. However, significant engineering judgement would have to be used to determine what safety factor should be applied to the outputs from the model. Dynamic simulations would provide additional information to enable the engineer to use tighter safety factors and increase the confidence in the capacity rating. Uncertainty and risk would be significant considerations in this model and should be discussed by all stakeholders.

UP Step 2: Data collection and reconciliation

In addition to the data requirements described in Section 7.2 step 2 to establish an appropriate estimate of sludge production and solids retention time, a wider range of the considerations for data collection and reconciliation must be considered in this case.

Of particular concern for nitrification are: (1) developing an accurate estimate of the influent TKN, including sidestreams; (2) determining if alkalinity will be limiting; (3) noting if there is any evidence of substances in the influent that might cause nitrification inhibition (e.g. industrial wastes or cyanide from incinerator scrubbers); (4) establishing suitable profiles for dissolved oxygen; and, (5) understanding the aeration system.

For denitrification, data is required to: (1) accurately characterise the influent carbon substrate concentrations; and (2) determine where and when in the process anoxic conditions occur, particularly if "simultaneous nitrification and denitrification" (SND) is suspected to be of significance. The latter consideration requires a comprehensive understanding of the reactor hydrodynamics and the aeration system.

UP Step 3: Plant model set-up

Any one of many biokinetic models could be used for this application. If the plant has clearly defined anoxic and aerobic zones, no significant SND, no accumulation of nitrite and no evidence of biological phosphorus removal then ASM1 or ASM3 are adequate models to describe nitrification and denitrification.

Note:

If biological phosphorus removal is possible (e.g. extended anoxic zones that may become anaerobic, and/or high concentrations of organic acids or other readily-biodegradable COD in the influent), then a biokinetic model with biological phosphorus removal processes should be chosen so that the competition for substrate is adequately modelled.

If the plant has significant SND or nitrite accumulation a modified biokinetic model that includes these capabilities (e.g. two-step nitrification and denitrification) should be considered.

As noted in the Data Collection and Reconciliation section, both the reactor hydrodynamics and aeration system operation are important factors in nitrogen removal systems. Sufficient detail should be included in the model to enable these aspects to be calibrated.

UP Step 4: Calibration and validation

Calibration and validation for this model starts by establishing the sludge production and clarifier performance. Once completed, calibration will extend to the nitrification and denitrification processes, and finally the oxygen transfer model (Figure 7.9).

Nitrification

Parameters for nitrification are well documented in the literature. Some of them may need to be measured, or deduced from the plant's performance data. Calibrating nitrification parameters also requires verifying that neither the pH (in most models through modelling alkalinity) nor inhibition are limiting factors.

Aeration has a big impact on modelling nitrification (and denitrification). In order to separate the calibration of the biokinetic model from the calibration of the oxygen transfer model, it is recommended that the DO concentration be set to a constant value or controlled to its measured values at this stage. It is important to have knowledge of the DO profile through the plant and the concentrations in the model should be set accordingly.

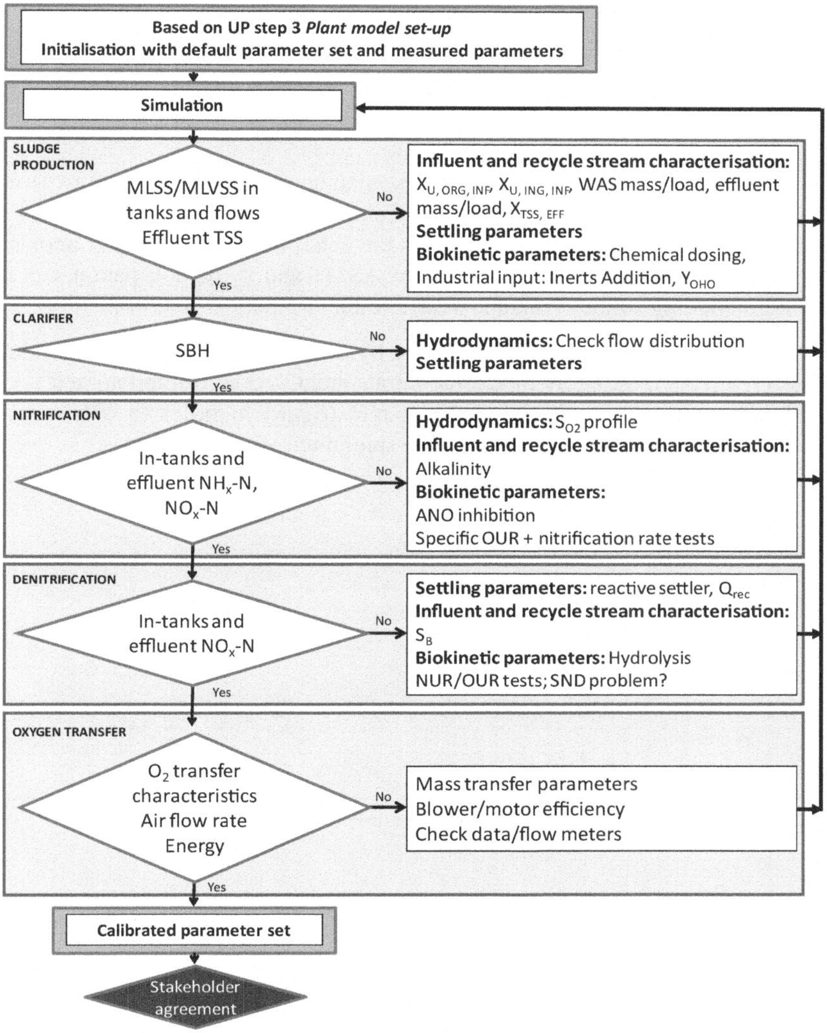

Figure 7.9 Calibration of a model for nitrogen removal.

Figure 7.10 illustrates that calibrating nitrification parameters may well be done using the effluent ammonium concentration, although using the nitrogen profiles in the mixed liquor may give more reliable results. Such profiles often catch more dynamics than the effluent data. Ammonia measurements in the effluent of a fully nitrifying plant typically provide no measure for calibrating nitrification dynamics as the concentrations often are at or near zero and insensitive to parameter changes.

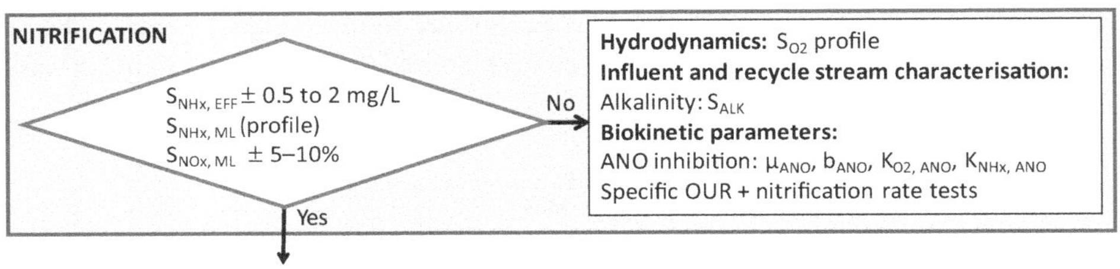

Figure 7.10 Calibration of nitrification.

Figure 7.11 is an example that shows the N profiles in an aeration tank obtained experimentally. Both calibrated and uncalibrated model results are shown. Calibration in this case consisted of modifying the autotrophic biomass growth and decay rates as well as the heterotrophic yield under anoxic conditions.

This example highlights two main modifications of ASM1 and its default parameters that have been suggested in several modelling studies (Hauduc *et al.* 2011):

(1) The need to change the ASM1 model structure by introducing a heterotrophic yield under anoxic conditions ($Y_{OHO,Ax}$) to properly model the nitrate and COD consumption, and
(2) The need to increase the autotrophic decay rate (b_{ANO}) in order to obtain stable maximum autotrophic growth rates ($\mu_{ANO,Max}$) when the sludge retention time varies.

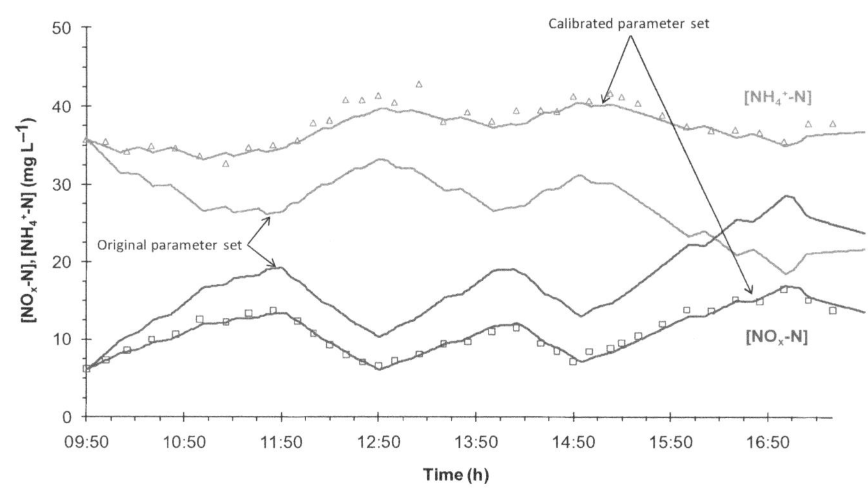

Figure 7.11 Observed and simulated nitrogen concentrations in an aeration tank during three aerobic/anoxic periods–ASM1 original and calibrated parameter sets (Choubert *et al.* 2005).

Denitrification

Before adjusting biokinetic model parameters, one has to check whether denitrification occurs in the settler due to a significant mass of sludge and anoxic conditions. In this case, an appropriate model including biological reactions should be used to simulate the denitrification occurring in the sludge blanket. This requires going back to UP step 3 *Plant Model Set-up*.

The effluent nitrate concentration can be used as a target variable for calibrating the denitrification processes, but in-tank measurements (N-profiles in anoxic zones or during anoxic periods and in the RAS stream) tend to be more useful for describing the dynamics of the system.

Denitrification is highly susceptible to available carbon so calibration of the denitrification process typically involves the carbon processes. Carbon availability is adjusted by manipulating the influent composition (ratio $S_B/(S_B + X_B)$) or by adjusting its release in tanks (due to hydrolysis and/or fermentation parameters, half saturation constants, and others). This procedure is illustrated in Figure 7.12.

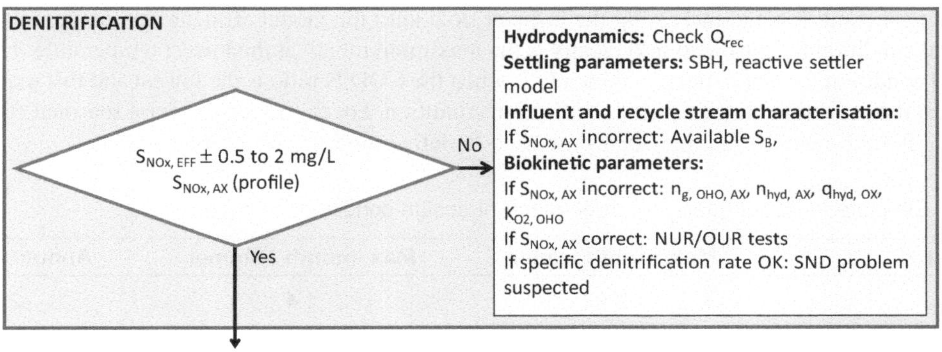

Figure 7.12 Calibration of denitrification.

When all other parameters have been checked, the occurrence of simultaneous nitrification/denitrification (SND) may be the last option to explain a mismatch between simulated and observed data. The requirement for modelling SND is often caused by over-simplification of the hydraulic model neglecting anoxic zones or phases.

Oxygen transfer

The oxygen transfer model is not part of the biokinetic model and the model type depends on the simulator implementation. However, oxygen transfer is one of the essential processes in biological wastewater treatment. In order to separate the calibration of the biokinetic model from the calibration of the oxygen transfer model, it is recommended that the DO concentration be set to a constant value or controlled to its measured values at first. It is important to have knowledge of the DO profile through the plant and the concentrations in the model should be set accordingly.

The following procedure to calibrate the oxygen transfer model is suggested:

- In a first step, the measured DO concentrations should be retraced with the model (e.g. with a simple proportional controller). This allows to calibrate the plant model (biokinetic, hydraulic, influent model, etc.) without interference with the oxygen transfer model.
- When the plant model is calibrated, the DO control loops can be implemented in the model.
- In a last step, the oxygen transfer model is calibrated by changing the α-value or the oxygen transfer rate (OTR) in order to fit the measured DO concentrations. One parameter, the αF-value (oxygen transfer correction factor to account for the difference between clean water and activated sludge (α), including potential diffuser fouling (F)) or the OTR (Oxygen Transfer Rate) should be fixed to a reliable value to simplify the calibration.

If independent measurements for the oxygen transfer rate (OTR) and the α-value are available (e.g. through off-gas measurements and clean water tests), these values should be used. If the results are not in accordance with the measurements, this can be due to concentration gradients, measuring errors, changing influent composition (e.g. surfactants, which decrease α-values, Gillot & Héduit, 2008) or diffuser aging processes (Rosso *et al.* 2008).

UP Step 5: Simulation and result interpretation

The calibrated model for this application is used to determine the current plant capacity by running the model under a critical condition and increasing the influent flow until the model effluent hits the permitted limit. The critical condition for nitrification typically is the maximum month at the lowest temperature. However, the critical condition for denitrification typically is when the COD:N ratio is the lowest and this may or may not occur at the same time as the critical nitrification condition. For this reason, several seasonal conditions should be run in the model and the worst flow used to define the capacity (Table 7.1).

Table 7.1 Simulated effluent quality for three different design conditions in mg/L.

Parameter	Max. month winter	Max. month summer	Annual average
BOD_5	2.4	1.4	2.1
TSS	4.6	4.3	3.9
Total Nitrogen	6.7	5.6	4.4
Ammonia (as N)	0.7	0.6	0.9
Nitrate + Nitrite (as N)	4.3	3.3	2.2

Rather than simply running simulations repeatedly and adjusting flows and other controlled parameters until a flow is determined, it can be more instructive to use a sensitivity analysis to assess the plant capacity. Figure 7.13 shows an example. In this particular case the analysis shows that a BOD/TKN of 6 or more would maximise nitrogen removal.

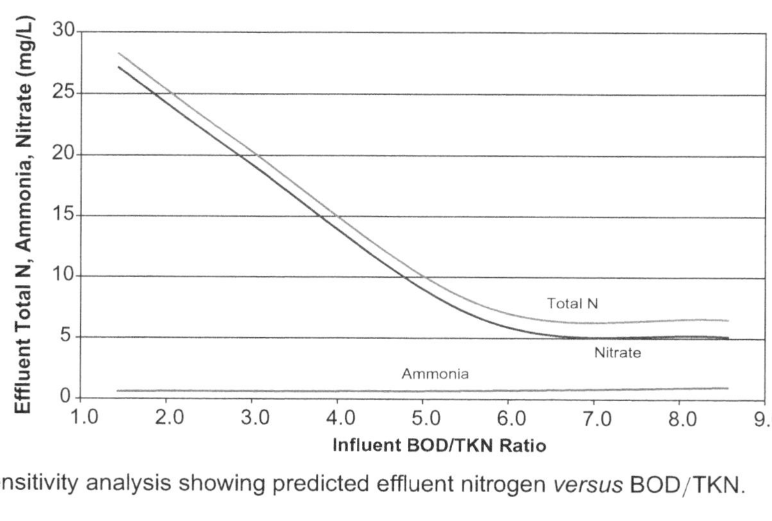

Figure 7.13 Sensitivity analysis showing predicted effluent nitrogen *versus* BOD/TKN.

Another useful approach is to plot nutrient profiles under different design conditions. Figure 7.14 shows an example. In this case the ammonia profile in a plug flow reactor shows that the ammonia concentration is less than 1 mg/L in reactor 4. Because of this, reactor 5 constitutes additional reaction volume in excess of what the model predicts is required. This additional volume provides some buffer capacity or "safety factor" and allows for uncertainties in the model predictions. A higher flow could presumably be found that results in an ammonia concentration just under 1 mg/L in the final reactor; however, this would be considered a more aggressive design with no buffer capacity and its appropriateness would have to be assessed through a risk assessment on the uncertainties in the model predictions.

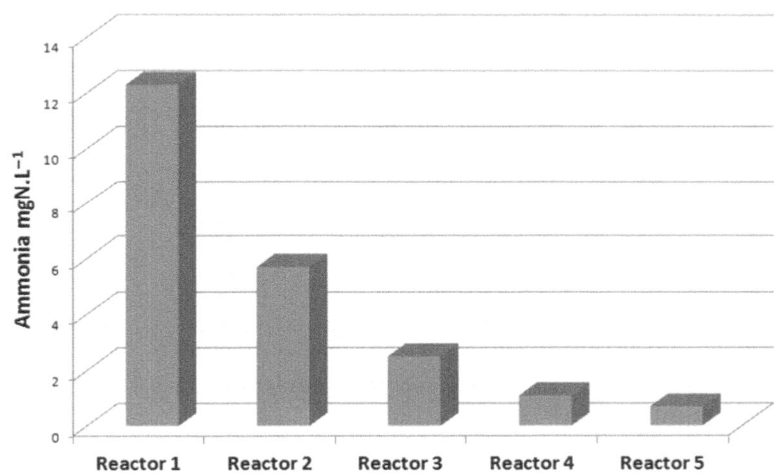

Figure 7.14 Representation of ammonia profile in a plug flow reactor operating at steady-state.

7.4 DEVELOP A SITE SPECIFIC MODEL FOR OPERATOR TRAINING

UP Step 1: Project definition

In this case it is assumed that a nitrogen and phosphorus removal facility with effluent limits of 10 mg/L total nitrogen and 1 mg/L total phosphorus based on monthly averages has been commissioned. A model is required to train operators, engineers and/or other personnel on the impact of making process changes to their specific plant. The model and presentation material will include some interactive "hands-on" work and model demonstration runs.

This example requires a great deal of focus and attention at the project definition stage in order to set reasonable expectations for the modelling quality and number of scenarios covered. The model outputs must match plant data with reasonable accuracy or the operators will not trust that the model is useful. At the same time, expectations should be set. Outputs may not match the plant data exactly but should be indicative of the likely performance as influent or operational parameters are changed. Careful consideration must also be given to scoping out what simulations will be run with the model. It is possible to create a long list of scenarios and sensitivity analyses that could be run. Many of these may be of interest to the designer but not to the operators or *vice versa*. The scenarios should be realistic and useful for developing an understanding of the plant.

UP Step 2: Data collection and reconciliation

In addition to the considerations discussed for modelling sludge production and nitrogen removal, this example adds two more degrees of sophistication to the modelling effort that require more data. Firstly the model is dynamic and so good time-series data is required for all pertinent model outputs that will be used to calibrate and validate the model. Secondly, because biological phosphorus removal is being modelled, the influent characterisation must consider the fractionation of the soluble substrate into volatile fatty acids (VFA) and other organics that are used as state variables in the kinetic model.

UP Step 3: Plant model set-up

Several biokinetic models can be used to simulate nitrogen and phosphorus removal. Table 5.3.5 describes the features of the different biokinetic models.

Hydrodynamics are especially important when carrying out dynamic simulations as the peaks and troughs of diurnal variations and other rapid changes will be accentuated if the flow is more plug-flow and attenuated if there is a higher degree of back-mixing.

Biological nitrogen and phosphorus processes are sensitive to dissolved oxygen concentrations which means that the aeration system must be modelled with reasonable accuracy to show the system dynamics.

UP Step 4: Calibration and validation

The overall procedure to calibrate and validate a full-BNR facility is shown graphically in Figure 5.4.3. The details for sludge production and nitrogen removal were described in the text for the previous examples. Details of the phosphorus removal calibration and oxygen transfer are given below.

Phosphorus removal

Biological phosphorus removal consists of both the assimilation of phosphorus by all biomass in the system and the growth of phosphorus accumulating organisms (PAO) that will provide enhanced biological phosphorus removal (EBPR or bio-P) when conducive conditions exist. The predicted phosphorus assimilation by all organisms depends on the model parameters describing the P content of the biomass. These parameters do not change significantly so for P assimilation it is normally sufficient to match the measured sludge production. Calibrating bio-P processes on the other hand is a complex task and needs an understanding of both the processes and the model structure.

Three different approaches exist for modelling bio-P:

- ASM2d-type models (e.g. ASM2d or ASM3 + EAWAG Bio-P Module)
- UCT-type models (e.g. Barker & Dold model or UCTPHO+)
- Metabolic models (e.g. TU Delft model)

In the *ASM2d-type models*, the storage of polyhydroxy-alkanoates (PHA) and polyphosphates and the growth of PAOs are modelled as separate processes but are highly interconnected. The user should be extremely careful in changing from the default parameters (however, note that for ASM2d no widely accepted default parameter set has been published) since the ratio between polyphosphate storage and growth should remain constant. Experimental results show that oxidation of stored organic compounds provides energy for both PAO growth and polyphosphate storage (Wentzel *et al.* 1989). Consequently, the PAO growth and poly-phosphate storage yields are linked and depend on the oxidation of stored organic components.

It is suggested that the biokinetic parameters should not be changed, but rather the model behaviour should be varied through the manipulation of the availability of volatile fatty acids (VFAs) instead. This can be done in a number of ways including the influent fractionation, the hydrolysis/fermentation rates, or the mass of nitrate entering through the RAS stream which has a significant influence and should be calibrated first. Figure 7.15 illustrates the calibration of bio-P processes.

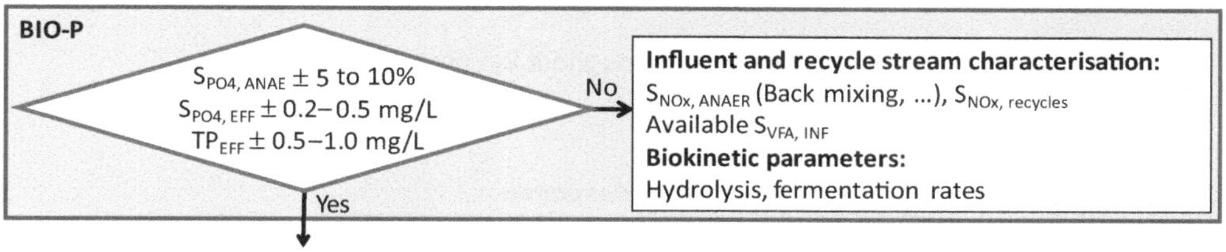

Figure 7.15 Calibration of biological phosphorus removal.

In the *UCT-type models*, the polyphosphate storage and biomass growth are described together as one model process. This simplifies the calibration procedure; however, the availability of VFAs and the amount of nitrate entering the anaerobic zones should be altered first before adjusting kinetic parameters.

In the *metabolic models*, both the growth and storage yields are linked to energy production. Again, the availability of VFAs and the amount of nitrate entering the anaerobic zones should be calibrated first.

Chemical precipitation is not the focus of this STR and only simplified empirical equations are presented in this report (Appendix A). The main goal of these precipitation models is to predict the chemical sludge production and chemical dosage is therefore considered in the first process calibrated (*Sludge production*). It is suggested that the ASM2d precipitation processes be calibrated by changing precipitation and redissolution efficiency to match the chemical P-removal efficiency (Figure 7.16). Co-precipitation processes are not represented in the models discussed in this report but might have a significant impact on the results.

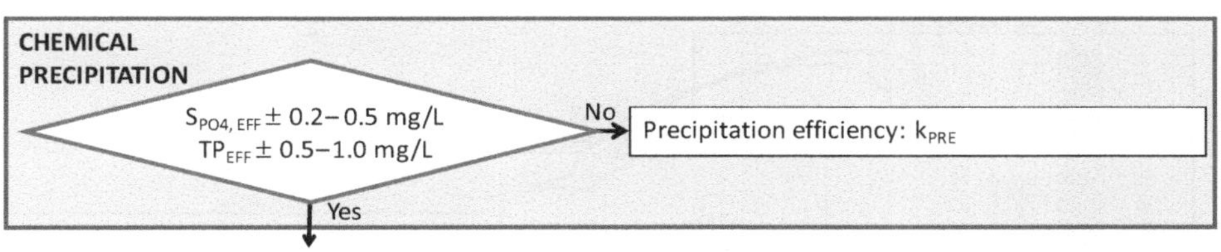

Figure 7.16 Calibration of chemical phosphorus removal.

UP Step 5: Simulation and result interpretation

This model is to be used for training. This can be done most effectively using dynamic simulations with dynamic outputs so that the trainee can "see" that adjusting operational parameters or input data has an impact on the outputs. Figures 7.17 to 7.20 and Table 7.2 show examples of outputs that can be used for this purpose.

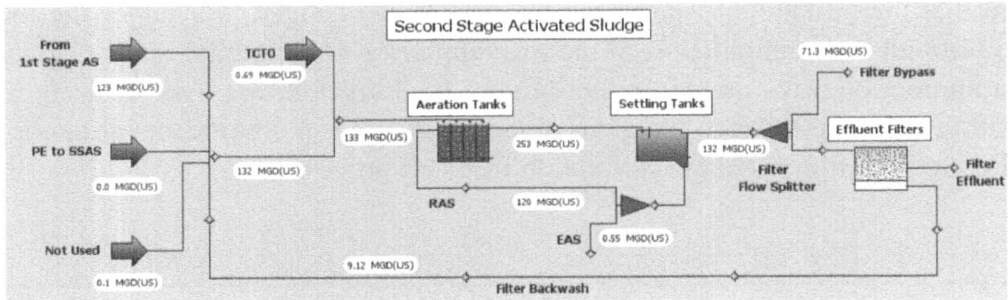

Figure 7.17 Process flow diagram showing flows throughout the process.

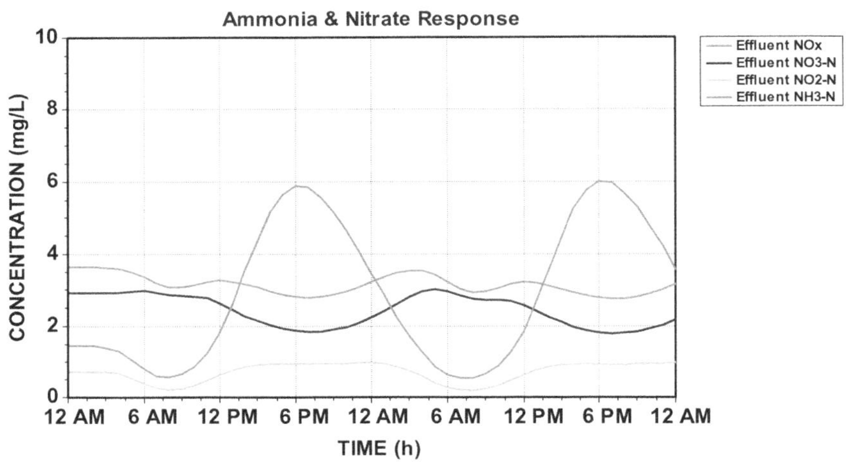

Figure 7.18 Graph showing effluent nitrogen species concentrations over 2 days.

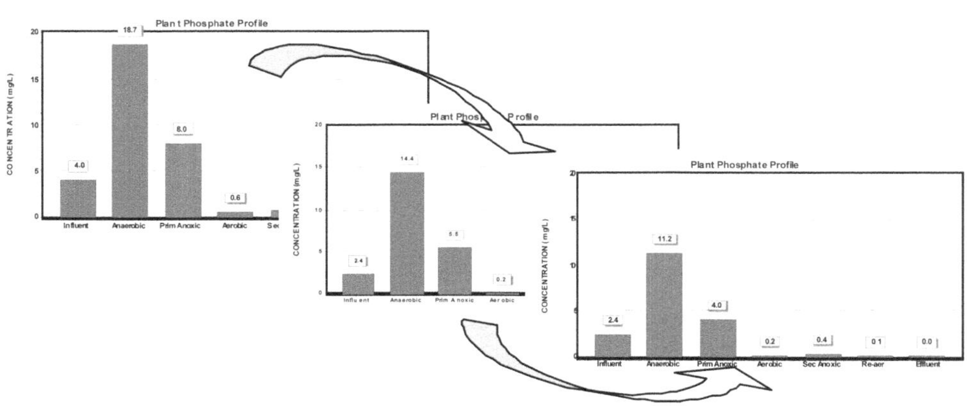

Figure 7.19 Bar graphs showing phosphate profile changing with time (each graph is viewed at a different simulated point in time).

Table 7.2 Outputs from dynamic simulation including summary statistics (note that "..." denotes data removed to keep table short for this publication).

Time (days)	$BOD_{5,EFF}$ (mg/L)	TSS_{EFF} (mg/L)	$NH_x{-}N_{EFF}$ (mg/L)	$NO_x{-}N_{EFF}$ (mg/L)	$N_{tot,\ EFF}$ (mg/L)	$P_{tot,\ EFF}$ (mg/L)	$PO_4{-}P_{EFF}$ (mg/L)	MLSS (mg/L)	Q_{EFF} m^3/d
0.00	2.20	8.46	0.26	2.93	4.75	0.41	0.17	2192	9930
0.05	1.99	7.31	0.25	2.92	4.67	0.37	0.17	2192	8336
–	–	–	–	–	–	–	–	–	–
0.95	2.18	7.82	0.30	2.80	4.63	0.32	0.10	2210	9085
1.00	2.24	8.21	0.26	2.68	4.49	0.33	0.10	2207	9628
Average	***2.21***	***8.49***	***0.41***	***3.22***	***5.19***	***0.44***	***0.20***	***2189***	***9957***
Peak	*2.69*	*11.21*	*1.00*	*3.89*	*6.57*	*0.61*	*0.32*	*2213*	*13,454*
Minimum	*1.77*	*6.25*	*0.20*	*2.68*	*4.49*	*0.32*	*0.10*	*2155*	*7099*
SD	*0.30*	*1.70*	*0.26*	*0.38*	*0.68*	*0.10*	*0.07*	*16*	*2199*

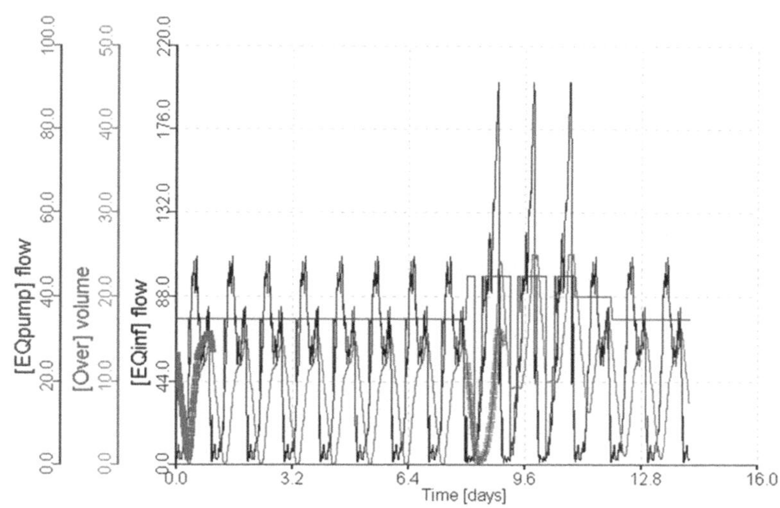

Figure 7.20 Example input data for a simulation investigating 3 peak flow days, following 7 typical diurnal flow days.

Uncertainty and sensitivity

Appendix H lists several sources of uncertainty in modelling. Several methods exist to help the modeller assess the sensitivity of the model to parameters for which there is quantifiable uncertainty. Methods include:

(1) Running different load conditions across the range of expected loads from minimum to maximum loads.
(2) Running extreme conditions such as very high and very low N/COD or P/COD ratios
(3) Carrying out sensitivity analyses of individual critical parameters (example in Figure 7.21)

(4) Monte Carlo simulations using frequency distributions for inputs and parameters in order to produce frequency distributions for the outputs that can then be displayed graphically as cumulative distributions (Figure 7.22), box and whisker plots or other graphical formats.

More details on dealing with uncertainties in modelling can be found in the Scientific and Technical Report of the IWA Task Group on Design and Operations Uncertainty (Belia *et al.* in preparation).

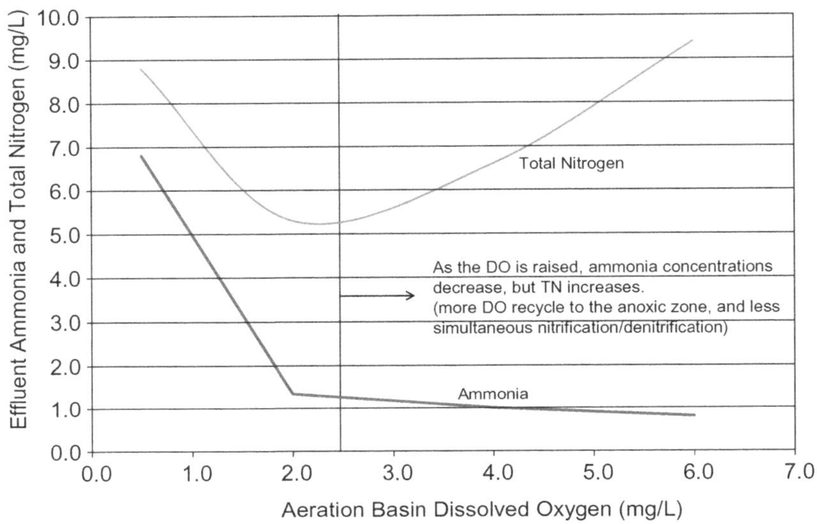

Figure 7.21 Example sensitivity plot varying one critical parameter (in this case, dissolved oxygen).

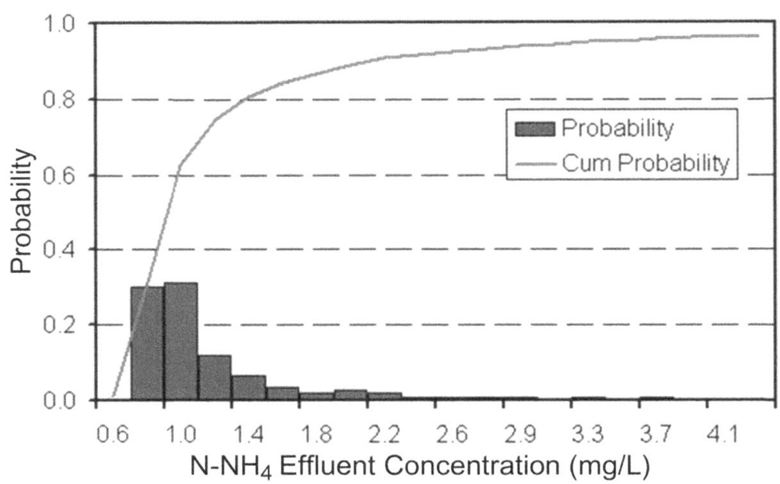

Figure 7.22 Example output from a Monte Carlo analysis (Martin *et al.* 2010).

7.5 A PLANT WIDE PROCESS MODEL FOR BEENYUP WWTP DESIGN UPGRADE

This text is adapted from Third *et al.* (2007).

UP Step 1: Project definition

Background

The Beenyup wastewater treatment plant (WWTP) consists of screening, grit removal, primary treatment using rectangular primary settling tanks (PST's) and secondary treatment using nitrifying/denitrifying activated sludge and circular secondary settling tanks (SST's). Primary sludge is thickened in the PST's and waste activated sludge (WAS) is thickened using dissolved air flotation (DAF) units. Mesophilic digestion is used to stabilise the combined sludge which is then dewatered using centrifuges. The model was commissioned to help plan an upgrade to the treatment capacity to 135,000 m^3/d initially and ultimately to 150,000 m^3/d.

Modelling objectives

The primary objective of the modelling project was to assess the unit process capacities under different operating conditions and to identify which unit processes were critical bottlenecks. It was reasoned that a model would illustrate the interactions between the liquid stream and solids processing units in a plant-wide mass balance.

Requirements

Upgrading the sludge treatment processes was a major issue so the accurate prediction of current and future sludge quantities was a focus of the modelling effort. It was expected that the model would estimate primary sludge and waste activated sludge (WAS) loads to within $\pm 10\%$ based on annual or monthly average values. Assessment of the treatment capacity of individual unit processes and the plant as a whole was carried out using conventional engineering loading criteria. As such a whole plant model was used to assess the loadings on each unit process but was not used to predict expected performance at excessive loadings beyond their design values. The model was also used to assess the aeration capacity of the secondary treatment and the ability of the activated sludge plant to meet target effluent total nitrogen loads. In order to assess the aeration capacity dynamic simulations with peak hourly air demands, weekend versus weekday loads, and seasonal variations were used.

UP Step 2: Data collection and reconciliation

As noted throughout this STR it is crucial that suitable data be used in the model. The assessment and analysis of the data used to develop the model for Beenyup was a challenge. Various issues are summarised here.

Influent flow validation

According to the available plant data, the average daily flow to the Beenyup WWTP was 95 ML/d as calculated by a flow totaliser on the inlet flumes. A comparison of instantaneous flows compared to the totalised flows showed that the PLC calculating the totalised flow contained an error. The error resulted in a 25% underestimation of the total plant inflow. This was verified by assessing the instantaneous flows measured at various flow measurement points across the plant. The analysis concluded that the actual daily average inflow to the plant in 2006 was 122 ML/d on week days, and as much as 135 ML/d on the weekend. For dynamic simulations, a representative week of diurnal profiles was used.

Influent loading rate

An analysis of the historical plant raw influent data was carried out for the period May 1999 – January 2007 to establish loads. Various changes over the years caused variations in the BOD and total suspended solids (TSS) measurements. The changes are summarised in Figure 7.23. The BOD sampling method was improved in June 2006 to analyze samples more rapidly which resulted in a 20% increase in reported BOD. The implementation of liquid polymer dosage for sludge dewatering in August 2006 improved the solids capture in the centrifuges and decreased the TSS concentration in the centrate. Finally, the raw influent sampling point was moved in December 2005 to a more representative sample location.

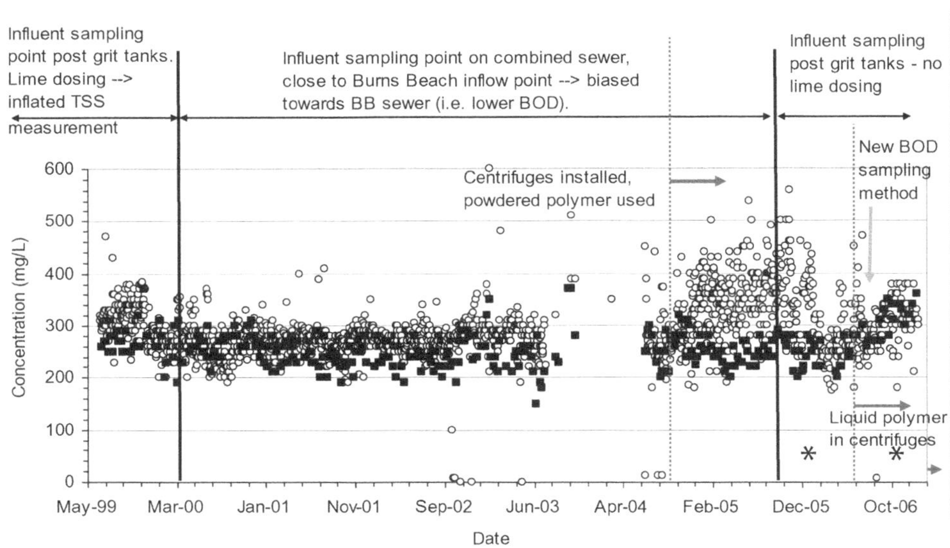

Figure 7.23 Summary of operational changes over the period 1999–2006 and their effect on BOD and TSS concentrations in the raw influent. Asterisks on the X-axis indicate special sampling periods.

The above information showed that process modelling and design should be based only on data during the period October 2006–January 2007. Data prior to this was deemed either unreliable or not representative of recent operation. The average concentrations were used for the steady-state modelling. The raw influent measurements contain the centrate flow and so the process model was used to estimate the centrate loads. The raw wastewater was then estimated by subtracting the estimated centrate loads from the measured raw influent values. The derived wastewater characteristics were used to develop the future loads.

Influent characterisation

Appropriate characterisation of the wastewater is one of the most important factors driving the model predicted behaviour. To produce reliable wastewater characterisation data for Beenyup WWTP, two series of "special sampling" runs were performed in 2006. Flow-paced, time-paced and composite samples were taken over several days. The first data set contained measurements of the raw influent and primary effluent simultaneously, and was therefore used as the first step in developing the fractionation.

The second data set contained primary effluent data only, and was used together with more recent influent data to refine the fractionation. In developing an appropriate fractionation for the raw wastewater composition, particular attention was paid to adjusting the readily biodegradable COD fraction, biodegradable particulate COD, and unbiodegradable particulate COD fractions in order to match the BOD/VSS and VSS/TSS ratios in the raw and primary effluent. The nitrogen and phosphorous fractions were similarly adjusted.

UP Step 3: Plant model set-up

In defining the model structure, particular attention was paid to several things including the level of complexity, the modelled processes, the model stability as well as the speed of the simulations. Secondary treatment at Beenyup consists of 4 "modules", with flow splits of 40% and 60% between Modules 1 & 2 and Modules 3 & 4, respectively. Modules 1 & 2 consist of four step-feed bioreactors, while Modules 3 & 4 consist of 5 bioreactors in a conventional single-pass anoxic-aerobic configuration, with a single recycle from the aerobic zone to the anoxic zone.

An important step in developing the model structure was determining the appropriate number of tanks in series to properly characterise the hydraulics of the main reactors. The formula proposed by Fujie *et al.* (1983) was used to estimate the number of tanks in series. Figure 7.24 illustrates the model setup.

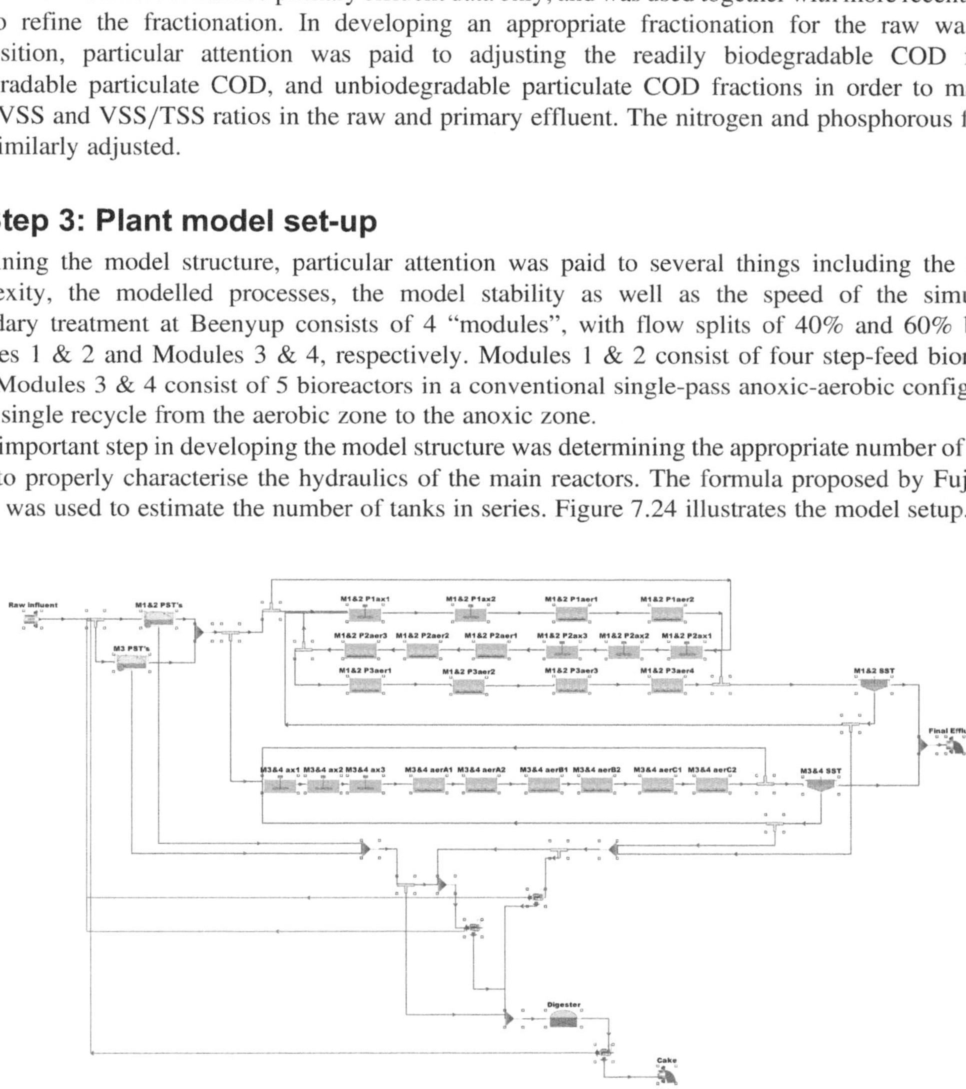

Figure 7.24 Schematic overview of the whole plant process model for Beenyup WWTP.

The physical characteristics, operational conditions (i.e. flow splits, recycles, etc.) and wastewater fractionation was entered into the model layout.

A spreadsheet for the results was developed. Actual plant data was also entered alongside the simulated outputs to enable a direct comparison between simulated data and plant performance. The original design capacities of each process unit were entered into the spreadsheet for comparison with the modelled outputs and maximum operating capacities. Steady-state simulations were carried out and the generated outputs were entered into the spreadsheet. The model structure and initial outputs appeared to be suitable for the purposes of the modelling; the next step was to calibrate and validate the model based on plant recorded data.

UP Step 4: Calibration and validation

Plant data recorded over the reliable sampling period of September 2006–January 2007 was used to calibrate the model. The following four general steps were taken:

(1) *Solids Balance*: The first step was to derive an accurate match between the solids balance generated in the model and the solids loads measured across the plant. The initial simulated numbers differed significantly (>10%) from the plant data, which triggered an extensive analysis of the plant data to determine the origin of the discrepancy. During this analysis, a number of measurement inaccuracies in the plant came to light. For example, the flow totaliser on the thickened WAS stream from the DAF's contained a similar PLC calculation error to the inlet flumes, resulting in a mismatch between the WAS and the total sludge mass.

(2) *Aeration system*: Aeration model parameters need to be individually derived for each particular treatment plant as they are system-specific, and depend on the depth, diffuser density and type of system. Reasonable estimates for these parameters were developed using the methodology developed by Johnson (1993). Once calibrated, the plant model indicated that the required air flows were significantly higher (>20%) than the air flows being recorded from the flow transmitters at the plant. Further investigations indicated problems with the air flow meters and a programme was subsequently implemented to replace or improve flow measurements at the plant. This highlights the value of modelling for uncovering plant deficiencies.

(3) *Nitrification*: Once the solids balance and aeration system were calibrated, nitrification was calibrated. Default nitrification kinetic parameters were not changed as there was no cause to believe that any form of inhibition was occurring nor was alkalinity an issue. Steady-state simulations indicated that full nitrification should be achieved; however, actual plant data showed an average effluent NH_x-N concentration of 2.3 mg N/L. This discrepancy was noted for further investigation during dynamic simulations.

(4) *Denitrification*: A significant discrepancy was found between the measured and predicted denitrification. While the model predicted nitrate concentrations as high as 20 mg N/L, measured nitrate was on average as low as 11 mg N/L. This led to the conclusion that significant denitrification by means of simultaneous nitrification and denitrification is occurring in the plant

A significant problem encountered during model calibration was the high variability in plant data. This variability made it difficult to establish an "average" operating condition for steady-state modelling. The plant was operating in a transient state during the 5-month period of plant data and this further complicated the calibration of the steady-state model.

Despite these issues, it was possible to match the model output to the measured data to within ±10% for most parameters.

UP Step 5: Simulation and result interpretation

The calibrated model was used to meet the main modelling objective, which was to identify current process bottlenecks and determine the required staging of design upgrades. Steady-state modelling was used for determining the major process bottlenecks. Dynamic simulation was used to check the plant's ability to treat peak loads. The outcome of the steady-state modelling is summarised in Figure 7.25. The overview in this figure provides a clear picture of the process units that are over their maximum design capacity and require upgrading.

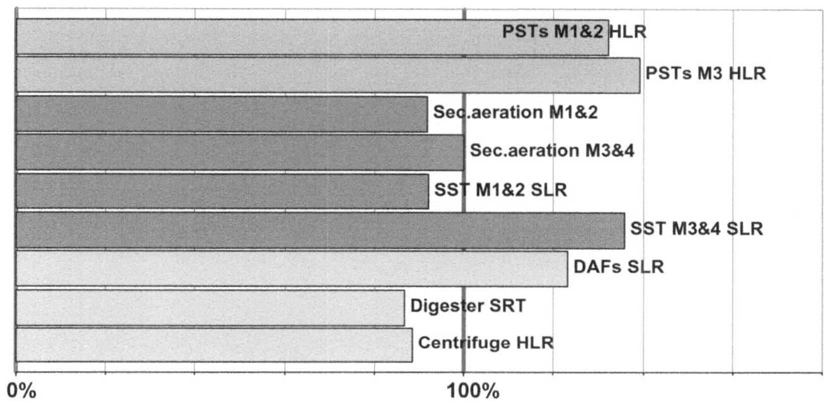

Figure 7.25 Overview of process unit capacities at Beenyup WWTP (M1 & 2 = modules 1 & 2; M3 & 4 = modules 3 & 4; PST = primary sedimentation tanks; HLR = hydraulic loading rate; SLR = solids loading rate; SST = secondary sedimentation tank; SRT = sludge retention time).

The main objective of the dynamic modelling was to assess the aeration system and the ability of the blowers and the diffusers to deliver peak air demand. During the steady-state modelling it became apparent that the plant should be able to nitrify completely (i.e. effluent NH_x-N < 0.1 mg N/L) assuming no oxygen limitation. However, the plant data showed an average effluent NH_x-N concentration of 2.3 mg N/L. One possible explanation for this would be an oxygen limitation under operating conditions. Further investigation of the air distribution showed that the air requirement for complete nitrification is not effectively being delivered into each aeration zone due to imbalances in the air distribution system and problems with control elements. The front-end aerobic zones required approximately 15% more air flow than the current system could deliver.

Conclusions

The plant wide process model for the Beenyup WWTP was calibrated and proved to be a useful tool in assessing unit capacities, identifying bottlenecks, and enabling the design engineers to stage plant upgrades and expansion. The GMP Unified Protocol provided a systematic framework for the development of the model, by providing a logical approach to developing the process model. Not unexpectedly, there was a strong link between the calibration and data validation steps.

Chapter 8

Use of activated sludge models for industrial wastewater

Short summary

In order to promote the use of activated sludge models in industrial wastewater treatment, it is beneficial to address industrial wastewater treatment modelling in its own section. The model application steps are the same in both municipal and industrial treatment and, for the most part, the *Unified Protocol* as detailed in Chapter 5 can be followed directly. One key difference in industrial wastewater modelling is the availability of detailed information on the wastewater source. Understanding the upstream production processes and characterising the major influent streams individually are the key factors for efficient data collection, model development, and effective scenario analysis. Another large difference is that extension and modification of the process model is often required in many industrial applications because of the unique nature of the wastewater components and plant operating conditions. These modifications require a fundamental understanding of process modelling. This chapter describes the limitations of existing activated sludge models when applied to industrial wastewaters, possible model modifications, typical pitfalls, and suggestions.

8.1 INTRODUCTION

Although the STR focuses on activated sludge (AS) models applied to municipal treatment plants, this chapter has been included because activated sludge is used to treat industrial wastewater in a wide range of industries. In industrial activated sludge plants, practitioners face very similar design and operational problems as they do in municipal plants. This chapter should also raise awareness of the typical pitfalls encountered with municipal plants that receive significant contributions of industrial wastewater. The GMP Task Group believes that transferring modelling experience from municipal to industrial wastewater applications and *vice-versa* is highly beneficial for all practitioners.

Municipal and industrial plants share many modelling objectives, and several examples of successful model applications to industrial WWTPs can be found in the literature (e.g. Bury *et al.* 2002; Eremektar *et al.* 2002; Ky *et al.* 2001; Orhon *et al.* 2009). The general steps of the *Unified Protocol* can be applied to the industrial sector with few adaptations. However, this chapter will focus on specific aspects related to modelling industrial wastewater treatment.

A key aspect of industrial wastewater treatment modelling is understanding a model's limitations. The risks of inappropriate model use in industrial applications is higher than in municipal applications because the standard AS models have not been developed for industrial activated sludge plants (Henze *et al.* 2000). It should be noted that many industrial applications require specific model modifications. The focus of this chapter is on modelling pure industrial wastewater, but similar considerations apply in a municipal application where a significant industrial contribution is identified.

This chapter discusses special characteristics of some industrial wastewaters, the impact on population dynamics caused by the potentially high variability of industrial influents, and operational conditions. A list of possible pitfalls and suggestions is also provided.

8.2 LINKS TO UNIFIED PROTOCOL STEPS

Industrial wastewater modelling studies should follow the general *Unified Protocol* steps. This section highlights some additional concerns and suggestions.

For Step 1, *Project Definition*, a more detailed agreement on the scope of influent modelling will be required due to the characteristics of industrial wastewater. For example, if the influent source can be controlled, this could have a large effect on the project objectives.

In Step 2, *Data Collection and Reconciliation*, the data quality is of the same importance and similar methods for data quality evaluation and reconciliation can be used. Additional data needs and respirogram analysis are discussed.

Step 3, *Plant Model Set-up,* is usually the same as often similar process units are used. When multiple influent streams are considered, the influent models often need to be expanded to describe the different stream characteristics. Model modification and expansion at this step could be the main challenge when modifications are required. Targeted production system information and additional data can be used to determine which processes and components need to be expanded and modified in the model.

In Step 4, *Calibration and Validation,* the calibration process follows the same general rules, though typically there is less experience available on parameter values. Therefore, parameter identification will require additional effort especially for expanded and modified process models.

During Step 5, *Simulation and Results Interpretation,* often special attention is given to scenario options which include influent source control.

Table 8.1 summarises specific aspects of industrial wastewater modelling with respect to the *Unified Protocol* steps.

Table 8.1 Unified Protocol steps and industrial application aspects.

Protocol steps	Specific aspects for industrial applications
(1) Project Definition	(1) Production plant information – Product, raw material, chemical use, water use – Site information (single or complex plant) – Production plant operation mode (continuous or batch) – Water collection scheme (major streams) – Typical upsets, incidents – Future expansion plan (2) pH and temperature range and variation (3) Effluent quality variation (unbiodegradable or slowly degradable component) (4) Acclimation concerns (plant shutdown and start-up)

(Continued)

Table 8.1 Unified Protocol steps and industrial application aspects (*Continued*).

Protocol steps	Specific aspects for industrial applications
	(5) Inhibition related incidents (6) High particulate content (hydrolysable or unbiodegradable) (7) Physico-chemical aspects (precipitation, VOC, ORP)
(2) Data Collection and Reconciliation	(1) Production plant information (2) Efficient data collection – Raw material analysis – Raw material recovery ratio – Water use information – Replacement with simple measurement (EC, UV/vis) (3) Laboratory measurement – Continuous treatability test – Soluble unbiodegradability test – Respirometry with change of soluble COD in time – COD, N and P fractionation – Yield evaluation – Denitrification test – Dose response test for inhibition and toxicity – VOC purge test
(3) Plant Model Set-up	(1) Influent consisting of multiple streams (2) Model modifications – Additional states (soluble, particulate) and processes – Number of components from respirograms – Acclimation (multiple heterotrophs and autotrophs) – Inhibition or toxicity – pH and temperature dependency – Macro- and micronutrient requirements – Physico-chemical model – Precipitation – Chemical redox reactions – VOC fate
(4) Calibration and Validation	(1) Fitting respirometry model to respirogram data – Hydrolysis rate and saturation constants – Inhibition or toxicity parameters (2) Estimated end of pipe respirogram validation (3) Estimated effluent soluble unbiodegradable validation
(5) Simulation and Result Interpretation	(1) Effect of specific stream – Variability, future expansion, future shutdown (2) Possibility of production plant suspension

8.3 WASTEWATER SOURCES

Understanding the production process, which generates the wastewater, is paramount in industrial applications. Detailed information on the wastewater generation may be available though this is not always the case. The following questions might be used to guide the industrial wastewater characterisation.

(1) What does the production plant manufacture?

Knowledge about the plant will help the modelling process. For instance, knowledge about what the plant produces, how many major production processes exist, where and how the water is used and disposed of, what kind of raw materials are used, and what chemicals are added will help the modeller with the characterisation of the incoming wastewater. This information provides insight into the wastewater characteristics including flow distribution, contaminant load distribution, biodegradability, and often shows the potential for source control.

(2) How does the production plant operate?

The mode of operation (e.g. continuous or intermittent) has a major impact on the variation in load to the wastewater treatment plant and this information should be the basis for planning the data collection.

Certain production plants operate continuously for months or years without any major break in operation. This is often the case for bulk chemical production plants. Typically in such cases the WWTP loading shows minimal variation.

On the other hand, certain production plants frequently operate intermittently within a relatively short time. Most food industry plants operate during the day with cleaning in the evening. The cleaning cycle generally means increased loading to the wastewater treatment plant. Such cycles can be hourly, daily, weekly, monthly or sometimes seasonally. Typical in chemical industries is to produce products in batches which leads to large variations in influent quality. Data collection schedules need to take into account these cycles to get representative data and to allow reproduction of wastewater characteristics in the model.

Unexpected wastewater discharge related to plant accidents is another source of variation. Production plants often discharge out-of-specification products, which can be the main source of large loading peaks. In this context, information on typical incidents also needs to be collected.

(3) How is the wastewater collected?

In addition to the piping and sewer system itself, there may be upstream reservoir tanks, final equalisation tanks, and active water flow control in the collection system. This information is critical as it could impact the decision of which streams need to be modelled. The ability to model the control of volumetric and mass loads may be important. If the equalisation capacity is sufficient to equalise all the variation, the project may be able to focus on the characterisation of equalised end-of-pipe wastewater. If this is not the case, the modeller needs to carefully decide how many streams need to be included in the collection system scheme.

8.3.1 Wastewater composition with few specific contaminants

Many processing plants produce a limited number of products, especially when the plant is producing bulk chemicals. The water volume and organic loading to this type of industrial facility can be very stable during the production period, and complete acclimation of the activated sludge can be achieved. However, a wastewater with a simple chemical composition can have a significant impact on treatment plant design and operation. The high strength of many industrial wastewaters may require special attention.

As an example, acetic acid is frequently used as a raw material and solvent in the chemical industry and it is not unusual that the wastewater organics at such a plant are largely composed of highly concentrated acetic acid. Acetic acid can be treated easily with activated sludge even at a high loading rate. The influent pH can be very low, but provided the reactor pH stays neutral, treatment can be achieved. If such a plant is overloaded (e.g. by insufficient oxygen supply), the residual organic acids can lower the reactor pH.

Once the pH is below the optimal range for biological growth, the biological activity is impeded causing an increase in acetic acid concentration and a further lowering of the pH. This is a typical positive feedback effect in organic acid treatment and can cause serious loss of performance. Load control or active pH control is essential for such an influent especially during the start-up period.

To be able to model this scenario, the effect of pH on biomass activity, the pH variation itself and potentially a pH controller need to be incorporated in the model. Acetic acid is often stored and not metabolised immediately by the biomass, which has a strong decoupling effect on the dynamics of oxygen requirement and soluble COD concentration in the reactor. In this case a storage process such as the one introduced in ASM3 may be required in the model.

8.3.2 Multiple wastewater sources of different nature

In the chemical industry, end-of-pipe influent characteristics can drastically change depending on the production lines in operation. This situation typically poses challenges related to acclimation of biomass and should be modelled if that condition exists.

Modelling lumped wastewater characteristics can be very difficult because production at the plant can produce a large number of wastewater combinations and each of these combinations can have different wastewater characteristics. However, modelling each wastewater source individually may provide the answer. Instead of sampling at the outlet, the combined stream variation may be estimated from the combination of upstream flows. Estimated end-of-pipe characteristics should be verified by actual measurements. Figure 8.1 illustrates this situation. Once a multiple source model is built, it is relatively easy to conduct scenario analyses for varying wastewater characteristics in combination with different treatment plant operating conditions.

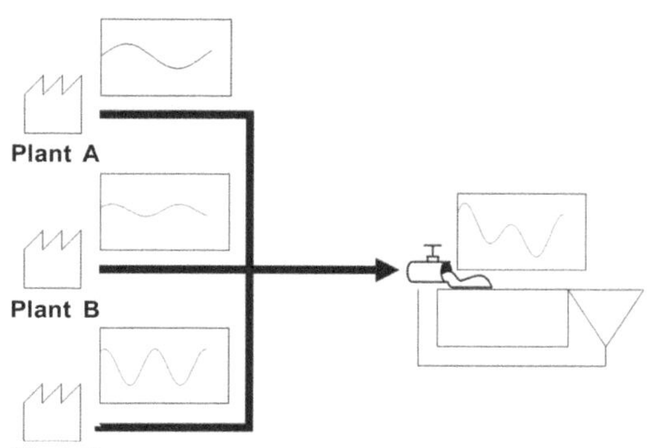

Figure 8.1 Variations of end-of-pipe wastewater characteristics based on stable multiple sources.

When the plant model consists of multiple influent streams, each of the components in these streams may have different characteristics. Standard AS models assume only one value for some critical variables. For example, the N and P fraction parameters (e.g. as introduced in ASM2d and ASM3) assume fixed nutrient ratios for the carbonaceous state variables and this can make the modelling of multiple streams difficult. An approach with lumped soluble and particulate nitrogen state variables (as in ASM1) may be more adaptable.

8.3.3 Source control

Another characteristic, which sets industrial models apart, is the possibility of having some degree of control over the wastewater sources. The wastewater plant manager may be able to identify the major wastewater sources which have a critical impact on final effluent quality. In an emergency situation an operator might be able to limit or suspend a production process to ensure that the treatment plant continues to meet its effluent quality criteria. For modelling purposes, it is important to understand the characteristics of the major wastewater sources and the possibilities for control.

8.4 INFLUENT COMPONENTS

The main difference between municipal and industrial wastewater is the potential for higher variability in industrial wastewater and the potential existence of specific components having an impact on degradability, sludge production and effluent concentrations. In this section different fractions and components are discussed which require special attention in terms of measurements and modelling.

8.4.1 Unbiodegradable fractions

The high variability of the unbiodegradable fractions in industrial wastewaters is an important aspect for wastewater treatment and modelling.

8.4.1.1 Soluble unbiodegradable fractions

Industrial wastewater can have a significant soluble unbiodegradable (inert) organic fraction (Germirli *et al.* 1991). When end-of-pipe wastewater characteristics are reasonably stable, the soluble unbiodegradable fraction can be estimated from effluent filtered COD measurements. Unfortunately, in industrial applications, stable characteristics are not common and this can lead to significant variations in the effluent COD. To properly predict soluble effluent quality, the soluble unbiodegradable content of major streams should be experimentally determined. Figure 8.2 shows an example of soluble unbiodegradable fractions in individual waste streams at a sugar production factory. Streams from some processes contain very low soluble unbiodegradable fractions while others contain significant soluble unbiodegradable material.

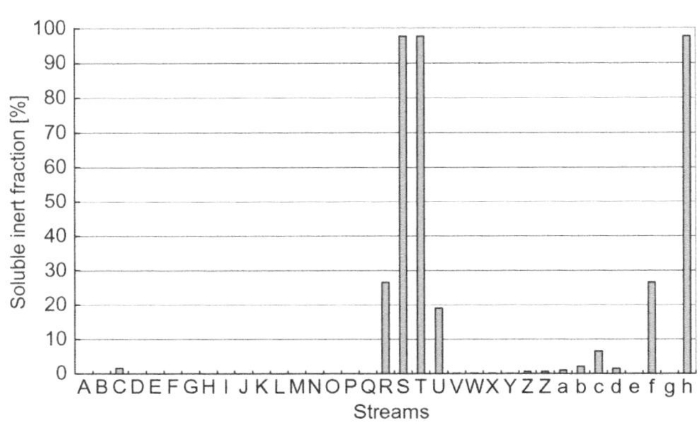

Figure 8.2 Example of soluble unbiodegradable fractions (% of total COD_{Mn}) of wastewater streams in a sugar production plant. COD_{Mn} (COD measured by the potassium permanganate method).

Modelling of soluble unbiodegradable fractions

Even though the production of soluble organic unbiodegradable components during the decomposition of influent organics and the decay process is documented (Boero *et al.* 1990), it is not typically included in standard activated sludge (AS) models. Industrial plants often operate using long SRT conditions, which could increase the importance of these processes typically ignored in standard AS models.

8.4.1.2 Particulate unbiodegradable fractions

Industrial plants often have to deal with significant organic or inorganic particulate unbiodegradable loads. This material has a large impact on sludge production. Occasionally, inorganic components such as calcium ions can form precipitates in the reactors. To be able to consider the potential formation of these precipitates, information on unbiodegradable components from the production plant operators and background on chemical speciation and precipitation will be required.

Modelling of particulate unbiodegradable fractions

In most cases, the existing unbiodegradable fraction in standard AS models can be used to model the particulate unbiodegradable fractions. If significant precipitation takes place, the related processes have to be identified and modelled.

8.4.2 Biodegradable organic fractions

In industrial wastewaters, the biodegradable fraction can vary considerably and is often characterised using a time dependent oxygen uptake rate (OUR) profile (i.e. a respirogram) (Eremektar *et al.* 2002; Orhon *et al.* 1999; Insel *et al.* 2003). The example respirograms in Figure 8.3 show a wide variety of biodegradation profiles. The oxygen uptake rate over time provides information on the contribution of major biodegradable components. An initial peak in the graph suggests that there will be a high demand for oxygen at the inlet of a plug-flow activated sludge reactor. Respirograms obtained from chemical process wastewaters suggest that multiple components with different degradation rates can exist in the same influent and as a result, the possibility for substrate storage may have to be considered.

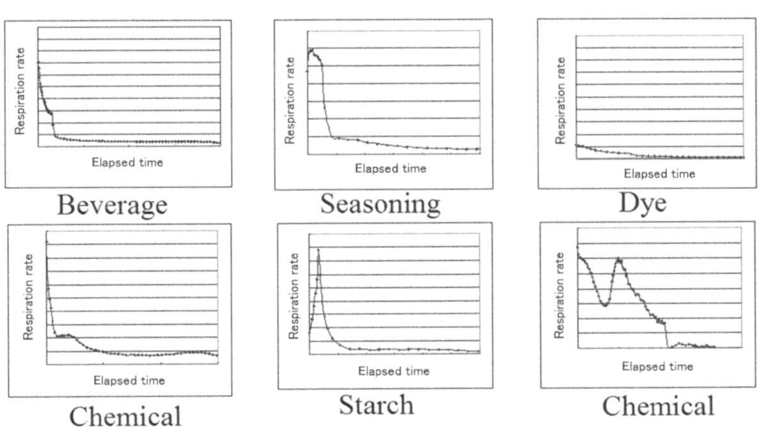

Figure 8.3 Respirogram examples of various wastewaters.

Because of the chemical characteristics of industrial wastes, it is not uncommon for soluble components in industrial wastewaters to produce respirogram patterns that are similar to municipal slowly degradable material (typically particulate and/or colloidal municipally) (Bury *et al.* 2002). If a soluble, slowly biodegradable component cannot be adsorbed to the sludge flocs or stored by the organisms, the potential exists for it to bleed through the reactor causing variations in effluent quality.

Modelling biodegradable fractions

In industrial applications, multiple biodegradable components with distinct kinetics may exist and therefore special attention should be given to identifying different fractions with specific biodegradability. The model needs to reflect these fractions, which may mean additional state variables and processes.

The addition of multiple soluble and particulate hydrolysable model components is occasionally necessary when the modelling of multiple streams is required. Model components such as soluble rapidly hydrolysable, slowly hydrolysable, adsorbable, particulate slowly biodegradable COD should be defined based on the degradation characteristics of the organics and the molecular size of the components as molecular size may play a role in the treatment technology (i.e. membrane bioreactor treatment). A conceptual model example is presented in Figure 8.4 and a multi-stream model is shown in Figure 8.5. In Figure 8.5, fermentable substrate and fermentation products are assumed to exist in all streams.

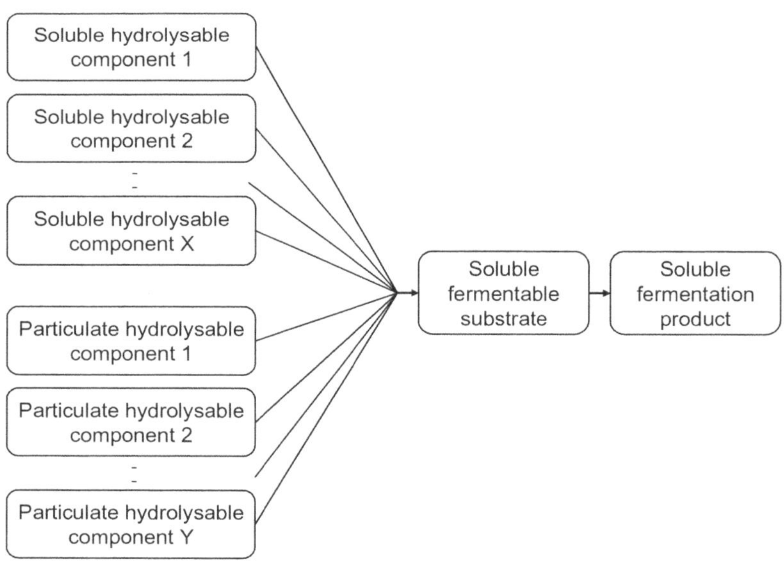

Figure 8.4 Conceptual model example for multiple slowly hydrolysable components.

The kinetic models used for the conversions of these components need to be carefully selected. Equation group 8.1 shows the most often used kinetic expressions; other expressions may include the ones shown in Equation group 8.2. Various batch tests, including respirogram measurements can be used to select the most suitable equations. Experimental setups typically include different substrate concentrations and different sludge concentrations as these will give information on the reactions and aid the model selection process.

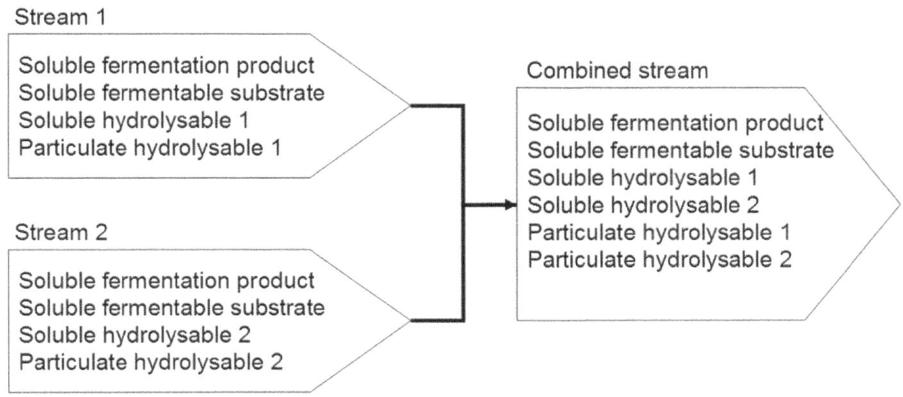

Figure 8.5 Conceptual model example for multiple streams.

For model selection examples see for example Goel *et al.* (1998) and Orhon *et al.* (1999).

$$\begin{aligned} & K_h \qquad (1) \\ & K_h \cdot X \qquad (2) \\ & K_h \cdot \frac{X_s}{K_x + X_s} X \qquad (3) \\ & K_h \cdot \frac{X_s/X}{K_x + X_s/X} X \qquad (4) \end{aligned} \tag{8.1}$$

where:

(1): Zero order
(2): First order
(3): Monod-type
(4): Surface limited

Substrate degradation, sorption and/or storage processes may need to be considered if estimation of soluble COD in the reactor and effluent is important. Baker (1994) included an adsorption step for the modelling of hydrocarbon components. In that conceptual model, hydrocarbons are quickly adsorbed to the activated sludge floc thereby moving the material from the liquid to the solids phase (Figure 8.6). The adsorbed "particulate" hydrocarbon is directly utilised for heterotrophic growth with surface limited model kinetics. This modelling approach combined with the storage process model was successfully applied to reduce the difference between the measured and estimated soluble COD.

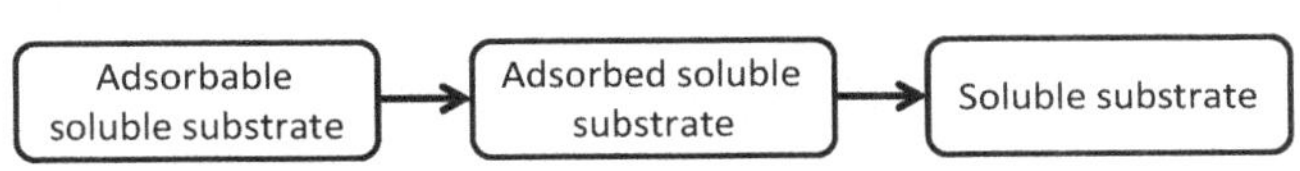

Figure 8.6 Adsorption concept model (Baker, 1994).

Figure 8.7 shows a conceptual model for particulate hydrolysable components. The kinetic and stoichiometric components to this type of model modification need to be selected based on experimental data. Hydrolysis is difficult to characterize, but soluble nitrogen release, which is easier to measure, may provide additional information for model calibration. Typically a surface limited expression similar to degradation kinetics of particulate substrate or a first order model proportional to active heterotrophic biomass can be chosen. In some cases nutrient fraction-based models can be difficult to calibrate so separate state variables representing the applicable nitrogen and phosphorus components (ASM1 approach) may provide the necessary flexibility for the model.

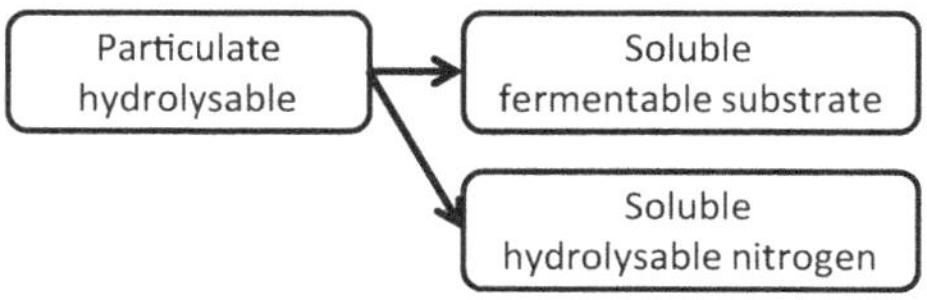

Figure 8.7 Conceptual model example for nitrogen fractions in a hydrolysable component.

Modelling soluble biodegradable components

Industrial wastewater can have several soluble biodegradable components, but soluble biodegradable matter in most standard AS models is modelled as one state variable. ASM2d provides volatile fatty acids (S_{VFA}) and readily biodegradable (fermentable) substrate (S_F), but even this may not provide enough flexibility for multiple soluble components in an industrial wastewater. If a soluble, slowly biodegradable component is modelled using the slowly biodegradable particulate state variable available in the AS models, significant inconsistencies can occur between measured and estimated soluble COD.

Modelling particulate hydrolysable components

In ASM1 and ASM2d, the biodegradable particulate component originating from the influent and the slowly biodegradable substrate generated in the decay processes are not differentiated from each other and are modelled as the same state variable X_B. In industrial applications, the characteristics of influent particulate components, their hydrolysis rate, and nitrogen content may be different from stream to stream and can be significantly different from the slowly biodegradable substrate generated by decay. In ASM3 endogenous respiration solves this problem as X_B is used exclusively for influent particulate slowly degradable matter. However, ASM3 suffers from the same limitations when it comes to modelling multiple streams with different components that have different hydrolysis rates.

8.4.2.1 Particulate very slowly biodegradable fractions

Very slowly biodegradable particulate organics is a source of controversy in wastewater treatment modelling and this is especially applicable to certain industrial wastes. This component may be considered undegradable under typical municipal plant conditions. In the food industry, for example, many production plants use raw vegetables or grains. Typically the treatment approach in these plants is longer SRTs and higher MLSS concentrations to maximise the degradation of this very slowly biodegradable material.

Modelling of very slowly biodegradable fractions

The modelling of the hydrolysis of particulate components like this can be the key for the accurate estimation of sludge production, nitrogen removal, phosphorus removal and oxygen requirements. The nitrogen content and the organic/inorganic ratio of these components can be significantly different from other substrates. In many modelling projects where a specific particulate biodegradable component is involved, careful characterisation of such components may be important. It may not be possible to confirm hydrolysis rates experimentally, so plant records on sludge production may serve as a surrogate method and a detailed nitrogen analysis can provide additional information.

If this very slowly degradable fraction is not included, it is possible that the developed model will not be able to accurately predict the long-term hydrolysis of the particulate wastewater fraction and could potentially misrepresent the process kinetics and the sludge production. So again, the modeller will need to fully understand the influent wastewater characteristics in order to include the necessary processes within the model.

8.4.3 Nitrogen fractions

In industrial wastewaters, soluble organic nitrogen is not always linked to a specific organic component nor is the nutrient content in different streams always identical (Bury *et al.* 2002), and this makes the application of the standard AS models difficult.

Most standard AS models only include a simplified one-step nitrification process. In some situations, a more detailed process description may be required including multi-step nitrification and denitrification (e.g. to predict nitrite accumulation or nitrous oxide formation). This may require adding specific model state variables and processes.

8.4.4 Inhibitory and toxic components

In industrial applications, inhibitory chemicals (including salinity) can significantly deteriorate plant performance. There are numerous reports on the inhibitory effects of industrial wastewater and specific chemicals (metals or detergents) on nitrifiers (Wood *et al.* 1981). This issue was recognised in the WERF manual for wastewater characterisation (Melcer *et al.* 2003) which recommends the evaluation of the nitrifier growth rate before sizing the nitrification reactor volume especially for plants with significant industrial contributions. It should also be recognised that certain substances can accumulate in the sludge and cause chronic toxicity. If inhibition is suspected, then it should be investigated and modelled if necessary.

Phenol can be used as a case study. Phenol is a common raw material for plastic production and various phenolics are known to be generated in the coke process. Phenol's inhibitory effect on the activated sludge process is well described in the literature (Jones *et al.* 1973; Rozich & Gaudy, 1985). An important aspect of phenol inhibition is that it shows typical substrate inhibition characteristics. That is, it is inhibitory at high concentrations but at low concentrations it is a biodegradable substrate. If a plant is continuously loaded with phenol, and the load is degradable, the phenol concentration in the reactor will stay below its inhibitory concentration. On the other hand if the load exceeds what can be degraded, the phenol concentration in the reactor may increase and exert an inhibitory effect on biomass activity, which in turn will cause further accumulation of phenol. This is another example of a typical positive feedback cycle.

An inhibitory chemical may not kill the organisms but may reversibly reduce the activity of those same organisms. In the case of an inhibitory chemical, dilution can be an effective remedy. The dose-response relationship provides basic information on the critical concentration of the inhibitory effect, because it is generally considered that there is a threshold concentration for inhibition. This contrasts toxic chemicals which irreversibly reduce the activity of the population. After a toxic event, the reactors may have to be cleared of the chemical and the process may have to be restarted. Information on treatability by pre-processing or dilution is required to propose effective remedies.

Modelling inhibition and toxicity

It is interesting to note that when modelling inhibitory chemicals, the reactor configuration may need to be considered. In a plug-flow reactor it is possible that the concentration at the inlet may be high and may cause severe inhibition. Consideration of this situation may be critical for the estimation of the overall contaminant removal. These situations provide a powerful rationale for quantitative modelling of the inhibitory effect on the biodegradation performance. Table 8.2 contains a list of items that need to be considered when inhibition is the focus of the study.

Table 8.2 Inhibitory aspects to be considered when modelled.

Inhibition type	Inhibitory (reversible) or toxic (irreversible) discrimination determines the proposed remedies
Effect of concentration level	Dose-response relationship establishes the dilution effects
Biodegradability of inhibitory chemical	Biodegradability and specific population requirements can affect potential remedies
Treatability by pre-processing	Inhibitory effect reduction by pre-processing such as coagulation, Activated carbon or physical oxidation may need to be confirmed for remedy proposals.

The standard AS models do not include inhibition or toxicity terms, but the following sections outline a few options that might be used when extending/modifying a published model.

A *reversible inhibition effect* is typically modelled by reducing the organism's overall growth rate (Nowak *et al.* 1995). The following inhibition models are frequently used and are described in Equation group 8.3:

- *(1) Competitive inhibition* displays an increased half-saturation constant for substrate under higher inhibitor concentrations.
- *(2) Haldane inhibition* models are often used for high substrate concentration inhibition.
- *(3) Non-competitive inhibition* shows decreased growth rate under higher inhibitor concentration.

$$\mu \cdot \frac{S}{K_s \cdot (1 + I/K_I) + S} \quad (1)$$

$$\mu \cdot \frac{S}{K_S + S + \frac{S^2}{K_I}} \quad (2)$$

$$\mu \cdot \frac{Ki}{Ki + Si} \frac{S}{K_S + S} \quad (3) \qquad (8.2)$$

A *toxic (irreversible) effect* can be described in the model using an increased decay rate. Appropriate model selection and parameter identification can be guided by growth rate and half saturation evaluation using respirometry. The dose-response relationship should always be observed. Increased dose of a suspected inhibitor should always produce more inhibitory effect.

Nowak *et al.* (1995) discussed biodegradability of inhibitors based on the example of allylthiourea, and modelled it as non-competitive inhibition by an adsorbed biodegradable inhibitor. The inhibitory effect of phenol is typically modelled as a competitive biodegradable inhibitor using Haldane type equations.

8.4.5 Physico-chemical characteristics of specific chemicals

Precipitation of inorganic components will affect the sludge production and this can be pronounced in industrial treatment. Inorganic components that cannot be oxidised biologically may still be chemically oxidised and can have a significant impact on oxygen demand. Baker (1994) showed this when modelling the oxidation of sodium meta-bisulphite. If volatile organic compounds (VOCs) are included in the influent, volatilisation by aeration can reduce these components in parallel with the biological oxidation.

This section provides examples of how standard AS models have been extended to include physico-chemical characteristics.

Modelling of chemical and physico-chemical processes

Simplified phosphorus metal precipitation is included in ASM2d, but other physico-chemical processes are neglected in all standard AS models. Volatilisation and redox processes are ignored; however these processes can have a significant effect on industrial applications.

Depending on the application, the modeller may need to account for precipitation processes. An empirical phosphorus precipitation model is included in ASM2d and this can be expanded if necessary. For example, Ky *et al.* (2001) appended phosphorus precipitation with calcium to achieve a better precipitation estimate under high calcium concentrations and high pH based on the investigation of Maurer *et al.* (1999). de Haas *et al.* (2001) compared metal addition scenarios by discriminating between aluminium and ferrous precipitation stoichiometry. For more generic applications, equilibrium precipitation and dissolution models based on chemical speciation are available (Allison *et al.* 1991). These models provide information on which precipitation processes are likely to occur and can be linked to activated sludge models. The modeller will have to decide if the model modification will be based on an equilibrium or some other approach. To ensure that an appropriate dynamic model is developed, it is often necessary to combine knowledge from chemistry with engineering experience and observations on plant behaviour and data from laboratory experiments.

Many efforts to model VOCs exist (e.g. Melcer *et al.* 1994) and these can be combined with standard AS models. These models are typically based on the *a priori* information of the existence of these chemicals in the influent and literature information of volatility and biodegradability. Unfortunately, all the required information is not available for all chemicals. Another approach that might be taken is the evaluation of bulk volatile component concentration and transfer constants based on experimental or real plant data. Baker (1994) proposed a COD based evaluation of the overall volatilisation constant.

In an aerated reactor, stripping is considered a predominant process for VOC transfer to the gaseous phase (Melcer *et al.* 1993). In an open reactor, stripping can be modelled by the following first order equation:

$$r_V = Kv \cdot C \tag{8.3}$$

where:

r_V: Stripping rate $[ML^{-3}T^{-1}]$
K_V: Rate constant of volatilisation $[T^{-1}]$
C: Concentration of volatile organic compound $[ML^{-3}]$

The rate constant of volatilisation is actually a function of temperature and the aeration method employed. Information from production personnel can be the primary source of relevant information.

Alternatively, if volatilisation is suspected, the aeration of an influent sample without sludge or solids may lead to a drop in soluble COD and may provide evidence in support of VOC modelling.

For chemical oxidation processes, basic chemical redox reactions may be required. For example sulphurous acid is used in various food and chemical industries. It can be chemically oxidised in aeration tanks as shown by the following equation:

$$2Na_2SO_3 + O_2 \rightarrow 2Na_2SO_4 \tag{8.4}$$

The speed of any oxidation reactions will dictate the model equations used. OUR measurements without sludge addition (or with sludge inactivated with bactericide) will provide insights into the existence of such chemicals and their reaction rates.

8.4.6 Additional nutrient and essential metal limitations

It is not uncommon for an industrial wastewater to be deficient in nutrients and essential metals. Biomass can adapt to these deficiencies, but growth may be unbalanced. Viscous bulking and deterioration of COD removal efficiency are typical outcomes. The direct measurement of nutrients and essential metals in the influent or the low nutrient content of the sludge can be indicators of this situation. It is common practice to dose deficient nutrients and metals for stable treatment. Conversely, excessive addition of nutrients and metals may adversely impact the effluent and potentially cause an effluent violation. When modelling industrial treatment, deficiency analysis and control strategy development are commonplace.

Modelling nutrient and metal requirements

In certain cases, the model may need to be expanded to include nutrient and essential metal requirements. Typically simple switching functions can be used to describe the impact of these things on the modelled growth rate. Equation 8.5 is an example of such an implementation.

$$Growth\, rate = \mu \cdot \frac{S}{K_s + S} \cdot \frac{Nut}{K_{Nut} + Nut} \cdot \frac{Met}{K_{Met} + Met} \tag{8.5}$$

where S, Nut and Met, respectively indicate substrate, nutrient and metal, and K indicates a half saturation coefficient for each of these components.

Examples can be found in the literature: Nowak *et al.* (1996) added phosphorus requirements to ASM1 for the evaluation of a phosphorus deficiency in a wastewater from a rendering plant. Ky *et al.* (2001) included a magnesium limitation for the phosphorus removal process from cheese factory wastewater, as it is known to be an essential metal for phosphorus removal. Both cases are specific to these industrial applications. It should be noted that in reality the lack of these nutrients may not result in impaired growth as the biomass may be able to adapt to the growth conditions, however depending on the model structure, this adaptation may be difficult to reproduce mathematically.

8.5 IMPACT ON BIOMASS COMPOSITION

The organism population is mainly a result of the wastewater characteristics. In industrial treatment plants the influent is often highly variable which can cause population shifts. Specific components that are typical in industrial wastewaters frequently favour the growth of specialist organisms.

8.5.1 Varying biomass yields

In industrial wastewater treatment plants the true yield of heterotrophic biomass can be significantly different from municipal wastewater treatment. This has been attributed to two issues:

(a) Growth of specialist organisms: For instance, it is well known that methanol degraders, often called methylotrophs, have a significantly lower yield than ordinary heterotrophs (Wilkinson & Hamer, 1979; Dold *et al.* 2007).
(b) Growth on a specific component: Organisms (or functional groups of organisms) may have different yields when growing on different substrates.

Another aspect which has to be considered in relation to the yield is the substrate storage process. This model feature is incorporated into ASM3 (Gujer *et al.* 2000). In industrial applications it is not unusual for the influent organic loading to be short chain fatty acids or sugars. Karahan-Gül *et al.* (2002) showed that organisms growing on short chain fatty acids and sugars have different yields if the substrates were involved in a storage process. The difference is assumed to be related to the storage polymers, (i.e. poly-ß-hydroxyalkanoates (PHAs) or glycogen) and the difference in energy utilisation efficiency.

A priori information on the influent characteristics can be advantageous. Interviewing production and treatment plant operators is one strategy that can be used. Measuring the concentration of soluble COD over time simultaneously with a respirometric measurement is another. If the COD is quickly depleted but the corresponding respiration rate is not sufficient to close the COD balance, substrate storage is a possible explanation.

As another possibility, if the calculated yield is found to be very high, volatilisation may play a role. Volatile organic compounds (VOC) reduction by stripping during respirometric measurements can cause COD reduction without oxygen requirement. Lower oxygen requirement from the respirogram with the same apparent sample load results in erroneous high calculated yields. A VOC stripping test using aeration (Baker, 1994) can be used to test for the presence of VOCs.

Modelling of different yields

The modelled yield has a direct impact on the predicted oxygen requirement and sludge production. For example, because of the reduced yield on methanol, an activated sludge process treating methanol will have reduced sludge production and a higher electron acceptor demand per COD treated than a typical municipal plant. Evaluation of the specific yield value is recommended in most industrial applications due to its impact on these two important process indicators.

8.5.2 Acclimation and activity loss (decay)

When modelling, acclimation can be a major consideration in industrial applications. It has been postulated that the decomposition of specific chemicals is mediated by specialised organisms (Wilkinson & Hamer, 1979; Hamer, 1997; Dold *et al.* 2007) and the growth of these organisms may have to be modelled as a separate organism type. Methanol degradation is mediated by methylotrophic organisms for example and methyltrophs may require an acclimation period if methanol is not continuously present in an influent stream. Sulphur bacteria are another example as reduced sulphur can be rapidly oxidised with specific sulphur oxidising autotrophic bacteria (Buisman *et al.* 1990). If specialist population exist in the system, then they may have to be incorporated in the modelling applications.

On the flip side, the lack of a continuous supply of a specific chemical in a waste stream can cause the loss of a specialist population. Industrial plants are known to suspend operations between production cycles and this can have an adverse impact on the treatment plant performance. Depending on the waste, it is possible that the system performance will reflect the population dynamics caused by these variations in load. If acclimation to specific chemicals is the focus of the modelling objectives then the modelling of specific populations based on data from well designed experiments may be required.

Modelling acclimation processes

In standard AS models, the acclimation processes can only be modelled for a few specialist biomass populations like nitrifiers and phosphorus accumulating organisms. They do not describe specialist organism growth for example, for biological phenol degradation or sulphur oxidation.

If modelling of acclimation to specific chemicals is required, additional state variables for the specific chemical and related specialist biomass may need to be introduced. Literature on the biodegradation of specific chemicals should be the basis for model construction. Figure 8.8 shows an example of a conceptual model with two heterotrophic populations. In this case, the decay products of the two heterotrophic organisms are not differentiated.

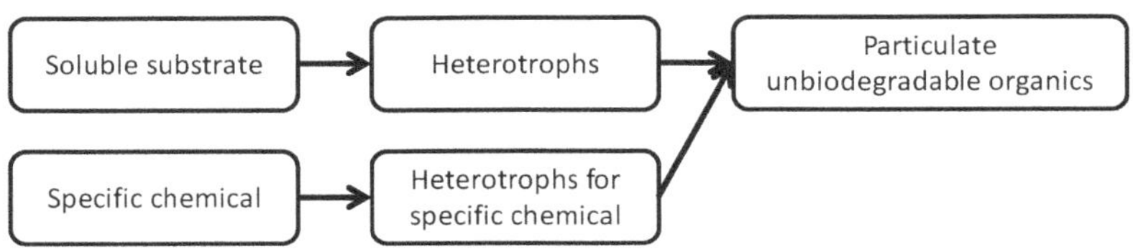

Figure 8.8 Conceptual acclimation model example by dual heterotrophs.

8.6 VARYING OPERATIONAL CONDITIONS

Industrial plants often operate at elevated temperatures due to the high temperature of the production process. This may require different kinetic parameter values for modelling. The process pH can also vary. Some processes have to deal with high inorganic salt concentrations. The activated sludge process can operate at high salt concentrations, but rapid changes in this concentration can cause problems. Some production plants utilise ion exchange resins for decolouring (e.g. sugar production) and these require high concentrations of acid and base for their regeneration. Other streams at the same facility may contain very low salt concentrations (e.g. washed sugar and water). In such a plant, a salt concentration model may be required to capture the plant behaviour.

8.6.1 Modelling temperature dependency

Some of the original AS model publications (e.g. ASM1) did not include the impact of temperature on kinetic parameters and so these models, as published, did not accurately reflect process changes caused by temperature. However, most commercial simulators have these models implemented with temperature dependencies. The extremes in industrial treatment temperatures may fall outside the scope of these models so care should be exercised when using these models outside their intended temperature ranges.

Arrhenius type temperature functions are widely used but are typically valid only in a limited temperature range. If the operating temperature is expected to exceed the valid range, which is often the case in industrial applications, the extrapolation of the Arrhenius equation is not appropriate and a suitable alternative may need to be explored. For example, it is known that nitrifiers have an upper temperature limit of approximately 40°C (Hellinga *et al.* 1998). In cases of extreme temperature, an empirical relationship between temperature and growth rate needs to be included in the model. One such example is to model a function similar to the pH dependency shown in Figure 8.9.

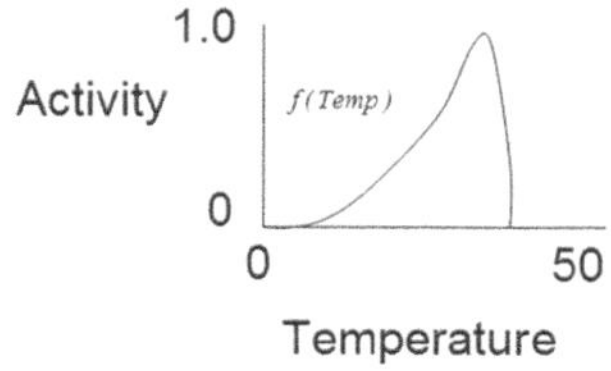

$$Growth\ rate = f(Temp) \cdot \mu \cdot \frac{S}{K_S + S}$$

Figure 8.9 Temperature dependency model including high temperature range.

The temperature dependency can be confirmed by maximum OUR measurements in a laboratory. Long term (adapted) and short term (temperature shock) relationships can be quite different so any supplemental experiments should be carefully designed. If the high temperature causes irreversible loss of activity, the net growth rate might be manipulated by increasing the decay rate of the biomass. It should be noted that temperature impacts are not completely understood and there could be other unknown impacts. For example it has been observed in industrial plants that the ability of the biomass to form flocs is impaired at higher temperatures so the sludge may not settle as well.

8.6.2 Modelling pH effects

ASM2d and later models relate alkalinity to biological activity, but in reality, the activity of micro-organisms depends, in part, not on alkalinity *per se* but rather on the pH environment. The effect of pH needs to be clearly distinguished from autotrophic substrate (CO_2) limitation effects. The calculation of pH is not included in any of the published AS models and when significant pH changes are expected, reproducing observed behaviour is difficult.

pH can have an impact on the process behaviour so pH may need to be considered when modelling industrial treatment. pH modelling is not trivial and any comprehensive pH model will include not only the components contributing to the pH but also a suitable solver and gas transfer modelling among other things.

Figure 8.10 illustrates an empirical model extension for pH dependency of a growth rate (Batstone *et al.* 2002). pH dependency can be confirmed by maximum oxygen uptake rate (OUR) measurements in a laboratory so if pH issues are suspected, supplemental experiments may be required. It should be noted that there are many unknown effects of a sub-optimal pH environment on various processes, which might not be observable in the experiments.

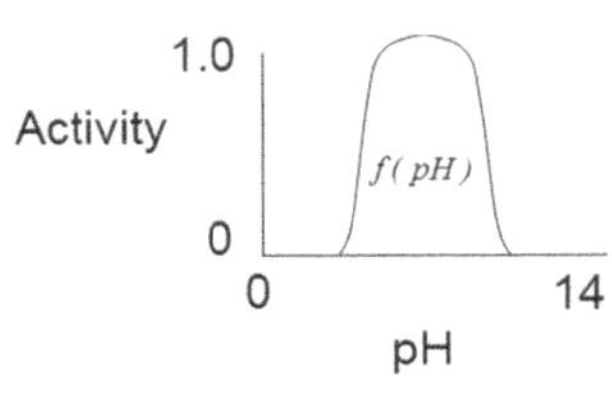

$$Growth\ rate = f(pH) \cdot \mu \cdot \frac{S}{K_S + S}$$

Figure 8.10 pH dependency growth rate based on experimental result.

8.7 EXPERIMENTAL METHODS FOR INDUSTRIAL APPLICATIONS

Table 8.3 presents a selection of experimental procedures that are useful in industrial applications.

Table 8.3 Experimental tests used for industrial model applications.

Test	Objective
Continuous treatability test	Achievable effluent quality Acclimation effect Floc formation Effluent suspended solids Sludge production
Soluble unbiodegradable evaluation	Effluent quality estimation Effect of upstream influent variation
Respirogram evaluation	Oxygen requirement Sludge yield Component changes Hydrolysis rate Inhibitory effect screening Inhibition model selection Storage and adsorption effect
Nitrifier growth rate	Nitrification tank sizing Inhibition effect
Inhibition test (dose response on maximum OUR)	Maximum acceptable concentration
Denitrification rate	Effect of influent organics combined with various electron donors
pH dependency	Performance estimation at different pH environments
Temperature dependency	Performance estimation under different temperatures
VOC evaluation	VOC variation Stripping rate

8.8 PITFALLS AND SUGGESTIONS IN INDUSTRIAL APPLICATIONS

This section describes potential problems and pitfalls when applying standard AS models to an industrial facility.

Upstream influent characterisation is preferable in most applications

Characterisation of the main incoming streams is always recommended and is preferred over end of pipe type characterisation. Many industrial applications depend on the effect of specific streams and the individual characterisation of streams should provide insight on their individual effects. Upstream characterisations can sometimes be validated using data from the combined stream.

Influent data collection

In industrial applications, often the major source of information is the production site. The cost for additional data collection tends to limit the scope of most projects and so it is important to minimise costs based on information provided by the production site. A few examples are listed below:

- Product analysis, information on recovery ratios and plant cleaning schedules may substitute for repeated end of pipe analyses.
- If influent composition is known to be stable, after initial calibration, the following measurements may substitute or alleviate inflated analytical effort and cost:
  - Electric conductivity
  - Spectroscopy
  - Refraction meter (for sugar, etc.)

pH can be assumed as an input in many applications

pH modelling is complex and often not enough information is available to construct a reliable and robust model. In industrial treatment, the installation of pH controllers is standard practice when the influent pH varies. In these cases, pH can be treated as a model input parameter. This approach combined with empirical models on pH dependency may reduce the overall modelling effort.

Applicability of phosphorus removal models

Compared to biological carbon and nitrogen removal, biological phosphorus removal is not well established in the industrial field. Before applying a model for excess biological phosphorus removal to industrial applications, it should be confirmed that the specific influent characteristics are similar to municipal wastewater and that the influent is free of inhibiting components. Good practice is to first prove the occurrence of excess biological phosphorus removal by intensive data collection and then use a model to confirm the kinetics of the process.

Fraction-based model considerations

The *fraction-based* models ASM2d and ASM3 assume a stable influent characterisation with respect to the nitrogen and phosphorus associated with various COD components. When the influent changes (i.e. due to a production change) changes to the wastewater characteristics may be required. This is often the case in industrial applications so the *component-based* modelling approach used in ASM1 may be a better choice.

FURTHER READING

The references in this section provide examples and possible model expansions for various industrial applications.

Orhon *et al.* (2009) proposed steps for industrial model development and calibration. Combining the *Unified Protocol* proposed in this report with their industrial model development steps could form the basis for an industrial treatment simulation protocol.

Baker (1994) developed models for the petrochemical and refinery wastewater treatment process. The author includes a detailed description of influent characteristics and the model features soluble component hydrolysis, sulfur oxidation, phenol inhibition, and stripping.

Bury *et al.* (2002) provides a general discussion on the requirements for industrial modelling applications.

Melcer *et al.* (2003) contains an industrial application chapter, providing a discussion on industrial wastewater characterisations and related model expansions. This reference also provides four industrial application examples.

Chapter 9

Frequently asked questions

How does the model compare to traditional design methods?

Different countries and regions in the world have different design methods, so the question could also be asked: How do traditional methods compare to each other? There is some inherent variation between particular design methods and different models with different model parameter sets. However the difference is not large or unreasonable. At 12°C and 3 days of SRT there are no methods or models (with reasonable parameter values) that will predict nitrification, while in a 7 day SRT system under the same conditions they all will.

Why do the models use COD?

A fundamental aspect of the models (structured models) in use today is that they are based on mass balances. This is trivial for phosphorus, relatively easy for nitrogen, but not possible for organic material if it is measured in VSS, TSS or BOD terms. COD is a measure of organic material in terms of oxygen units, and can conveniently link influent loading, sludge production and required oxygen demand on a mass balance basis.

Are the models applicable to industrial wastewater treatment?

The methods and the model structure used in the ASM-type models are applicable to a certain extent to industrial wastewater treatment, but the default parameters often are not. Modellers should take special care to ensure the model parameters being used are applicable. Experimental measurements are often required before modelling to establish fractions, kinetic and stoichiometric parameters for a specific industrial wastewater.

When should I use a simplified clarifier model (ideal or point settler) vs. a flux-based (one-dimensional) or a CFD model?

Simplified models can be chosen if average effluent and return solids concentrations are sufficient considering the modelling objectives. If dynamics in effluent solids, sludge blanket height or RAS concentration are important, flux-based models will provide more accurate predictions. Neither of these models describes clarifier geometry very well, and cannot be used for detailed design, baffle placement or similar objectives. For these tasks CFD models are required.

Does the model predict foaming/bulking/filaments/deflocculation?

No, these processes are not described in the standard activated sludge models. However, model extensions that attempt to describe some of these processes are available.

Which model should I use in my project?

There is no simple answer to this question – many aspects, including model features and capabilities, the user's familiarity with the model or software, project scope and objectives, budget and available time all have to be considered. Narrowing the question to the technical perspective, it is important to ascertain that the model describes the processes that are required to cover the project scope (e.g. bio-P model for a bio-P project), and that the modeller is familiar with the model, its performance and default parameter values.

When is dynamic modelling needed and when is steady-state enough (or preferred)?

Steady-state modelling is a good approach if the project satisfies two conditions: only long term behaviour is investigated, and the plant does not have significant inherent dynamics (SBRs or intermittently aerated reactors are dynamic processes that are not well represented with steady-state models). Steady-state models are much simpler to set up and run compared to dynamic models. Steady-state models are obviously not applicable when dynamics play an important role, such as the evaluation of peak aeration demand (blower sizing), diurnal behaviour or handling of a storm. Chapter 5.5 provides more detailed examples.

How much does modelling and sampling cost?

Small projects with well defined, limited objectives that do not require additional data collection can be successfully finished in a few days, while a complete process evaluation of a large plant may take many person-months and significant laboratory and testing costs for what may be hundreds of additional samples. Different example applications and their efforts are given in the Application Matrix (Chapter 6).

How will the model benefit me/the project? Will it save us money?

Modelling does cost money (typically in effort and additional laboratory costs), but it provides more accurate information for process design, lowers the risk of process failures and it results in better optimised operating procedures. If these benefits can be successfully converted to savings by good engineering and operating practices, the model will save money (as it should) for the plant.

How close do I need to get to say the model is "calibrated"?

There is a large variability in plant data in part due to sampling, sample processing and analytical issues. The model will not be more accurate than the plant data. Generally, ± 5 to 15% difference from measured data (or for effluent components with low concentrations, 10–100%, up to one mg/L difference) is considered acceptable.

Should I be changing the biological kinetics or the influent characteristics?

Influent characteristics are absolutely imperative to determine due to their impact on the process and due to the fact that there is considerable variation in influent composition between different countries and regions. Kinetic parameters for municipal wastewaters are reasonably well known, and their measurement is required only if the plant is in an unstable, marginal situation (overloaded) or inhibitory or toxic components are suspected in the influent. During setting up a model influent characteristics should be changed prior to the kinetic parameters.

Have the models been verified and validated? How accurate are the model predictions?

ASM-type models have been calibrated and validated – some to a much higher degree than others. Good model predictions should be within 5–15% of high quality data, except at very low effluent concentrations where the acceptable margin can be higher.

What is the difference between the software packages?

All packages are built on the same basic functionality (solving differential equations), but are very different from the user's standpoint. The available models, the user interface, the ability to customise and technical support are some of the key issues to consider in choosing the appropriate package.

How long does it take to learn to use models?

Learning to use a model for a quick solids production calculation can be done with a few days of effort. On the other hand, learning to configure and calibrate a complex plant typically requires considerable experience. Gaining this experience may take several months or years to obtain.

Why do I need more reactors in my model than I have in the plant?

You may have one plug-flow reactor in the plant which is represented as 4 or 6 reactors (Completely Stirred Tank Reactors, CSTRs) in series in the model. This is necessary to describe concentration profiles (DO, NH_x, PO_4, etc.) along the real reactor.

Can we model 24 clarifiers/aeration basins as one representative big basin? If we have to do that, should we add the surface areas and volumes?

If the flow and load distribution is relatively uniform (or the flow splits are unknown) to the 24 trains, it is a good modelling practice to represent the total as one large train. In this case the total reactor volume and total clarifier surface area have to be used.

What are the most important influent characteristics?

Depending on the modelling objectives:

(1) COD and its different fractions. These should be measured (Chapter 5.2) or inferred from BOD, VSS and TSS measurements
(2) TKN and ammonia
(3) Total phosphorus and orthophosphate
(4) For special modelling purposes other measurements may be necessary

My special sampling data is different than my regular sampling. Why is this and how do I handle it?

This is typically due to the different sampling methods. Regular sampling may be manually performed while an automatic sampler may have been set up for the sampling campaign (or the other way around). The sampling location, time, sample storage, flow or time weighting will influence the results. It is important to collect representative samples on a flow weighted basis, and process them as soon as possible to prevent sample deterioration.

What is the number of influent samples and the duration of influent characterisation required?

The principle is the more the better, and if possible the use of (properly maintained) online monitors should be considered. However due to practical limitations, a minimum of 2 weeks of daily composite sampling and two days of two-hourly samples (all together $14 + 24 = 38$ samples) is known to provide a reasonable initial data set. However, one has to keep in mind that loads and diurnal variations on weekends might differ considerably from weekdays.

Would COD fractions vary between scenarios? Annual average, peak loads, etc.

Yes, wastewater fractions can vary if the circumstances are different.

Since I calibrated my model the influent loads have changed. Is my model valid outside the calibration range?

Models have some ability to extrapolate within their validity field, but the further the new operating conditions are from the calibration situation, the less confidence can be attached to the results.

Glossary

Absorption: Incorporation of molecules into the physical structure of a solid or liquid without chemical reaction.

Accuracy: Closeness of computations, estimates or measurements to the exact or true values. Accuracy can be further differentiated into *Trueness* and *Precision*.

Activated sludge: The biologically active solids maintained in concentrated and suspended form in the reactors of wastewater treatment plants.

Adsorption: Attachment of molecules onto the physical surface of a solid without chemical reaction.

Advanced wastewater treatment: Treatment processes designed to remove pollutants that are not adequately removed by conventional secondary treatment processes.

Aerated SRT: *Sludge Retention Time* within the aerated portion of an activated sludge tank. See solids residence time.

Aerobe: An organism that requires free oxygen for respiration.

Aerobic condition: Aerobic conditions are characterised by the presence of free oxygen.

Algorithm: A precise rule (or set of rules) for solving a problem.

Alkalinity: The ability of water to neutralise an acid. Alkalinity is usually given in the unit meq/L (milli-equivalent per litre). Sometimes alkalinity is given in the unit mg $CaCO_3$/L.

Alum: Common name for aluminium sulphate (spelled aluminum sulfate in the US), frequently used as a coagulant in water and wastewater treatment. Chemical formula: $Al_2(SO_4)_3 \cdot 14H_2O$.

Aluminum sulfate: See *Alum*.

Anaerobic condition: Anaerobic conditions are characterised by the absence of free oxygen, nitrite and nitrate.

Anaerobic digestion: Sludge stabilisation process operated specifically without oxygen in which part of the organics are converted to methane and carbon dioxide.

Analytical models: Models that can be solved mathematically in closed form. For example, some model algorithms that are based on relatively simple differential equations can be solved analytically to provide a single solution.

Anoxic condition: Anoxic conditions are characterised by the absence of free oxygen and the presence of nitrate.

Application Matrix: A representative set of examples for model application proposed by this report.

Ash: The non-volatile inorganic solids that remain after incineration.

Autotroph: Plants and bacteria which can synthesise organic compounds from inorganic nutrients.

Average daily flow: The total flow past a physical point over a period of time divided by the number of days in that period.

Average day, maximum month: The average day flow or constituent mass during the maximum month for that constituent.

Average flow: The arithmetic average of flows measured at a given point.

Averaging periods: Averaging periods are the unit of time over which a measurement is taken and then subsequently averaged.

Batch reactor: A reactor in which the contents are completely mixed and flow is neither entering nor leaving the reactor vessel during the period of interest.

Bias: Systematic deviation between a measured (i.e. observed) or computed value and its "true" value. Bias is affected by faulty instrument calibration and other measurement errors, systematic errors during data collection, and sampling errors such as incomplete spatial randomisation during the design of sampling programs.

Biochemical oxygen demand (BOD_5): A standard measure of wastewater strength that quantifies the oxygen consumed in a stated period of time, (typically 5 days as signified by the subscript) at 20°C. Also BOD and cBOD.

Biodegradable: Term used to describe organic matter that can undergo biological decomposition.

Biological process: The process by which the metabolic activities of bacteria and other microorganisms transform organic materials.

Biomass: The mass of living biological material contained in a system.

Bio-P: See *Enhanced biological phosphorus removal (EBPR)*.

Bioreactor: A vessel containing activated sludge promoting biological transformations of contaminants in wastewater.

Biosolids: Particulate organic matter recovered from municipal wastewater treatment that can be beneficially used, especially as a fertiliser. Biosolids are solids that have been stabilised within the treatment process, whereas sludge has not.

Black box model: See *Statistical Model.*

Boundary conditions: In mathematics used for differential equation systems: restraints on values the state variables can assume (i.e. negative concentrations are outside the boundary). Sets of values for state variables and their rates along problem domain boundaries, sufficient to determine the state of the system within the problem domain.

Calibration: The process of adjusting model parameters within physically defensible ranges until the resulting prediction gives the best possible fit to the observed data. In some disciplines, calibration is also referred to as "parameter estimation".

Calibration, dynamic: The calibration of a model to specific influent and/or environmental conditions that vary dynamically with time.

Calibration, steady state: The calibration of a model to a fixed set of influent, operating and environmental conditions.

Carbonaceous biochemical oxygen demand: The portion of biochemical oxygen demand whereby oxygen consumption is the direct result of carbon oxidation; typically measured after a sample has been incubated for 5 days ($cBOD_5$). Typically, $cBOD_x$ is measured in the same way as BOD_x but the sample has been inhibited to prevent nitrification.

Chemical equilibrium: The condition that exists when there is no net transfer of mass or energy between the components of a system. This is the condition in a reversible chemical reaction when the rate of the forward reaction equals the rate of the reverse reaction.

Chemical oxygen demand (COD): Measurement of the oxidation potential of a water or wastewater using a chemical oxidising agent. For modelling purposes, this measure is used to describe the concentration of organic matter (mg/L) in a wastewater or sludge sample and is the unit of measure used most commonly in modelling programs that simulate biochemical treatment systems.

Code: Instructions, written in the syntax of a programming language, which provide the computer with a logical process. Code may also be referred to as a computer program. The term code describes the fact that computer languages use a different vocabulary and syntax than algorithms that may be written in standard language.

Code verification: Substantiation that a model code is in some sense a true representation of a conceptual model within certain specified limits or ranges of application and corresponding ranges of accuracy.

Coefficient: A numerical quantity, interposed in a formula that expresses the relationship between two or more variables to include the effect of special conditions or to correct a theoretical relationship to one found by experiment or actual practice.

Combined variable: See *Composite variable*.

Completely-mixed, stirred-tank reactor (CSTR): An ideal reactor in which the concentrations are uniform throughout. The effluent concentration equals the reactor concentration.

Complexity: The opposite of simplicity. Complex systems tend to have a large number of variables, multiple parts, mathematical equations of a higher order, and are more difficult to solve. In relation to computer models, complexity generally refers to the level in difficulty in solving mathematically posed problems as measured by the time, number of steps or arithmetic operations, or memory space required (called time complexity, computational complexity, and space complexity, respectively).

Composite sample: Multiple samples (time, volume or flow weighted) typically over a defined time interval combined into one sample for analytical processing.

Composite variable: A combination of state variables usually to form variables that can actually be measured in the plant (e.g. BOD_5, Total COD, TKN, Total Phosphorus, TSS, VSS). Also called combined variable.

Conceptual model: A description of reality in terms of verbal descriptions, equations, governing relationships or 'natural laws' that describes reality. This is the user's perception of the study area and the corresponding simplifications and numerical accuracy limits that are assumed acceptable in order to achieve the purpose of the modelling.

Constant: A quantity with a fixed value (e.g. the speed of light or gravitational force) representing known physical, biological or ecological activities.

Continuity equation: Equation describing the conservation of mass, energy, charge, etc.

Design criteria: (1) Engineering guidelines specifying construction details and materials.

(2) Objectives, results, or limits that must be met by a facility, structure, or process in the performance of its intended functions.

Design standards: Standards established for design of equipment and structures. The standards may or may not be mandatory.

Detention time: The theoretical time required to displace the contents of a tank or unit at a given rate of discharge.

Deterministic Model: A model that provides a single solution for the state variables. Changes in model outputs are solely due to changes in model components (i.e. model inputs).

Dissolved solids: Solids in solution that cannot be removed by filtration.

Diurnal: A daily fluctuation in flow or composition.

Diurnal pattern: A repeating daily fluctuation in flow or composition that is of similar pattern from one 24 –hour period to another.

Domain boundaries (spatial and temporal): The spatial and temporal domains of a model are the limits of extent and resolution with respect to time and space for which the model has been developed and over which it should be evaluated.

Domestic wastewater: Wastewater originating from sanitary conveniences in residential dwellings, office buildings, and institutions. Also called sanitary or municipal wastewater (in difference to industrial wastewater).

Dry weather flow: The flow of wastewater in a sanitary sewer during dry weather; the sum of wastewater and dry weather infiltration.

Dynamic model: Model in which time is an independent variable.

Dynamic simulation: Time-varying solution of the ODE system: usually a simulation in which the inputs to the model and outputs from the model vary with time.

Empirical model: A model where the structure is derived, not from knowledge on the physical, biological or ecological processes, but from mathematical analyses of relationships among observed data.

Endogenous respiration: Bacterial growth phase during which microbes metabolise their own protoplasm without replacement.

Enhanced biological phosphorus removal (EBPR): The biological removal of phosphorus from a wastewater through the cultivation and wasting of bacteria that retain phosphorus in excess of their minimum biologically required amount. Also called bio-P.

Equalisation: The process of dampening hydraulic or organic variations so that nearly constant conditions can be achieved in a flow.

Extrapolation: Extrapolation is a process that uses assumptions about fundamental causes underlying the observed phenomena in order to project beyond the range of available data. In general, extrapolation is not considered a reliable process for prediction.

Final effluent: The effluent from the final unit of a treatment process at a wastewater treatment plant.

Flow rate: The volume or mass of a gas, liquid, or solid material that passes a point in a stated period of time.

Flux: Flow rate per unit area.

For membranes: The volumetric filtration rate for a given area of a membrane. A typical unit of flux is litres per square meter of membrane area per day.

Forcing/driving Variables:: External or exogenous (outside the model framework) factors that influence the state variables calculated within the model.

Function: A mathematical relationship between variables.

Grab sample: A single water or wastewater sample taken at a time and place.

Gujer Matrix: Tabular representation of state variables and their interactions within the model. Formerly known as a *Petersen matrix*.

Half-saturation coefficient: The half-saturation coefficient is a parameter in the Monod term defining the shape of the resulting curves and corresponds to the substrate (or other component) concentration at which the value of the Monod saturation/inhibition function is 0.5.

Half-saturation concentration: The concentration at which the process rate is half of its maximum rate.

Henry's law: Defines the equilibrium concentration of a component in solution and in air.

Heterotrophic bacteria: A type of bacteria that derives its cell carbon from organic carbon; most pathogenic bacteria are heterotrophic bacteria.

Hydraulic loading: Total volume of liquid applied per unit of time to a specific tank or treatment process.

Hydraulic retention time: Vessel volume divided by the liquid throughput rate, typically expressed in minutes, hours or days depending on the situation.

Inerts: Constituents that are assumed not to react in the model. Inerts may be soluble or particulate, organic or inorganic.

Influent characterisation: See *Influent fractionation.*

Influent fractionation: Proportioning of the influent constituents into state variables. Also termed "influent stoichiometry" and "influent characterisation".

Influent stoichiometry: See *Influent fractionation.*

Inorganic matter: Substances of mineral origin, not containing organic carbon, and not subject to decay.

Input variable: For modelling purposes, variables may be divided into state variables (or internal variables), input variables and output variables.

Insoluble: A compound that cannot be dissolved in a solvent.

Integrated model: Model that includes several domains, for example, collection system – wastewater treatment – receiving water body.

Interface model: A model that describes how output variables are passed from a model of one type to a model that uses different variables for its inputs.

Jar test: A laboratory test procedure for evaluating coagulation, flocculation, and sedimentation in a series of parallel comparisons.

Lumped model: A model that considers several reactors (or process trains) as one unit without explicitly accounting for the spatial variability of their characteristics. The parameters are considered to be valid for the system taken as a whole.

Lysis: The rupture of a cell that results in loss of its contents.

Mass balance: Balance of material flows, including input, output, production or loss using the principle of mass conservation.

Mass transfer: Movement of atoms or molecules by diffusion or convection from high to low concentration regions.

Material balance: See mass balance.

Measurement error: Errors in the observed data that are a function of human or instrumental errors.

Mechanistic model: A model that has a structure that explicitly represents an understanding of physical, chemical, and/or biological processes. Mechanistic models quantitatively describe the relationship between some phenomenon and underlying first principles of cause. Hence, in theory, they are robust and useful for inferring solutions outside of the domain that the initial data was collected and used to parameterise the mechanisms.

Metabolic models: Models based on metabolic processes of biological treatment using the transformations of intermediate compounds.

Mixed liquor: The mixture of wastewater and activated sludge in the treatment reactor.

Mixed liquor suspended solids (MLSS): Suspended solids in the mixture of wastewater and activated sludge.

Mixed liquor volatile suspended solids (MLVSS): The organic fraction of the mixed liquor suspended solids as measured by flash incineration at 550°C.

Model: A representation of the behaviour of an object or process, often in mathematical or statistical terms. Models can also be physical or conceptual.

Model calibration: See *Calibration*.

Model error: Difference between observed and simulated variables; can be calculated in different ways as cumulative, absolute, quadratic, and so on.

Model prediction accuracy: A measure of how closely the model matches observed data.

Model set-up: Procedure to put together the required sub-models during the construction of a plant model.

Model testing: Comparing model predictions to independent data.

Model validation: See *Validation*.

Modeller: A specialist who undertakes the technical modelling activity.

Modular modelling approach: The coupling of different unit process models. The output data are usually transferred between the different components or sub-models. Often a modelling interface facilitates the interactions between the different types of models. See also *Interface model*.

Monod term: A mathematical function commonly used in the models to describe the variation in the kinetics of biological growth as a function of a model component (i.e. substrate, nutrient, pH…). Identical in form to the Michaelis-Menten equation often referred to in industrial applications. Also called *Switching function*.

Monte-Carlo simulation: A simulation technique used to approximate the probability of certain outcomes that involves running multiple simulations, with randomised inputs.

Net yield: Net mass of solids produced in a biological process divided by mass of substrate removed, typically expressed in BOD or COD units. Is equal to the synthesis yield minus decay.

Noise: Inherent variability in the data that the model does not characterise (see definition for variability).

Numerical solver: Mathematical solving routine included in a simulator used to solve the differential equations in a model.

Nutrient: A substance, element or compound, organic or inorganic, necessary for the growth and development of organisms. Carbon, oxygen, nitrogen, potassium and phosphorus are important examples of nutrients needed by living things.

Objective function: A function to quantify the deviations between model outputs and observations.

Observations: Measurements at a real plant.

ODE: Ordinary Differential Equation – the derivative of a variable depends on the variable itself.

Organic loading: The amount of organic matter fed to a treatment process.

Oxygen uptake rate (OUR): The oxygen used during biological oxidation, typically expressed as mg O_2/L/h in the activated sludge process.

Parameters: Terms in the model that are fixed during a model run or simulation but can be changed in different runs as a method for conducting sensitivity analysis or to achieve calibration goals.

Particulate: Typically considered to be a solid particle larger than 0.45 μm or large enough to be removed by filtration.

Petersen matrix: See *Gujer Matrix*.

Plant model: Combination of sub-models to simulate a plant or a part of it. A typical activated sludge plant model consists at least of an input, a transport or hydraulic model, a biokinetic model, an aeration model and a clarifier model.

Plant-wide model: Term often used in Europe. See *Whole plant model*.

Plug flow: Flow condition in which a fluid package passes through a tank without longitudinal mixing. The packages are discharged in the same sequence as they enter. Opposite of completely mixed, where the reactor concentration equals the effluent concentration. Ideal plug-flow does not exist and the actual hydraulic behaviour is somewhere in between ideal plug-flow and completely mixed.

Precision: For measurements: Term for random errors. Opposite of trueness as a term for lack of systematic errors.

For model predictions: Precision of simulation results is the degree to which several simulation results are similar to each other.

Process model: A model describing the behaviour of a certain unit process (e.g. activated sludge reactors).

Quality assurance (QA): Quality assurance (QA) is the procedural and operational framework used to build consensus among the organisations concerned in its implementation. QA assures technically and scientifically adequate execution of all tasks included in the study, and that all modelling-based analysis is reproducible and justifiable.

Quality control (QC): Quality control (QC) is part of the overall QA procedure and emphasises testing of model results.

Reliability: The confidence that (potential) users have in a model and in the information derived from the model such that they are willing to use the model and the derived information. Specifically, reliability is a function of the performance record of a model and its conformance to best available, proven science.

Residence time: The period of time that a component remains in a tank or system.

Sensitivity: The degree to which the model outputs are affected by changes in selected parameters.

Sensitivity analysis: Objective examination of the effect of changes in parameters of a model on the output of the model.

Sidestream: In plant returns of flows and pollutants from solids processing or odour control facilities to the main liquid stream treatment facilities.

Simulation: A model run providing outputs based on model inputs.

Simulator: Software used to run a model – usually with interactive inputs.

Sludge: A wastewater residual produced by primary, secondary, or advanced treatment that has not undergone any process of stabilisation or pathogen reduction. The term sludge is generally used only up to the point where stabilisation occurs and in conjunction with a process descriptor, such as in the phrases primary sludge, waste activated sludge, and secondary sludge.

Sludge age: See *Sludge retention time.*

Sludge retention time (SRT): The average time solids spend in the system. Also called *Sludge Age* or *Solids residence time.*

Solids residence time: See *Sludge retention time.*

Solubility: The amount of a substance that can dissolve in a solvent under a given set of conditions.

Stakeholders: A person or group that has an investment, share, or interest in something, as a business or industry.

Standard Methods: An assembly of analytical techniques and descriptions commonly accepted in water and wastewater treatment (i.e. Standard Methods for the Examination of Water and Wastewater) published jointly by the American Public Health Association, the American Water Works Association, and the Water Environment Federation.

State variable: Fundamental components (e.g. ammonia) in a model. Variables may be divided into state variables (or internal variables), input variables and output variables.

Statistical model: A model where the structure is derived, not from knowledge on the physical, biological or ecological processes, but from analyses of relationship among observed data. Often called black-box model.

Steady-state model: A model that is not dynamic, that is a model where everything is considered stationary in time.

Stochastic model: A model that includes variability in model parameters and variables. The solutions obtained by the model are therefore a function of both the deterministic input and model structure and the random variability.

Stoichiometric coefficients: Factors used to convert mass units between different state variables. Typically obtained from the stoichiometric (mass balanced) chemical equations or empirical observations that describes the transformations. Not to be confused with *Influent stoichiometry*.

Sub-model: A model used to describe a particular aspect of a WWTP within the plant model for the unit process.

Substrate: Constituents that are used for biological growth.

Supermodel: A model describing the whole wastewater treatment plant with one common set of state variables (as opposed to interface model). Also called "integrated plant model".

Suspended growth process: Biological wastewater treatment process in which the micro-organisms and substrate are maintained in suspension within the liquid.

Suspended solids: Solids captured by filtration through a glass fibre filter or 0.45 μm filter membrane.

Switching function: Switching functions are used to turn process rate equations on and off, depending on the environmental conditions. Monod kinetic equations are generally used; however other switching functions are also available (e.g. Haldane, Andrews).

System SRT: Solids residence time within a processing system. In activated sludge the system SRT includes the mass of solids in anaerobic, anoxic, and aerobic portions of aeration tanks plus solids in the secondary clarifiers.

Time series: Temporal sequence of consecutive data.

Time step (of a given model): Unit interval of time used by a discrete model for time series simulations (frequently variable).

Tracer test: Tracer experiments are carried out by injecting one pulse (or several) of an inert tracer or increase its concentration for a specified time at the inlet of the system under study. The tracer substance should not be degradable and not adsorb to the sludge. The time series of recovered tracer is measured at specified locations within the system and at its outlet.

Trueness: Term for systematic errors: the closeness of agreement between a measurement and an accepted reference value. The measure of trueness is usually expressed in terms of bias.

Ultimate biochemical oxygen demand (BOD_∞): The amount of oxygen required to completely satisfy carbonaceous and nitrogenous biochemical oxygen demand.

Uncertainty: The term used in this guidance to describe lack of knowledge about models, parameters, constants, data, and beliefs. There are many sources of uncertainty, including: the science underlying a model, uncertainty in model parameters and input data, observation error, and code uncertainty. Additional study and collecting more information allows error that stems from uncertainty to be minimised/reduced (or eliminated). In contrast, *Variability* is irreducible but can be better characterised or represented with further study.

Underflow: The concentrated solids removed from the bottom of a tank or basin.

Unified Protocol: Recommended steps by this report for a consistent and repeatable modelling study, job or exercise.

Validation: Substantiation that a model has sufficient accuracy consistent with the intended application of the model.

Variability: Spread in a variable or a probability distribution.

Variable: A quantity that varies in time. Variables in the model may be divided into state variables (or internal variables), input variables and output variables.

Velocity gradient: A measurement (G value) of the degree of mixing imparted to water or wastewater during flocculation.

Verification (code): Examination of the algorithms and numerical technique in the computer code to ascertain that they truly represent the conceptual model and that there are no inherent numerical problems with obtaining a solution.

Volatile solids: Volatile matter measured as the fraction (such as the percent) of solids material that is ignitable at 550°C. Typically this is used to represent the organic fraction of the sludge or other solids material.

Volatile suspended solids: Organic content of suspended solids in a water or wastewater. Determined from weight loss after a filtered sample is heated to 550°C.

Waste activated sludge: Excess activated sludge that is discharged from an activated sludge treatment process.

Whole plant model: A model used to describe all unit processes and their interconnections in a plant mass balance. In Europe often called *Plant-wide model.*

Appendix A

Sub-model descriptions

A.1 HYDRAULIC AND TRANSPORT MODELS

A.1.1 Reactor models

Computational requirements dictate that most simulators only provide *completely stirred tank reactor* (CSTR) models or plug flow tank models consisting of a series of CSTRs. The following types can be distinguished:

- Fixed volume: $Q_{OUT} = Q_{INF}$
- Variable volume:
  - Overflow: No flow until tank is filled, then $Q_{OUT} = Q_{INF}$
  - Variable overflow: $Q_{OUT} = f(Q_{INF})$, for example, using a weir equation where the outflow is depending on the water level.
  - Pumped flow: The outflow is set by a pump rate but limited by a minimum and maximum tank volume and input flow.

More complex reactor models include two-dimensional (2-D) or three-dimensional (3-D) flow/transport behaviour using *computational fluid dynamics* (CFD). CFD models are outside the scope of this report. The following paragraphs explain how CSTRs are used to mimic the hydraulic behaviour of various reactor types.

A.1.2 Flow scheme

Due to the required simulation time for 2-D or 3-D models, in most simulation programs the hydraulic behaviour is modelled as a series of CSTRs. Since the hydraulic behaviour depends on flow regime, basin geometry, baffling, etc., the number and combination of tanks has to be calibrated based on experience, measured performance data, empirical equations (e.g. Fujie *et al.* 1983), tracer experiments or CFD models.

Estimation equations

If no tracer test results are available several equations exist to estimate the required number of tanks in series. A spreadsheet with different estimation formulae is provided on the GMP WaterWiki website.

Tracer experiments

In a tracer experiment (Figure A.1) a non-reactive and easy to measure component (e.g. bromide, lithium, potassium chloride, rhodamine) is dosed into the influent of the plant section under evaluation (e.g. whole plant, biological stage, clarifier, etc.). The tracer concentration is measured at sensible locations until it leaves the evaluated plant section (including recycle streams). Beside the number of required CSTRs for the model, information can be gained about possible back flows and short circuits, dead zones, and the uneven distribution of flows to parallel lanes.

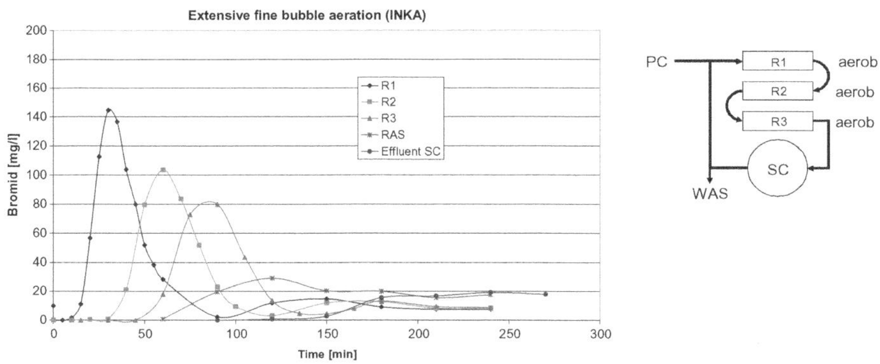

Figure A.1 Tracer experiment showing plug flow behaviour at WWTP Glatt (CH) (Sample locations at the end of each reactor).

A.1.2.1 Return Activated Sludge (RAS) and Internal Recycle (IR) flows

The RAS and IR lines can be modelled as a pumped flow (m^3/d) or as a fraction of a total flow (e.g. $f_{IR} = 0.1$).

A.1.2.2 Waste activated sludge (WAS) flow

The wastage rate is one of the most important settings in an AS plant model because it determines the sludge retention time (SRT) of the system and so special care should be put into the way the wastage is modelled. Typically three main options are used:

- A certain flow rate is "pumped" from the RAS line (following measured flow or calculated flow based on pump speed),
- The SRT is calculated first (e.g. based on a TSS or better on a P or Fe mass balance) and the mass of sludge to be wasted is removed accordingly from the last bioreactor or the RAS line,
- Based on the calculated SRT, the WAS is continuously removed from all bioreactors according to the dilution rate (1/SRT).

Whereas the first option is probably most often used, it has the disadvantage that only the measured flow rate is used as a model input and not the mass removal. If the modelled MLSS concentration and/or the loss of particles through the effluent are not correct, the SRT will carry an error as well.

The second option allows using the SRT obtained from a detailed analysis of the sludge inventory. The sludge is normally removed from the last tank in the model and not from the RAS stream with the advantage of separating SRT from the secondary clarifier efficiency. The authors suggest calculating the SRT based on a phosphorus mass balance or at least after checking a MLSS mass balance around the secondary clarifier.

The last option integrates the dilution rate into the set of differential equations. This option is mathematically elegant but makes it difficult to include the clarifier performance into the model.

A.1.2.3 Flow splitter

Different models are available to simulate flow splitters.

(a) *Ratio*: The input flow is split according to a pre-defined ratio.
(b) *Fraction*: One part of the incoming flow is withdrawn.
(c) *Pumped flow*: A certain volume is pumped (m^3/d) from the passing flow. The model has to make sure that this pumped flow does not exceed the inflow (if the splitter is simulated as a point object without volume).
(d) *Bypass*: A bypass is a splitter that will have no flow in the bypass below a certain flow level in the input. Above this flow level the excess flow is diverted to the bypass and the main flow remains constant at the maximum.
(e) A *flow router* is a splitter that can be modelled to switch between the two outlets according to a timer or other signals (e.g. send flow for 2 hours to the left, then switch for 2 hours to the right, etc.).
(f) *Flow pacing*: The flow is split according to another flow measurement (e.g. RAS flow controlled based on influent flow).

A.2 CLARIFIER MODELS

A.2.1 Overview

The following clarifier models are typically used:

(a) *Point clarifier*: This is the simplest model with a complete split of particulate and soluble state variables and no volume. No settling behaviour is modelled. In the simplest implementation all particles go to the RAS line. However, most clarifier models in commercial simulators include loss of particles through the effluent by an additional term. The effluent particulate material is modelled as a fraction of the incoming or RAS TSS concentration. Another approach is to model a predefined effluent TSS concentration, which allows the measured TSS concentrations to be followed for calibration purposes.
(b) *Ideal clarifier with volume*: This models a simplified hydraulic behaviour and sludge storage capacity of the clarifier, using a completely mixed reactor (but without biological reactions). Most often only the sludge blanket is modelled but not a separate clear water zone. An additional clear water zone (or several reactors) allows (simplified) modelling of the hydraulic delay of the effluent.
(c) *Layered clarifier models (also called 1-D or flux models)*: Several CSTRs in series are introduced to mimic one-dimensional settling behaviour (clarification, thickening and sludge storage). Different approaches are available (e.g. Takács *et al.* 1991; Otterpohl & Freund, 1992; Wett, 2002; Plósz *et al.* 2007; Burger *et al.* 2011) using different numbers of layers and settling function approaches.
(d) *CFD clarifier models*: Two- or three-dimensional Computational Fluid Dynamics (CFD) models have been applied to investigate settling problems. They model the flux of particles in different directions in high resolution. Because of the required simulation time, it might be helpful to use more-dimensional models to analyse the system and then derive a simplified model within the WWTP simulator and for the foreseen working range.
(e) *Reactive clarifier models*: A biokinetic model is added to the clarifier models, which include volumes, i.e. models (b), (c) and (d) as described above.

A.2.2 Selection of clarifier models

The selection of the clarifier model depends upon the objectives of the simulation study. Very simple clarifier models (e.g. ideal clarifier) are applicable if the hydraulics and the suspended solids load are relatively constant. If rain events or other significant dynamic perturbations will lead to considerable sludge shifts between activated sludge tanks and clarifier, more complex models have to be applied that take the settling behaviour into account.

In general, ideal clarifier models are sufficient for most typical objectives. More complex layer models have limitations in their prediction of effluent solids concentration. Even if the effluent solids is calibrated at dry weather flow, these models may not predict effluent solids correctly at high hydraulic loads, although the simulation of a rising sludge blanket indicates a potential failure and wash-out of solids. In some models the number of horizontal layers used in the model has a significant impact on the clarifier performance.

It should also be noted that detailed data to calibrate a layered model is not readily available and has to be obtained by additional measuring campaigns. Currently, CFD models are mainly used in research and therefore not yet part of standard WWTP simulators.

A.2.3 Reactive clarifier models

In addition to the hydraulic behaviour of the clarifier, the sludge blanket is a reactive zone in which biological reactions occur such as endogenous denitrification or secondary phosphorous release.

The sludge mass in the sludge blanket can be estimated using sludge level meters. However, this is a rather coarse estimate due to the normally unknown sludge concentration gradients. Therefore, the sludge blanket volume should be regarded as a calibration factor to fit the measured nitrate concentration in the return sludge. Coupling clarifier models with a biokinetic model increases the computational load which might be reduced by decreasing the number of reactive compartments or layers, respectively.

In a simplified approach denitrification can be modelled in a virtual sludge blanket compartment in the return sludge line decoupled from clarifier modelling.

A.3 BIOKINETIC MODELS

A great number of models or model extensions has been published or is implemented in simulators. Due to space constraints, the reader is referred to the original publications or software manuals.

A.3.1 Temperature dependency of parameters (Arrhenius equation)

Several biokinetic parameters are temperature dependent. Typically the Arrhenius equation is used to calculate the actual parameter values.

Note that two different implementations can be found:

$$k_T = k_{20} \cdot \Theta_{\mathrm{pow}}^{(T-20)} \quad \text{(e.g. ASM2, Henze } et\ al. \text{ 2000)} \tag{A.1}$$

$$k_T = k_{20} \cdot e^{\Theta_{\mathrm{exp}} \cdot (T-20)} \quad \text{(e.g. ASM3, Gujer } et\ al. \text{ 2000)} \tag{A.2}$$

where:

k_T: Kinetic parameter k at a given temperature T
k_{20}: Kinetic parameter k at 20°Celsius
Θ_{pow}: Temperature correction coefficient theta in power law implementation

Θ_{exp}: Temperature correction coefficient theta in exponential function implementation
T: Temperature [°C]

The two temperature correction coefficients can be easily converted from one to the other by means of the following expression:

$$\Theta_{pow} = e^{(\Theta_{exp})} \quad (A.3)$$

A.4 INPUT MODELS

Input models convert measurements into model state variables or other variables (e.g. energy consumption) or constants (e.g. set-points) the plant model requires as input.

A.4.1 Influent model

Wastewater characterisation is used to fractionate measurements (typically for COD, N and P) into model state variables. A practical approach is to measure, calculate or estimate average fractions required by the biokinetic model and to superimpose these ratios on the total compounds measured. The assumption is that the ratios between the measured total compounds and their fractions are stable over time (e.g. the readily biodegradable substrate concentration is proportional to COD_{tot}). It should be kept in mind that this is a simplification and that these ratios may vary significantly over time due to changes in the wastewater characteristics.

A.4.2 Concepts for influent fractionation

A.4.2.1 COD fractions

All ASM-type models are based on COD as this allows mass balancing of organic material and electron demand (oxygen). The two basic steps in fractionating a wastewater are to (i) split the total measurement into particulate and soluble (and in some models also colloidal) components and (ii) define the biodegradability. Typically, the degree of biodegradability is given as unbiodegradable, slowly biodegradable and readily biodegradable groupings, however, more complex models will have more substrate fractions. Biomass is defined as a separate biodegradable fraction. Figure A.2 shows an example for a COD influent fractionation.

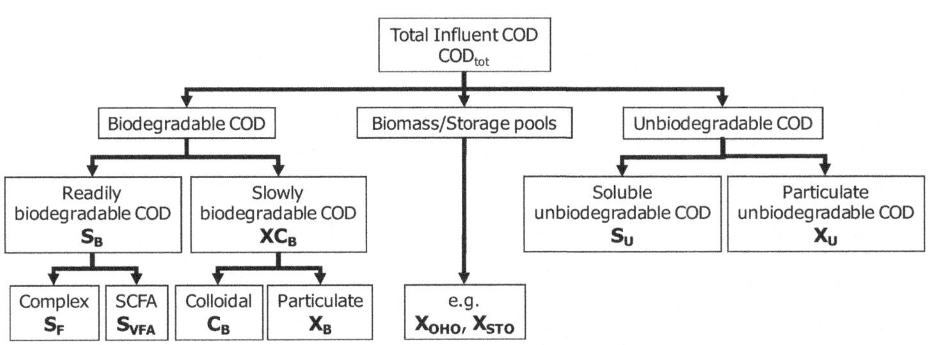

Figure A.2 Example for a COD influent fractionation (notation according to Corominas *et al.* 2010; reprint Appendix D).

Typical influent COD fractions are:

- Soluble unbiodegradable (S_U): S_U cannot be removed by bioprocesses or with the sludge and therefore leaves the system with the effluent. Its value limits the minimum soluble effluent COD concentration.
- Particulate unbiodegradable (X_U): X_U accumulates in the sludge and therefore has a major influence on the sludge production and thus on the MLSS concentration and the clarifier performance.
- Soluble readily biodegradable (S_B): S_B influences denitrification performance, the required anoxic tank size and the P removal performance
- Slowly biodegradable substrate (particulate plus colloidal!) (XC_B): XC_B needs to be hydrolyzed before it can be taken up by the heterotrophic biomass but influences the denitrification potential of a plant.
- Active biomass: X_{Bio} is often neglected in the influent but studies (e.g. Sperandio & Paul, 2000) showed a biomass content of up to 15% (or more) of the total COD.
- S_B, XC_B and X_{Bio} also determine the carbonaceous oxygen demand of the influent.

COD fractionation procedures

The selection of a fractionation method depends on the model used and the data available. The fractionation method has an impact on the model results; however, no standard procedure has arisen yet. A list with published fractionation methods is given in Tables 5.2.2 and 5.2.3 in Chapter 5.

As an example, a procedure for ASM1 is given:

- Measure $\mathrm{COD_{tot}}$, $\mathrm{COD_{fil}}$, $\mathrm{COD_{sol}}$ (section 5.2; $\mathrm{COD_{sol}}$ after flocculation step. Compare Melcer *et al.* 2003).
- $\mathrm{COD_{sol}}$ has to be split into the two soluble model components S_B and S_U.
- S_U can be estimated by measuring $\mathrm{COD_{sol}}$ in the effluent with the assumption that all biodegradable matter is degraded during the treatment process. Some methods (e.g. STOWA: Roeleveld & van Loosdrecht, 2002) assume some biodegradable soluble fractions in the effluent and correct S_U (STOWA: 90% of $\mathrm{COD_{sol,EFF}}$).
- The difference between $\mathrm{COD_{sol}}$ and S_U is assumed to be readily biodegradable S_B.
- The colloidal COD ($C_B = \mathrm{COD_{fil}} - \mathrm{COD_{sol}}$) is assumed to be part of the slowly biodegradable matter XC_B.
- Information on the total biodegradable fraction (T_B; e.g. from respirometry or estimation of total BOD based on BOD measurements) is used to calculate the remaining particulate slowly biodegradable fraction ($X_B = \mathrm{T}_B - S_B - \mathrm{C}_B$).
- The ASM1 and ASM2d fraction for slowly biodegradable COD (XC_B) is the sum of X_B and C_B.
- The remaining COD fractions are unbiodegradable COD (X_U) and biomass (X_{OHO} and X_{ANO}). Some methods assume no biomass in the influent, however, several cases were published where a significant concentration of heterotrophic biomass was measured. When no measurements are available, one value should be fixed to a typical value and the last COD fraction is calculated by difference.

A.4.2.2 Nitrogen and phosphorus fractions

In addition to COD components, most models contain nitrogen and some also phosphorus components. Nitrogen and phosphorus fractions consist of inorganic (e.g. NO_x-N, NH_x-N, PO_4-P) and organic fractions. There are two main modelling approaches to include organic nitrogen and phosphorus: (i) *component-based* and ii) *fraction-based*. The first approach uses specific state variables (Figure A.(3) and the second one models organic N and P as fixed ratios of the COD state variables (Figures A.4 and

A.5). Irrespective of the model approach used the measured total *N* and *P* has to be properly converted into model inputs.

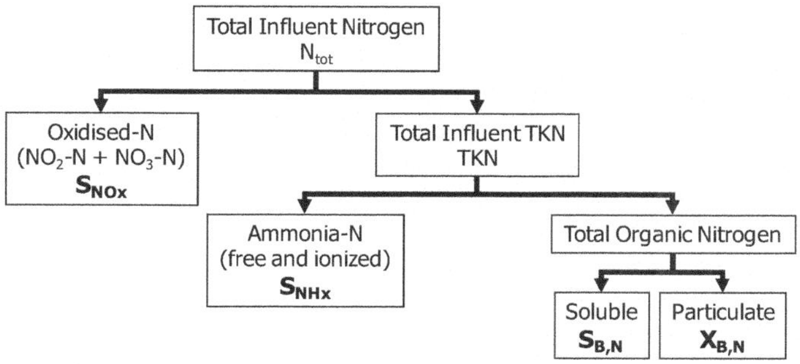

Figure A.3 Example for component-based model approach for nitrogen.

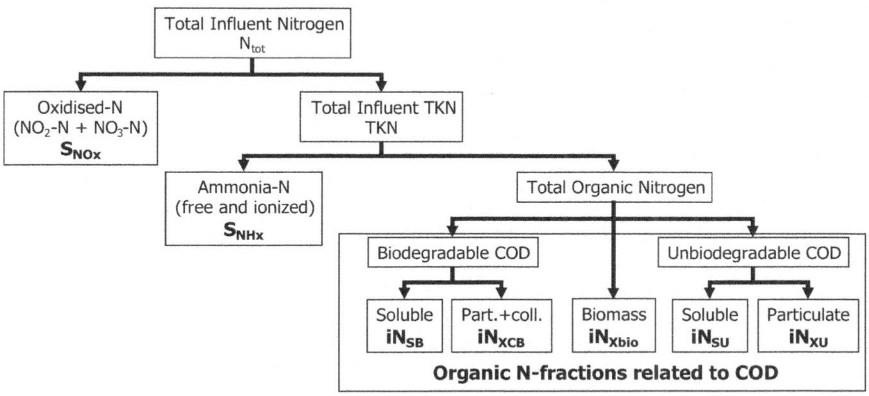

Figure A.4 Example for fraction-based model approach for nitrogen.

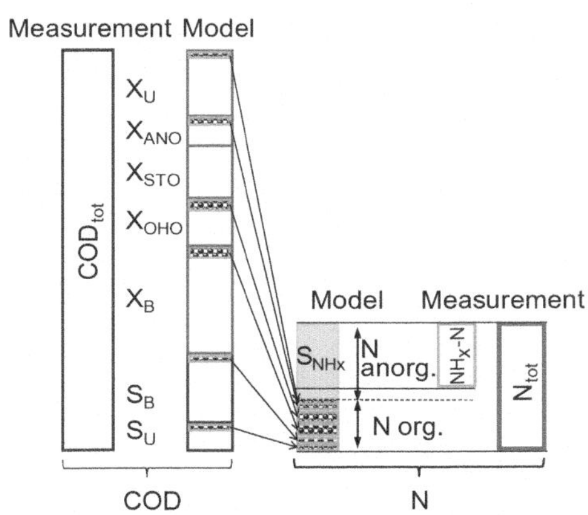

Figure A.5 Relation between measured and modelled COD and nitrogen in the influent for fraction-based models (ASM3).

Examples for the *component-based* approach are: ASM1, Barker & Dold, UCTPHO+. Models using a *fraction-based* approach are: ASM2d, ASM3, ASM3 + Bio-P and TU Delft.

The two model approaches are based on different levels of simplification of the real system but are both justified by accepted scientific and engineering principles. However, from the standpoint of the practitioner, there are a few very real differences that need to be considered when using a set of influent measurements to calculate influent values to different models as discussed below.

Most organic state variables as well as *N* and *P* fractions cannot be directly measured in the influent and have to be deduced (maintaining total COD, *N*, *P* concentrations) from measurements that account for a combination of those. This process is complex, partly based on experience and therefore error-prone.

N and *P* fractionation procedures

In the ***component-based*** models, the *organic nitrogen* is typically split into a soluble and a particulate fraction. The particulate fraction undergoes a hydrolysis step to soluble matter before it is transformed into ammonia in an ammonification process. For *organic phosphorous*, most models hydrolyse the particulate organic fraction directly to phosphates.

The split between particulate and soluble fractions in the influent can be based on measurements or used as a calibration parameter.

The ***fraction-based*** models (Figure A.5) need more attention during calibration as the same *N* and *P* fractions are used in the influent characterisation as in the MLSS. Therefore both locations have to be checked during calibration. Two approaches can be used for setting up the fraction-based influent:

- Calibrating the fraction parameters (i_{N_XB}, i_{P_XU}, etc.) throughout the system as noted above.
- Keeping the fraction parameters as defaults and use the ammonia and phosphate influent as the main calibration parameters in order to maintain total nitrogen/phosphorus input into the model.

The first approach can be dangerous when changing the parameters outside reasonable ranges. It is important to check the resulting *N* and *P* content of the sludge in the aeration tanks after setting up the influent model.

The second approach is less critical but it means that the measured ammonia/phosphates concentration cannot be used as a direct model input. For nitrogen, the procedure is to take the TKN measurements and subtract the organic nitrogen as calculated by the model (and nitrite + nitrate if total nitrogen is measured). The remaining nitrogen fraction is assumed to be the ammonia load and is used as model input. The difference between measured and calculated ammonia should be small, however, the critical point is to maintain the total nitrogen mass balances. Accordingly, the phosphates concentration should be calculated out of the measured total phosphorous concentration.

Note:

By changing the COD fractionation of a fraction-based model the total *N* and *P* mass is changed also. Subsequently the whole influent characterisation procedure has to be repeated including comparisons of the MLSS, *N*, and *P* masses. From Figures A.3 and A.4 it is evident that this is not necessary for component-based models.

A.4.2.3 Suspended solids fractions

An important variable in practice is the concentration of the total suspended solids (X_{TSS} in the model notation). It consists of a volatile part (volatile suspended solids, VSS) and an inorganic part (inorganic suspended solids ISS = TSS – VSS). There are several modelling approaches for the calculation of TSS:

- *COD based*: TSS is calculated based on the total concentration of particulate COD state variables and a factor according to the measured TSS/COD ratio. This approach is for example used in some ASM1 implementations.
- X_{TSS} as a state variable: X_{TSS} is modelled as a state variable, which is integrated through all simulation steps. This approach is for instance used in ASM2d and 3.
- X_{ISS} as a state variable: This approach introduces X_{ISS} as state variable and calculates TSS as a combined variable based on measured COD, TSS, VSS, and other concentrations in the influent. ISS accumulates in the system according to the sludge age. Some models also include increase of ISS due to biomass decay.

Note that most simulators introduced an external TSS calculation based on ISS in the influent and that the original method published with the biokinetic model might not be used.

The approach where X_{TSS} is introduced as a state variable is not completely described in the model publications and needs careful set-up within the model, especially the influent model. Every COD state variable is linked to a certain amount of organic (VSS) and inorganic material (ISS). Figure A.6 shows the TSS fractionation for ASM3, where each particulate model state variable includes a defined percentage of VSS and ISS.

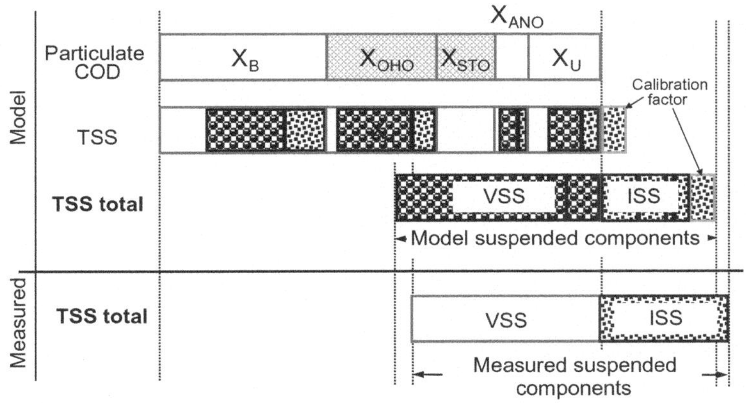

Figure A.6 TSS fractionation in ASM3.

As an example, Figure A.7 shows the calculation of the TSS content of biomass in ASM3. In the published biokinetic model, a COD/VSS ratio of 1.48 and a VSS/TSS ratio of 0.75 is assumed. In the model only an overall parameter $i_{TSS,XBio}$ is given with the default value of 0.9. In reality this parameter is SRT-dependent, which may introduce errors into the model.

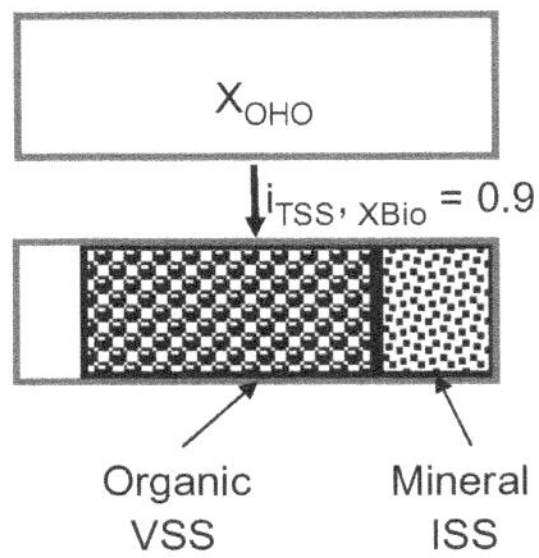

$$\frac{[\text{g TSS}]}{[\text{g COD}]} = \frac{1}{1.48 * 0.75} \qquad \frac{1}{[\text{g COD/g VSS}] * [\text{g VSS/g TSS}]}$$

Figure A.7 Calculation of TSS content of biomass in ASM3.

***Typical pitfalls**:* This simplified approach can lead to serious problems if:

- *The measured TSS in the influent is used directly as model input*: In typical engineering practice 0.45/0.7 µm filters are used, which let colloidal matter pass. This results in continuous underestimation of the influent TSS and therefore the TSS accumulation is too low in the activated sludge tanks in relation to the COD content. To avoid this effect the colloidal fraction has to be measured or the model TSS calculations, which are based on COD, should be used as an input for the X_{TSS} state variable.
- *X_{TSS} is used as target variable to fit sludge production*: The models are COD based and therefore the first step should be to calibrate the model to fit total COD. In a second step the COD/TSS ratio needs to be fitted.

A.5 pH AND ALKALINITY

Few models include pH calculations. Instead, alkalinity is introduced in several models to predict possible pH changes and to guarantee the continuity in ionic charge of the biological processes. Alkalinity is usually measured in equivalent concentration of HCO_3^- or in concentration of $CaCO_3$ (1 meq HCO_3^-/L = 50 mg $CaCO_3$/L). A low alkalinity concentration may lead to unstable pH, which could reach inhibiting levels. Three ways to deal with alkalinity have been proposed in activated sludge models:

- Alkalinity is not taken into account in the model at all (e.g. Barker & Dold and UCTPHO + models);
- Alkalinity is taken into account in the stoichiometry but does not limit the kinetic rates (e.g. ASM1);
- Alkalinity is taken into account in both stoichiometry and kinetic rates (e.g. ASM2d, ASM3, ASM3 + Bio-P and ASM2d + TUD).

A.6 OUTPUT MODELS

Output models convert model state variables into combined (composite) variables, which can be compared with measurements. Examples of typical composite variables are:

- Total suspended solids (TSS or MLSS)
- Total or soluble COD
- Total nitrogen or TKN
- Total phosphorus
- BOD

Some state variables can be directly compared with measurements (e.g. ammonia, nitrate, etc.)

A.7 AERATION MODELS

Diffused aeration: Aeration models can be split into a model for oxygen transfer and a model to link the oxygen transfer to an air flow rate. The oxygen transfer model calculates the resulting dissolved oxygen (DO) concentration of a bioreactor typically based on an oxygen mass transfer coefficient K_La. The second model takes specific information like diffuser density and tank geometry into account to link K_La to an air flow rate.

Surface aeration: Surface aeration models link energy consumption directly to oxygen transfer or calculate an oxygen transfer coefficient similar to K_La.

Depending on the aim of the simulation project, there are two main applications of aeration models (Figure A.8):

- A DO concentration is set and the model is used to calculate the required air flow (e.g. for blower and equipment sizing).
- Prediction of DO profiles over time and reactor length for given air flow rates (e.g. for design of diffuser distribution (tapering) or design and optimisation of aeration control system)

Note that a special case is when the detailed aeration system is of no interest. Then a simplified model could control K_La directly without linking to an air flow rate.

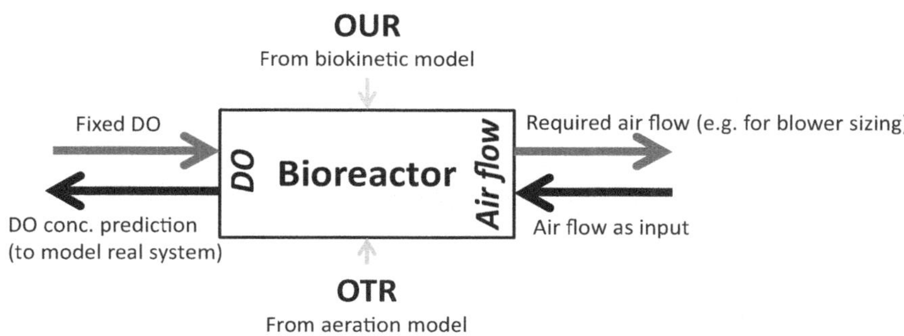

Figure A.8 Typical objectives of aeration models.

A.7.1 Oxygen transfer model

The oxygen transfer model is typically not part of the published biokinetic model and should be regarded as a separate model. However, it is often implemented as part of the biokinetic model matrix. Oxygen transfer models are based on the following basic equation:

$$r_{O_2,T} = k_La(S_{O_2,sat} - S_{O_2}) \qquad \text{(A.4)}$$

where:

S_{O_2}: dissolved oxygen concentration [mg DO/L]
$S_{O_2,sat}$: saturation concentration of dissolved oxygen [mg DO/L]
k_La: oxygen mass transfer coefficient [1/h]

Depending on the objectives of the project, the value of the mass transfer coefficient and of the DO saturation value may be important and linked to other operating parameters (air flow rate, power …).

If the goal is to provide sufficient oxygen to the model, the DO concentration can be set to a fixed value (e.g. 2 mg DO/L) without an aeration model (the oxygen state variable has always the same concentration or might be removed from the variable list). Another approach is to model the oxygen transfer and use a controller (e.g. a proportional or PI controller with a high proportional factor) and control the k_La value to obtain a specified DO set-point (e.g. 2 mg/l).

For calibration purposes, the DO set-point can be replaced by measured DO concentrations to separate calibration of the biokinetic and the oxygen transfer model.

The highest demands in terms of model accuracy are required if the DO concentration has to be predicted based on a specific air flow rate. A more complex aeration model is required taking the oxygen transfer efficiency into account to calculate k_La with typical inputs being energy (surface aerators) or air flow (diffused aeration).

Several oxygen transfer models are available in the different simulators. The reader is referred to the software manuals for a detailed description of the model used.

A.7.2 Aeration control model

A.7.2.1 DO control loops

If a goal of the study is to investigate the aeration control system, the control loop has to be modelled according to the real conditions. Although often the real controllers are implemented in a discrete way, in the plant model a continuous control loop might be used for reasons of calculation speed. The important point is to model the same time constants as measured at the real plant. In some cases even the response times of sensors (Rieger *et al.* 2003) and actuators (Rieger *et al.* 2006) have to be considered.

A.7.2.2 Reactor definition

A critical part of an activated sludge model is to model the correct DO concentrations in accordance with measured oxygen gradients in length and depth. Often a compromise is necessary between the optimal representation of the hydrodynamic behaviour (e.g. based on a tracer experiment) and the oxygen profiles.

As an example, for WWTPs with longitudinal flow the number of modelled reactors can be selected according to zones with equally distributed diffusers (Figure A.9). For surface aeration of an oxidation ditch type circuit a more complex compartmentalisation may be required to model the hydrodynamic behaviour as well as the oxygen profiles and the connected simultaneous nitrification/denitrification (SND).

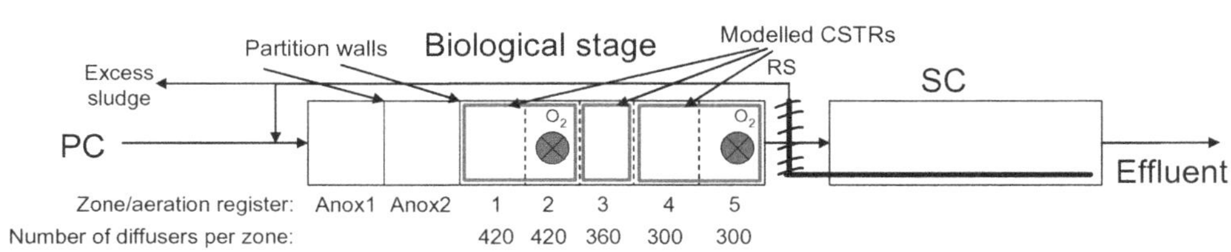

Figure A.9 Process scheme and aeration register of WWTP Werdhoelzli (CH).

A.7.2.3 Location of DO sensors

Beside the reactor definition and the air distribution the location of the DO sensors has a major influence on the simulation results. If there is only one DO sensor per lane in the real plant that information is available only at this one location and is not providing insights into the existing DO profile. In the model a good approximation of the DO profile is required, which might also necessitate a change in the measured DO time series if the DO sensor cannot be placed at a reasonable location in the model (e.g. due to a limited number of CSTRs). Gradients in length and depth should be measured in advance to define and calibrate a suitable model.

A.7.3 Detailed aeration system model

In some special cases it may be necessary to model further elements of the aeration system including the blower system, pipes, valves and diffusers (Alex *et al.* 2002). The level of detail again depends on the objectives of the study. CFD models have been used to predict the flow behaviour as well as clean water oxygen transfer in aeration tanks (Fayolle *et al.* 2007).

A.8 PHOSPHORUS PRECIPITATION MODEL

Describing models for chemical phosphorus precipitation in detail is outside the scope of this report. However, chemical phosphorus precipitation can significantly contribute to the sludge production of a plant. Even if the precipitation itself is not modelled, at least an estimation of the chemical sludge produced should be used before comparing model results with the measured sludge production. The chemical sludge production will depend largely on the mass of chemical (usually iron or aluminium salt) added, which itself depends mainly on the required effluent phosphorus limits and the technology used (e.g. pre-, post- or simultaneous precipitation).

As an example: To reach a 1 mg P/L effluent limit it can be assumed that 1.4–1.8 molar dose (Fe/P ratio, "r") is required in iron, and a somewhat lower dose in aluminium. For significantly lower effluent P concentrations (e.g. 0.1 mg P/L or less) the required dose could double. A simple calculation can then be used to estimate the chemical sludge formed:

- The assumed hydroxo-phosphate precipitate can be represented as $Me_rH_2PO_4(OH)_{(3r-1)}$; the molecular masses are Fe = 55.8, Al = 27, H = 1, P = 31, O = 16.
- To reach 1 mg P/L effluent phosphate concentration (assuming $r_{AL} = 1.2$ and $r_{Fe} = 1.6$), for each kg of P removed 5.6 kg aluminium sludge or 8.1 kg iron sludge will be formed.
- To reach 0.1 mg P/L effluent phosphate concentration (assuming $r_{AL} = 2.4$ and $r_{Fe} = 3.2$), for each kg of P removed 8.6 kg aluminium sludge or 13.6 kg iron sludge will be formed.
- This mass has to be added to the sludge stream (or influent as convenient) as unbiodegradable inorganic suspended solids (ISS).

REFERENCES

Alex J., To B. and Hartwig P. (2002). Improved design and optimization of aeration control for WWTPs by dynamic simulation. *Water Science and Technology*, **45**(4–5), 365–372.

Burger R., Diehl S. and Nopens I. (2011). A consistent modelling methodology for secondary settling tanks in wastewater treatment. *Water Research*, **45**(6), 2247–2260.

Corominas L., Rieger L., Takács I., Ekama G., Hauduc H., Vanrolleghem P.A., Oehmen A., Gernaey K. V., van Loosdrecht M. C. M. and Comeau Y. (2010). New framework for standardized notation in wastewater treatment modelling. *Water Science and Technology*, **61**(4), 841–857.

Fayolle Y., Cockx A., Gillot S., Roustan M. and Heduit A. (2007). Oxygen transfer prediction in aeration tanks using CFD. *Chemical Engineering Science*, **62**(24), 7163–7171.

Fujie K., Sekizawa T. and Kubota H. (1983). Liquid mixing in activated sludge aeration tank. *Journal of Fermentation Technology*, **61**(3), 295–304.

Gujer W., Henze M., Mino T. and van Loosdrecht M. C. M. (2000). Activated Sludge Model No. 3. In: Activated sludge models ASM1, ASM2, ASM2d and ASM3. M. Henze, W. Gujer, T. Mino and M. C. M. van Loosdrecht, (eds), *Scientific and Technical Report No. 9*, IWA Publishing, London, UK.

Henze M., Gujer W., Mino T., Matsuo T., Wentzel M. C. and Marais, G. V. R. (2000). Activated sludge model No. 2. In: Activated sludge models ASM1, ASM2, ASM2d and ASM3. M. Henze, W. Gujer, T. Mino and M. C. M. van Loosdrecht, (eds), *Scientific and Technical Report No. 9*, IWA Publishing, London, UK.

Melcer H., Dold P. L., Jones R. M., Bye C. M., Takacs I., Stensel H. D., Wilson A. W., Sun P. and Bury S. (2003). Methods for wastewater characterisation in activated sludge modelling. *Water Environment Research Foundation* (WERF), Alexandria, VA, USA.

Otterpohl R. and Freund M. (1992). Dynamic models for clarifiers of activated sludge plants with dry and wet weather flows. *Water Science and Technology*, **26**(5–6), 1391–1400.

Plósz B. G., Weiss M., Printemps C., Essemiani K. and Meinhold J. (2007). One-dimensional modelling of the secondary clarifier – factors affecting simulation in the clarification zone and the assessment of the thickening flow dependence. *Water Research*, **41**(15), 3359–3371.

Rieger L., Alex J., Winkler S., Böhler M., Thomann M. and Siegrist H. (2003). Progress in sensor technology – progress in process control? Part I: Sensor property investigation and classification. *Water Science and Technology*, **47**(2), 103–112.

Rieger L., Alex J., Gujer W. and Siegrist H. (2006). Modelling of Aeration Systems at Wastewater Treatment Plants. *Water Science and Technology*, **53**(4–5), 439–447.

Roeleveld P. J. and van Loosdrecht M. C. M. (2002). Experience with guidelines for wastewater characterisation in The Netherlands. *Water Science and Technology*, **45**(6), 77–87.

Sperandio M. and Paul E. (2000). Estimation of wastewater biodegradable COD fractions by combining respirometric experiments in various SO/XO ratios. *Water Research*, **34**(4), 1233–1246.

Takács I., Patry G. G. and Nolasco D. (1991). A dynamic model of the thickening/clarification process. *Water Research*, **25**(10), 1263–1271.

Wett B. (2002). A straight interpretation of the solids flux theory for a 3 layer sedimentation model. *Water Research*, **36** (12), 2949–2958.

Appendix B

Representation of biokinetic models – the Gujer Matrix

B.1 INTRODUCTION

Biokinetic models describe ***bio***logically mediated time-dependent (***kinetic***) transformations in activated sludge. ASM-type models are "structured models", which distinguish groups of organisms, substrates and other compounds of interest instead of relying on lumped constituents such as TSS or BOD. Structured models are more fundamentally based and have a wider range of applicability than lumped variable models. However, many state variables and parameters cannot be directly or even indirectly measured. Also, structured models require more information for wastewater fractions through site-specific wastewater characterisation. This wastewater characterisation step is specific for the different biokinetic models as it translates measurements into model state variables.

Biokinetic models consist of:

- *State variables* representing relevant components within the context of the model,
- *Kinetic rates* that define the rate of conversions for each process acting on the state variables.
- *Stoichiometric coefficients* that describe the conversion of state variables to others in each process
- *Elemental composition* (e.g. N or P content), *COD content* or *charge* of each state variable in the composition matrix to allow mass balance continuity. For models purely expressed in one unit (e.g. COD) this matrix may be missing but is inherently included in the state variable unit definitions.

This appendix describes the standard way of representing biokinetic matrices, called a *Gujer Matrix*. Seven published biokinetic models for AS systems are discussed in this report. Appendix E contains a detailed article on the model verification as published in Hauduc *et al.* (2010). The models have been carefully checked for consistency and continuity, and typos in the original publications were corrected. The corrected Gujer Matrices for the seven models are provided on the GMP WaterWiki website.

B.2 MATRIX FORMAT

One of the main steps towards the widespread use of activated sludge modelling was the introduction of a standardised model matrix notation by the IAWPRC (now IWA) Task Group on *Mathematical Modelling for Design and Operation of Biological Wastewater Treatment*. Since ASM1 (Henze *et al.* 1987), biokinetic models have been presented in this matrix format (the Gujer Matrix) allowing a concise representation of a large equation system in one easy and intuitive format.

For a full description of the matrix notation, the reader is referred to the original publication (Henze *et al.*, 1987) and Gujer and Larsen (1995). A brief summary of the key features is given below.

B.2.1 The structure of the Gujer Matrix

The Gujer Matrix organises the state variables, rate equations, stoichiometry and state variable composition information into a specific matrix format consisting of three main parts (Figure B.1):

(1) The *stoichiometry matrix* for all process rates and state variables,
(2) The *composition matrix* for all state variables and their constituents considered, and
(3) The *kinetic rate expressions* for all processes

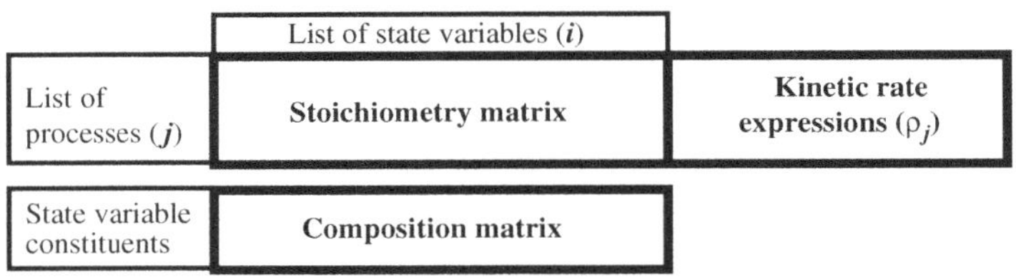

Figure B.1 Main parts of the gujer matrix.

Columns in the matrix (index i) contain the state variables (such as heterotrophic biomass or oxygen), except the last column which contains the rate expressions (ρ_j). Rows (index j) contain processes acting on the state variables with their relevant stoichiometry in the stoichiometry matrix part, and state variable compositions in the composition matrix part.

In addition, the list of stoichiometric parameters, found in the stoichiometry and composition matrix, and the list of kinetic parameters found in the kinetic rate expressions with their proper default values form an important part of any ASM-type model. The combination of the model structure (described in the Gujer Matrix) and the stoichiometric and kinetic parameters implemented in a simulator make it possible to run the model.

Stoichiometry and composition

As an example, let's consider the process of aerobic growth of heterotrophic organisms on one unit of substrate, expressed in COD. One unit of substrate generates Y (yield) unit of biomass (in COD) and 1-Y unit of electron acceptor, in this case oxygen, is used in the process (Figure B.2).

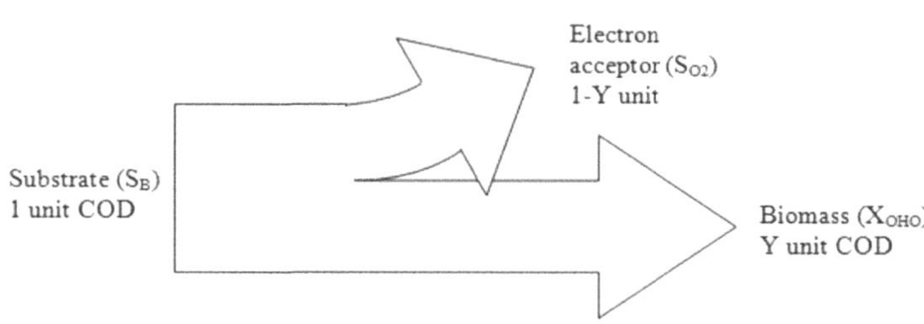

Figure B.2 Substrate, biomass and electron acceptor balance.

This stoichiometry is tabulated in the *Stoichiometry* part of the Gujer Matrix (Table B.1), usually (but not necessarily) normalised to one unit of biomass (all terms divided by Y).

Table B.1 Stoichiometry for aerobic growth of ordinary heterotrophic biomass (X_{OHO}).

	X_{OHO} g COD . m^{-3}	S_B g COD . m^{-3}	S_{O_2} - g COD . m^{-3}
Aerobic growth	1	$-1/y$	$-((1-y)/y)$

This process is balanced for electron acceptor in agreement with Figure B.2, since

$$1 - (1/y) + ((1-y)/y) = 0 \tag{B.1}$$

where the terms for X_{OHO} and S_B are multiplied by 1 (unit of COD) and the term for oxygen is multiplied by -1, that is negative COD.

Including ammonia nitrogen strictly for heterotrophic biomass synthesis demonstrates the use of the *Composition matrix* (Table B.2):

Table B.2 Stoichiometry and composition matrix for aerobic growth of X_{OHO}

Stoichiometry	X_{OHO} g COD . m^{-3}	S_B g COD . m^{-3}	S_{O_2} - g COD . m^{-3}	S_{NHx} g N . m^{-3}
Aerobic growth g COD · m^{-3} · d^{-1}	1	$-1/y$	$-((1-y)/y)$	$-i_N$
Composition				
COD	1	1	-1	
N	i_N			1

Empty cells signify a coefficient of zero. Heterotrophs in this model contain i_N amount of nitrogen per COD unit (usually 7.0–8.6% per COD), and they obtain that amount during growth from ammonia. Substrate contains no nitrogen in this model, and ammonia is measured in terms of its nitrogen content, therefore its nitrogen coefficient is 1 and COD coefficient is considered zero.

To obtain continuity for COD, each stoichiometric coefficient in the growth process must be multiplied with the relevant coefficient in the *Composition matrix*. This is equivalent to Equation B.1. For N continuity, the same principle is applied, resulting in Equation B.2:

$$1 \cdot i_N - i_N \cdot 1 = 0 \tag{B.2}$$

The same principle can be applied for all elemental components in more complex models (i.e. if they observe continuity for COD, C, O, H, N, P, S, Fe, charge, etc.).

An important feature of ASM-type models are electron equivalence conversions. This is trivial in case of oxygen and COD, because of their equivalence oxygen can also be considered as "negative COD", as shown in the example above. However, other electron acceptors such as nitrate, and in general the conversion of various nitrogen species to their electron equivalence (and therefore COD) requires specific consideration.

If we consider complete denitrification (nitrate to nitrogen gas), the following half reactions can be postulated (Equation B.3):

$$10e^- + 2NO_{3^-} + 12H^+ \longrightarrow N_2 + 6H_2O \tag{B.3}$$

that is the reduction of 1/5th mol of nitrate per one electron equivalence.

And, similarly in Equation B.4:

$$4e^- + O_{2^-} + 4H^+ \longrightarrow 2H_2O \tag{B.4}$$

that is the reduction of ¼ mol of oxygen per one electron equivalence.

Consequently, in electron equivalence units $1/5^{th}$ mol of nitrate is equivalent to ¼ mol of oxygen, that is, approximately 14/5 g N of nitrate-N is equivalent to approximately 32/4 g of oxygen, which is 2.86 g of oxygen per g of nitrate-N.

Table B.3 summarises the most typical electron equivalence values used in ASM models (in single and multiple step nitrification and denitrification processes) and their quick calculation method based on oxygen and nitrogen atomic masses.

Table B.3 Electron equivalence values for selected components.

to ↓	from →	NH_3	NH_2OH	N_2	N_2O	NO	NO_2	NO_3
NH_3			1.14	1.71	2.28	2.86	3.43	4.57
NH_2OH		1.14		0.57	1.14	1.71	2.28	3.43
N_2		1.71	0.57		0.57	1.14	**1.71**	**2.86**
N_2O		2.28	1.14	0.57		0.57	1.14	2.28
NO		2.86	1.71	1.14	0.57		0.57	1.71
NO_2		**3.43**	2.28	1.71	1.14	0.57		1.14
NO_3		**4.57**	3.43	2.86	2.28	1.71	**1.14**	
NH_3			O/N	3O/2N	20/N	5O/2N	30/N	40/N
NH_2OH		O/N		O/2N	O/N	30/2N	20/N	3O/N
N_2		3O/2N	O/2N		O/2N	20/2N	3O/2N	5O/2N
N_2O		2O/N	O/N	O/2N		O/2N	O/N	2O/N
NO		5O/2N	3O/2N	20/2N	O/2N		O/2N	3O/2N
NO_2		3O/N	2O/N	30/2N	O/N	O/2N		O/N
NO_3		40/N	3O/N	5O/2N	2O/N	3O/2N	O/N	

The atomic mass of oxygen (O) is 15.9994: ~16 g/mol.
The atomic mass of nitrogen (N) is 14.0067: ~14 g/mol.

Kinetic expressions

The Gujer Matrix provides the reaction rate (r_i) that is required in the mass balance equation for each state variable i. r_i is calculated by summing the products of the stoichiometric coefficients (ν_{ij}) and the process rate expression (ρ_j) over all processes j as follows:

$$r_i = \Sigma \nu_{ij} \rho_j \tag{B.5}$$

The process rates (ρ_j) are listed for every process in a separate column of the Gujer Matrix. The process rate expression is based on a maximum process rate (where μ stands for growth, b for decay, m for maintenance and q for other rates). This maximum rate is generally proportional to a biomass or component concentration (surface reactions), and the overall rate varies depending on reduction factors and a series of environmental factors such as pH, temperature, concentration of consumed components (substrates, nutrients, electron acceptor …) or inhibitors (e.g. presence of dissolved oxygen in a denitrification process). The general form is:

$$\rho_j = \mu_{\text{Max}} \cdot \eta \cdot \underbrace{\frac{S_B}{K_{\text{SB}} + S_B}}_{\textit{Saturation-function}} \cdot \underbrace{\frac{K_{\text{SI}}}{K_{\text{SI}} + S_I}}_{\textit{Inhibition-function}} \cdot X_{\text{BIO}} \tag{B.6}$$

where:

ρ_j = Kinetic expression for process j
μ_{Max} = Maximum rate
η = Reduction factor
X_{BIO} = Biomass concentration
S_B = Substrate concentration
S_I = Inhibitor concentration
$K_{\text{SB/SI}}$ = Half-saturation coefficient

A reduction factor reduces the overall rate due to for instance the fact that only a part of the modelled biomass is able to carry out this process or that the process runs with a reduced rate in the specified environmental conditions.

Saturation and inhibition functions (also called switching functions or Monod terms) are used to reduce process rates from the maximum rate according to the availability or non-availability of another component. Monod functions are generally used, however other functions are also available (e.g. Haldane, Andrews). The impact of the Monod-type saturation/inhibition functions is shown in Figure B.3. The half-saturation coefficient is a parameter defining the shape of the curves and corresponds to the substrate (or other component) concentration at which the value of the Monod saturation/inhibition function is 0.5.

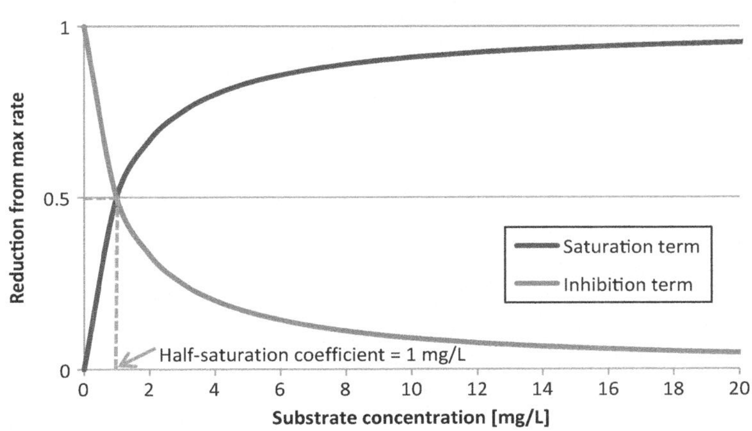

Figure B.3 Monod saturation/inhibition functions.

Special variants of the Monod-type functions are sometimes used to introduce a minimum component ratio (Equation B.7) or to model a maximum possible storage pool (Equation B.8). Note that the initial ratio of X_B/X_{BIO} has to be lower than $K_{\max}$ to prevent numerical problems.

$$\frac{X_B/X_{\mathrm{BIO}}}{K_{\mathrm{SB,OHO}}+(X_B/X_{\mathrm{BIO}})} \tag{B.7}$$

$$\frac{K_{\max}-X_B/X_{\mathrm{BIO}}}{K_{\mathrm{Stor}}+(K_{\max}-X_B/X_{\mathrm{BIO}})} \tag{B.8}$$

In the case where multiple substrates can be used in the same process, a term needs to be introduced making sure that the resulting rate is not higher than the specified maximum rate. An example for parallel consumption (without preference for one substrate) is described in Equation B.9).

$$\begin{array}{ll}\text{Process description for growth on substrate 1} & \dfrac{S_{B1}}{S_{B1}+S_{B2}} \\ \text{Process description for growth on substrate 2} & \dfrac{S_{B2}}{S_{B1}+S_{B2}}\end{array} \tag{B.9}$$

Equation B.10 is an example Monod equation for heterotrophic growth (assuming dissolved oxygen, nutrients, and alkalinity are not limiting). The heterotrophic growth rate μ_{OHO} is calculated by multiplying the maximum heterotrophic growth rate $\mu_{\mathrm{OHO,Max}}$ by a saturation function and finally by the heterotrophic biomass concentration.

$$\rho_{\mathrm{Growth-OHO}}=\mu_{\mathrm{OHO,Max}}\cdot\frac{S_{\mathrm{B}}}{K_{SB}+S_B}\cdot X_{\mathrm{OHO}} \tag{B.10}$$

where:

S_{B} = Substrate concentration
K_{SB} = Half-saturation coefficient of S_B for heterotrophic organisms

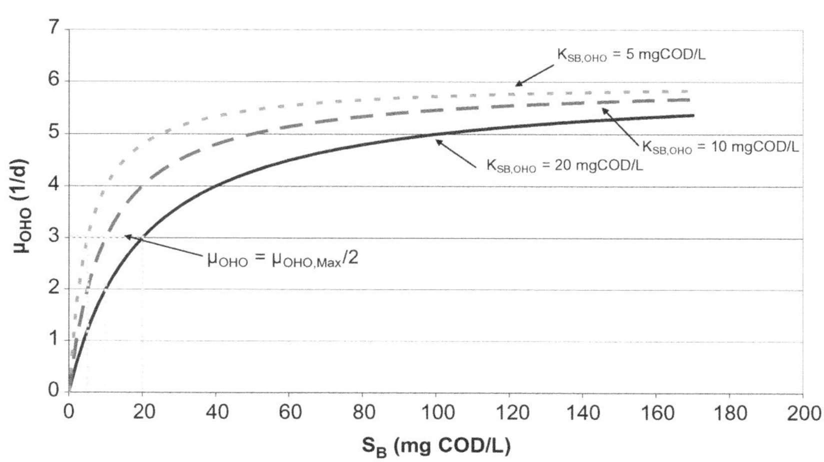

Figure B.4 Impact of $K_{\mathrm{SB,OHO}}$ on the heterotrophic growth rate μ_{OHO}.

Figure B.4 shows the heterotrophic growth rate μ_{OHO} as a function of the substrate concentration S_B for different half-saturation coefficients $K_{SB,OHO}$. In general, when S_B is close to zero the growth rate is also close to zero and when S_B is much greater than $K_{SB,OHO}$, the growth rate is close to the maximum growth rate (in this example 6 d^{-1}), that is, the Monod saturation function is close to 1. The value of $K_{SB,OHO}$ reflects the ability of the organisms to use low levels of substrate. A low $K_{SB,OHO}$ value implies a high affinity of the organisms to the substrate.

Typically, several saturation and inhibition functions and reduction factors are added to describe the multiple impacts on the maximum rate. The example in Equation B.11 shows the kinetic rate expression for heterotrophic growth with saturation functions for substrate, oxygen, and ammonia. Note that each term can stop the process, so even with enough substrate available, the lack of oxygen or ammonia (as nutrient) will reduce the overall rate to zero.

$$\rho_{\text{Growth-OHO}} = \mu_{\text{OHO,Max}} \cdot \frac{S_B}{K_{SB} + S_B} \cdot \frac{S_{O_2}}{K_{O_2} + S_{O_2}} \cdot \frac{S_{NHx}}{K_{NHx} + S_{NHx}} \cdot X_{OHO} \tag{B.11}$$

where:

S_B = Substrate concentration
K_{SB} = Half-saturation coefficient of S_B for heterotrophic organisms
S_{O_2} = Dissolved oxygen concentration
K_{O_2} = Half-saturation coefficient of S_{O_2} for heterotrophic organisms
S_{NHx} = Ammonia concentration
K_{NHx} = Half-saturation coefficient of S_{NHx} for heterotrophic organisms

REFERENCES

Gujer W. and Larsen T. A. (1995). The implementation of biokinetics and conservation principles in ASIM. *Water Science and Technology*, **31**(2), 257–266.

Hauduc H., Rieger L., Takács I., Héduit A., Vanrolleghem P. A. and Gillot S. (2010). Systematic approach for model verification – Application on seven published Activated Sludge Models. *Water Science and Technology*, **61**(4), 825–839.

Henze M., Grady C. P. L., Jr., Gujer W., Marais G. v. R. and Matsuo T. (1987). Activated Sludge Model No. 1. *IAWPRC Scientific and Technical Report No. 1*, IAWPRC, London, UK.

Appendix C

The numerical engine – solvers for beginners

The mathematical model of an activated sludge system usually consists of a large number of linked algebraic and differential equations which need to be solved efficiently under different conditions. These calculations are performed by various algorithms, or "solvers", which form part of the simulator's numerical "engine". The following brief review is intended to explain common terminology that is used to operate solvers for various types of simulations.

Four types of numerical algorithms are considered in this section and are summarised in Table C.1. Many solvers found in the mathematical and programming literature are not sufficiently efficient or robust enough to solve an activated sludge plant model. Solvers in practice are usually implemented with modifications that make them more suitable for the complex, nonlinear, discontinuous and stiff numerical environment they have to operate within.

Table C.1 Numerical solvers.

	Typical use	Typical features	Example algorithms
Steady-state solver	Finding the steady-state solution of a model for a certain condition	Iterative local or global search methods	Newton-Raphson, and so on.
Dynamic solver	Solving ordinary differential equations (ODEs) (e.g. the biokinetic model)	Fixed or variable step methods, with different error criteria and order	Euler, Runge Kutta, Gear's stiff, many others
Algebraic solver	Solving chemical equilibrium or fast kinetic variables, flow loops	Iterative search to solve algebraic loops or differential algebraic solver	DAE, dynamic delay
Optimisation algorithm	Finding an objective function (e.g. minimising effluent N_{tot})	Local or global search, convergence criteria	Nelder-Mead Simplex Gradient methods many other ...

C.1.1 STEADY-STATE SOLVERS

Steady-state can be calculated by running a long dynamic simulation with constant inputs. The simulations need to be long enough (~3 SRTs) so that the outputs stabilise to a steady-state value. Using an iterative solver is usually more efficient in terms of getting to the solution faster. There are many different solvers, and from the user's standpoint their parameters (if accessible at all) can be confusing. In general, if a steady-state solver cannot find a solution and is jumping around too much, reducing the step size (or related parameters) often helps.

C.1.2 DYNAMIC SOLVERS

Dynamic solvers are used to calculate state variable concentrations through their derivatives. Today's simulation software contains powerful, variable time step solvers. This means that at times where changes are large, for example starting up of blowers in intermittently aerated reactors, the solver will automatically reduce the time step to maintain a predetermined, small error. While at the same time, if the error is within the set tolerance, the solver will take larger time steps to minimise the time required for the simulation.

C.1.3 SIMULATION SPEED AND TIME STEPS

It is important to remember that "solver or integration time step" and "data or communication time step" are two different concepts. The former is the time step used in the integration calculation and the latter is the interval at which the solver provides results back to the user interface. Solvers use small integration time steps, typically in the order of seconds of simulated time. Simulation software collects results at a much larger data interval (e.g. one hour). Changing the "data or communication time step" has no effect on the results (except that a very small data or communication step may force the solver to reduce its step size, which impacts the simulation speed). It only affects the number of data points on an output graph for example. Changing the "solver or integration time step" will have an effect on results. If fixed time step solvers are used, usually very small time steps (e.g. 0.1 second) need to be used in order to avoid integration errors. In most cases, solvers use variable, self-adjusting time steps. Increasing time steps or increasing the error tolerance usually cannot be used to reduce the simulation time without a significant risk of calculating the wrong results. Certain solvers are more efficient for smooth simulations with slow dynamics, while others are better for simulating quick, rapid changes or stiff systems (a stiff system is one where the rate of change of state variables at certain times may differ by three orders of magnitude or more). If the model has to be used only a few times, usually it is best to use the default algorithms and settings. If a large number of simulations need to be performed with one specific model, it may be worthwhile to run a few tests with different solvers and chose the fastest option.

In some cases solver problems are related to problems with the setup of the plant model and so it is worthwhile checking if there is an error in the data or settings. In other cases solver problems indicate critical process conditions (e.g. near wash-out).

C.1.4 ALGEBRAIC SOLVERS

Some variables in a plant model will be solved algebraically. These are calculated directly instead of being part of the differential equation system. The reason is usually that the dynamics of these variables are orders of magnitude faster than that of the slower kinetic variables, so the intermediate dynamics can be ignored. Typical cases are flow loops or ionic speciation in equilibrium chemistry. Simulators solve these variables using either a dynamic lag approach (converting them to pseudo kinetic variables, with fast kinetic rates), or, preferably, with an iterative algebraic solver.

C.1.5 OPTIMISERS

Optimisation algorithms can be used for maximising, minimising or fitting any desired objective function. Typical applications are (i) calibration (fitting model to measured data), or (ii) optimising operations for example minimising effluent total nitrogen. In general, optimisers adjust a few user selected parameters to achieve the desired objective function. Discussion of optimiser settings is beyond the scope of this report.

Appendix D1

New framework for standardized notation in wastewater treatment modelling

Ll. Corominas[1], L. Rieger[1], I. Takács[2], G. Ekama[3], H. Hauduc[1,4], P.A. Vanrolleghem[1], A. Oehmen[5], K.V. Gernaey[6], M.C.M. van Loosdrecht[7] and Y. Comeau[8]

[1]modelEAU, Université Laval, Pavillon Pouliot, 1065 av. de la Médecine, Quebec (QC) G1V 0A6, Canada.
[2]EnviroSim Associates Ltd, 15 Impasse Fauré, Bordeaux 33000 France.
[3]Water Research Group, Department of Civil Engineering, University of Cape Town, Rondebosch, 7701, Cape Town, South Africa.
[4]Cemagref, EPURE, Parc de Tourvoie, BP44, F-92163, Antony Cedex, France.
[5]CQFB/REQUIMTE, Department of Chemistry, FCT, Universidade Nova de Lisboa, P-2829-516 Caparica, Portugal.
[6]Department of Chemical and Biochemical Engineering, Technical University of Denmark, Building 229, DK-2800 Kgs. Lyngby, Denmark.
[7]Department of Biotechnology, Delft University of Technology, Julianalaan 67, 2628 BC Delft, The Netherlands.
[8]Department of Civil, Geological and Mining Engineering, Ecole Polytechnique, P.O. box 6079, Station Centre-ville, Montreal (QC) H3C 3A7, Canada.

ABSTRACT

Many unit process models are available in the field of wastewater treatment. All of these models use their own notation, causing problems for documentation, implementation and connection of different models (using different sets of state variables). The main goal of this paper is to propose a new notational framework which allows unique and systematic naming of state variables and parameters of biokinetic

Previously published as: Corominas, L., Rieger, L., Takács, I., Ekama, G., Hauduc, H., Vanrolleghem, P.A., Oehmen, A., Gernaey, K.V., van Loosdrecht, M.C.M. and Comeau, Y. (2010). New framework for standardized notation in wastewater treatment modelling. *Water Science and Technology*, **61**(4), 841–857.

models in the wastewater treatment field. The symbols are based on one main letter that gives a general description of the state variable or parameter and several subscript levels that provide greater specification. Only those levels that make the name unique within the model context are needed in creating the symbol. The paper describes specific problems encountered with the currently used notation, presents the proposed framework and provides additional practical examples. The overall result is a framework that can be used in whole plant modelling, which consists of different fields such as activated sludge, anaerobic digestion, sidestream treatment, membrane bioreactors, metabolic approaches, fate of micropollutants and biofilm processes. The main objective of this consensus building paper is to establish a consistent set of rules that can be applied to existing and most importantly, future models. Applying the proposed notation should make it easier for everyone active in the wastewater treatment field to read, write and review documents describing modelling projects.

Keywords: Wastewater treatment, modeling, nomenclature, ASM, ADM

D1.1 INTRODUCTION

Mathematical modelling of wastewater treatment (WWT) processes has become a widely accepted tool in the past decade, and is used for research, plant design, optimization, training, and model-based development and testing of process control. Starting with the activated sludge system and now moving into whole plant modelling, the modelling community has produced a significant number of models describing the processes occurring in wastewater treatment plants (WWTPs). New models and model extensions are constantly being developed in response to changing requirements, for example stricter effluent limits, or new processes such as side-stream treatment.

One of the milestones in dynamic modelling of WWTPs was the research carried out by the University of Cape Town (Ekama & Marais, 1977; Dold *et al.* 1980). With this research a specific notation was introduced (further referred to as the "UCT system") and several research groups are still using this naming system (e.g. Barker & Dold, 1997, Lee *et al.* 2006). In 1987, a task group, chaired by Prof. Henze, of the International Association on Water Pollution Research and Control (IAWPRC originally, and since 2000 the International Water Association, IWA) introduced the first Activated Sludge Model for biological carbon and nitrogen removal (ASM1). ASM1 was based on the South African work but was presented in a new format (the Gujer or Petersen table, composed of a stoichiometric matrix, a rate vector and extra information as units and names) and with a new and standardized notation (in this paper referred to as the "IWA system"). The latter notation had its roots in the work of another IAWPRC/IUPAC task group, led by Prof. Grau (Grau *et al.* 1982a, 1982b, 1987).

The need to widen the model boundaries and to include other process units led to the development of several other models such as ADM1 for anaerobic treatment (Batstone *et al.* 2002), fixed biomass (Rittmann & McCarty, 1980; Wanner & Gujer, 1986; Horn *et al.* 2003) and membrane bioreactors (MBRs; Lu *et al.* 2001; Jiang *et al.* 2008). Nitrite as an intermediate compound is included in several models (Sin *et al.* 2008). Increased microbiological and biochemical insights led to the development of so-called metabolic models (e.g. Smolders *et al.* 1995; Murnleitner *et al.* 1997; Lavallée *et al.* 2008; Lopez-Vazquez *et al.* 2009). An emerging field is the modelling of the fate of micropollutants, where a number of models were proposed by several researchers (e.g. Joss *et al.* 2006; Schönerklee *et al.* 2009). All of these models are published with their own notation, sometimes using different names for the same compound or parameter, or the same name for different compounds/parameters.

D1.1.1 Motivation

The need for a common international notation standard in biological wastewater treatment was already highlighted in Henze *et al.* (1982), where examples were given of abuse of notation (e.g. double notation, double meaning, misdirection, etc.). It was concluded that notation is a common cause of confusion due to the absence of a universally-agreed system of terminology. At the same time a proposal for unifying the notation used in the description of biological wastewater treatment processes was presented by Grau *et al.* (1982a, 1982b, 1987). This proposal was presented by a Working Group set up by the IAWPRC and the Commission on Water Quality of the International Union of Pure and Applied Chemistry (IUPAC). In this report, several symbols are listed together with their description, dimensions and some specifications as footnotes. This notation standard has been followed for many years. However, the complexity of WWT models has significantly increased over the last 25 years (Gujer, 2006) and new modelling concepts have been introduced. Moreover, in the work of Grau *et al.* (1982a, 1982b, 1987) no clear framework was given for future developments.

The need to re-address this problem arose during the work of the IWA task group on "Good Modelling Practice – Guidelines for Use of Activated Sludge Models" (GMP TG 2008) as well as during the work on a book on Biological Wastewater Treatment (Henze *et al.* 2008). The IWA task group on "Benchmarking of Control Strategies for Wastewater Treatment Plants", implementing and describing the details of a plant-wide WWTP model, was confronted with the issue as well. A working group involving several experts was set up in connection with the 1st IWA/WEF Wastewater Treatment Modelling Seminar (WWTmod2008), where the state-of-the-art in wastewater treatment modelling was discussed. It was decided to address the following problems:

(i) The same state variables and parameters used in different models or different platforms are named differently.
(ii) Several specific pitfalls prevail in the existing notations (e.g. colloidal matter, see next section).
(iii) No internationally accepted framework is available to name new state variables and parameters.
(iv) Model documentation (including notation) is time consuming and can lead to implementation errors.
(v) Model exchange is a problematic issue especially for complex models (Gernaey *et al.* 2006).
(vi) Coupling different models is becoming common such as for plant-wide modelling (Grau *et al.* 2009) making the use of one notation indispensable.
(vii) Different notations in reporting and coding can cause implementation errors and make double-checking difficult.

Given all of the above, it appears that a new and extendable notational framework is needed, that should:

(a) be as similar as possible to the existing notations,
(b) be simple,
(c) be straightforward and easy to understand,
(d) provide unique names within a model context,
(e) describe physical, biological and chemical properties important within the model context,
(f) be extendable for future developments.

To present the new framework, this paper is organized as follows; first, the general objectives for the framework are laid out and general notation rules are introduced. Then, separate sections for state variables and parameters are presented. They include a discussion on problems currently encountered and a description of the proposed new framework with some examples. Finally, the contributions of the new framework and the conclusions are described.

D1.2 GENERAL FRAMEWORK

The proposed notation should be valid for the different subfields of WWT modelling, and is mainly focused on biokinetic models. Therefore, the new notation has been developed considering models for activated sludge, anaerobic digestion, sidestream treatment, membrane bioreactors, micropollutant fate and biofilm processes, and so on. In addition, the notation also considers metabolic modelling approaches. The main objective of this consensus building paper is, first to create a consistent set of notation rules that can be applied to existing and more importantly, future models and second, to promote the establishment of a consensus on variable/parameter names.

D1.2.1 Naming system established for the new notation

The main goal of the new notation is to provide a framework which allows unique naming of state variables (the compounds or components used in the model's mass balances) and parameters. The resulting name is kept as short and mnemonic as possible and previously accepted notation is applied whenever feasible. An important element of the new notation is that the symbols are consistently defined as a main symbol with different subscript levels, which accounts for the increasing complexity of the models. Elements in the subscript are separated by a comma or underscores. The authors suggest using only the subscript levels required to make the name unique within the model context (e.g. for understanding the behaviour/role of the compound or parameter, or to prevent misinterpretation). *Thus, all levels of subscripts are optional and only those required should be used in a given model context.* Additional levels can be added by the model developer if further specifications are required (e.g. compartment or units of expression). The naming system also allows the use of superscripts under certain conditions (e.g. biofilm modeling) if this makes the symbol more comprehensible or better readable. The resulting symbols have to be properly documented in each of the models used and they are always presented with their units and their values. A list of the proposed abbreviations and symbols is given in Table D1.4. In the preparation of the proposed list, existing abbreviations were favoured over new ones.

D1.3 STATE VARIABLES

While analysing current models, the most obvious problems were encountered with respect to the naming of state variables. New models, model extensions, the connection of models using different sets of state variables (e.g. in whole WWTP and other fields of integrated modelling) were driving forces to develop the new notational framework. The new notation should provide the information required in the context of the model used (e.g. on the physical, chemical and biological properties of the compounds).

D1.3.1 Specific problems encountered

Different naming systems: Looking at the most common models one can observe that there is no real consensus with respect to the use of standardized symbols (Table D1.1).

- Main letter: In the IWA system the main letter is used to differentiate between the particulate ("*X*") variables that will settle out of the bulk liquid and soluble ("*S*") variables that will remain dissolved. The former UCT system uses the main letter to differentiate between units of measurement where "*S*" represents substrate, "*Z*" volatile solids in COD units, "*X*" volatile solids in VSS units and "*N*" nitrogen (e.g. $S_{bs,c}$, Z_{BH}, N_{obs}).

Table D1.1 Selection of state variable symbols used in different models.

Description	ASM1 [1]	ASM2d [1]	ASM3 [1]	GenASDM [2]	UCTPHO [3]	UCTPHO+ [4]	TU Delft-P [5]	ADM1 [6]
Fermentable organic matter		S_F		S_{BSC}	$S_{bs,c}$	S_F	S_F	$S_{su} + S_{aa} + S_{fa}$
Volatile fatty acids		S_A		S_{BSA}		S_A	S_A	$S_{va} + S_{bu} + S_{pro} + S_{ac}$
Propionate				S_{BSP}				S_{pro}
Dissolved methane				S_{CH4}				S_{ch4}
Soluble inert organic matter	S_I	S_I	S_I	S_{US}	S_{US}	S_I	S_I	S_I
Dissolved oxygen	S_O	S_{O2}	S_O	DO	O	S_{O2}	S_O	
Particulate inert endogenous products	X_P	X_I	X_I	Z_E	Z_E	X_E	X_I	X_I
Particulate inert organic matter from influent	X_I			X_I	Z_I	X_I		
Soluble biodegradable organic N	S_{ND}			N_{OS}	N_{obs}			
Total ammonia	S_{NH}	S_{NH4}	S_{NH}	S_{NH3}	N_a	S_{NH4}	S_{NH}	S_{IN}
Total nitrite + nitrate	S_{NO}	S_{NO3}	S_{NO}	$S_{NO2} + S_{NO3}$	N_{O3}	S_{NO3}	S_{NO}	
Ordinary heterotrophic organisms	$X_{B,H}$	X_H	X_H	Z_{BH}	Z_{BH}	X_H	X_H	
Nitrifying organisms (NH_4 to NO_3)	$X_{B,A}$	X_{AUT}	X_A	$Z_{AOB} + Z_{NOB}$	Z_{BA}	X_{NIT}	X_A	
Stored poly-β-hydroxyalkanoates in PAO		X_{PHA}		S_{PHA}	X_{PHA}	X_{PHA}	X_{PHA}	

[1] IWA Activated Sludge Models, Henze *et al.* (2000), [2] "General Activated Sludge Digestion Model" proprietary model in simulation software BioWin 3 (ESA, 2007), [3] University of Cape Town model, Wentzel *et al.* (1992), [4] latest version of UCT model, Hu *et al.* (2007), [5] TU Delft phosphorus removal model, Meijer (2004), [6] IWA Anaerobic Digestion Model No.1, Batstone *et al.* (2002).

- Subscripts:
  - Degradability: In the UCT system "B" stands for biodegradable and "U" for unbiodegradable (e.g. S_{US}). In the IWA notation they are given as "S" (substrate) and "I" (inert) (e.g. S_I). Conversion processes that do not depend on biodegradation, such as precipitation, acid-base reactions or adsorption, lack a clear notational framework to deal with these "abiotic" (non biological) reactions.
  - Structured biomass: With the introduction of structured biomass models (e.g. Smolders *et al.* 1995; Wild *et al.* 1995; Lavallée *et al.* 2008) it becomes necessary to link cell-internal products (e.g. PHA, RNA) to a specific biomass group (e.g. X_{PHA} in ASM2d is not linked to an organism group).
- Specification system:
  - Subscripts are not standardized with respect to upper and lower case (e.g. S_{BC} in GenASDM and $S_{bs,c}$ in UCTPHO).
  - The order of specifications is not standardized.
- Units: Different models use different units for the same state variables. For example, ADM1 model uses kg COD·m^{-3} as the basis in addition to kmol C·m^{-3} (for HCO_3^- and CO_2) and kmol N·m^{-3} (for NH_4^+ and NH_3). ASM models use g COD·m^{-3}, g N·m^{-3} and mol HCO_3^-·m^{-3} for alkalinity.

Another example is S_{ND} in ASM1 and N_{OS} in GenASDM which represent soluble biodegradable organic nitrogen. Both symbols use "*S*" for soluble and "*N*" for nitrogen, but they are combined in a different way.

Different names used: ammonia, nitrate, oxygen, volatile fatty acids and other compounds have different symbols or abbreviations in different models (see Table D1.1). Moreover, biomass names are abbreviated differently (e.g. nitrifying organisms in Table D1.1).

Non-meaningful names: Some state variable names do not provide a clear and unique meaning (e.g. S_A in ASM2d does not provide a clear meaning, unless the letter "A" would be considered a standardized abbreviation for acetate. However, the letter "A" is also used in ASM2d to abbreviate autotrophic, e.g. Y_A).

Colloidal matter: The first structured activated sludge models divided the organic substrate according to degradability (readily [S_S] and slowly [X_S] biodegradable compounds). For these models there is an incoherent use of "*X*" and "*S*" that refer to the particle size. S_S is mostly soluble, while X_S contains both particulate and soluble (colloidal) compounds in these models. This *S-X* distinction is frequently confusing modellers as it cannot be directly converted to soluble and particulate compounds, which were required later for primary and secondary settler and whole plant modelling. Introducing a colloidal fraction (which is both soluble and slowly degradable) solves this problem. Filtrates containing both soluble and colloidal fractions can be properly distinguished. The WERF and STOWA influent fractionation methods (Melcer *et al.* 2003; Roeleveld & van Loosdrecht, 2002) contain a specific flocculation step to discriminate between a colloidal fraction and truly soluble compounds.

D1.3.2 Framework

In the proposed notational framework, the main symbol is related to the particle size and should always be given. In the subscript, four levels can be provided, each referring to different information:

(1) Degradability
(2) Organic/inorganic compound
(3) Name of compound or organism
(4) Additional specifications.

The main symbol is in upper case and italics, the different elements of the subscript are in upper case (or combined with lower case if needed to make the name clearer, for example AcCoA) and not italicised, as defined in Table D1.4. Figure D1.1 shows the proposed framework and some examples that illustrate the notational procedure.

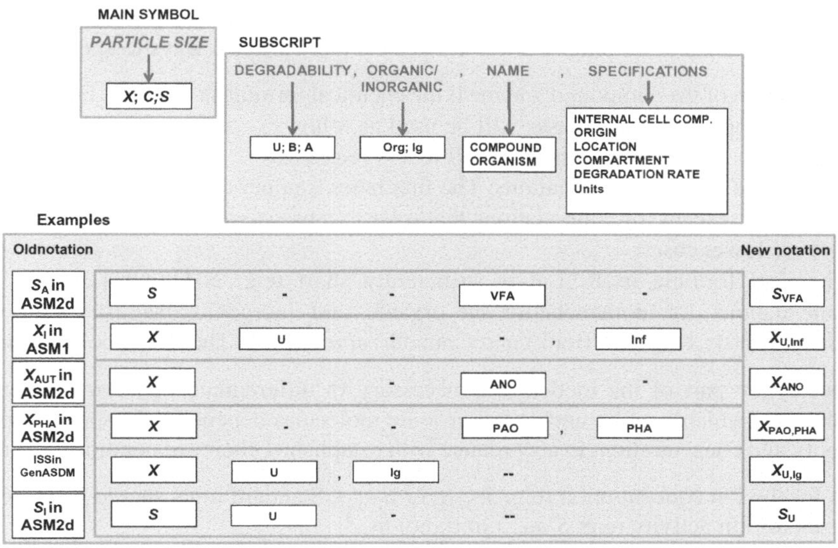

Figure D1.1 Description of the proposed notation for state variables.

In most cases, one or several of the subscript levels are not required (as illustrated in Figure D1.1), and therefore, are not included in the symbol. Generally, if the name of the compound is provided (e.g. Volatile Fatty Acids, abbreviated as VFA), it is not necessary to write the preceding levels (i.e. degradability or organic/inorganic). Finally, depending on the model or on the context for describing the model, it may be necessary to add specifications, as the final elements of the subscript. An example is X_I in the ASM1 model, which becomes $X_{U,Inf}$ when applying the proposed notation, with the subscript "Inf" referring to the fact that this fraction originates from the influent of the WWTP.

D1.3.3 Notational procedure

Particle size: The first upper case letter of the notation is related to the particle size. It is proposed to differentiate between soluble (S), particulate (X) and colloidal (C) matter. The novelty here is that the colloidal fraction is included explicitly, as was already proposed by Melcer *et al.* (2003). The filter size to distinguish between soluble, particulate and colloidal compounds cannot be specified at this stage, considering that MBR researchers need to adapt it according to the membrane pore size used. Therefore, the particle size used in a particular model (or study) should be specified and documented. Care should be taken not to confuse the use of "*C*" for colloidal and for total material concentration (as defined in Grau *et al.* 1987). It is proposed to use the symbol "*Tot*" for total material concentration.

Degradability: This is one of the most important aspects of WWT models. It is proposed to distinguish between undegradable (U), biodegradable (B) and abiotically convertible (A) compounds. The last symbol was already used in Howard *et al.* (1991) and refers to compounds that can be involved in conversion processes that are not related to the metabolism of an organism (e.g. photolysis, chemical reactions, adsorption, etc.).

Organic/inorganic: This differentiation is useful, notably, to distinguish between autotrophic and heterotrophic metabolism, where the carbon is obtained from inorganic (Ig) or organic (Org) compounds.

Name: For biomasses, all variables that contain an "*X*" (as an upper case letter) and a subscript that finishes with an "O", represent an organism (e.g. X_{OHO} for ordinary heterotrophic organisms).

For simple state variables describing specific molecules, there are two ways to specify names:

(a) An abbreviation of the compound's name if the chemical formula is too long. Examples are given in Table D1.4. Upper and lower cases will be used as follows:
 – For acronyms: Upper case is used for all letters (e.g. OHO)
 – For initial and syllabic abbreviations: The first letter is upper case and the rest is lower case (e.g. Inf, Org, Ig), except for abbreviations that refer to a process (e.g. hyd for hydrolysis) where all letters are lower case.

(b) The chemical formula itself if it is sufficiently short (e.g. NH4, CH4, NO2). The IUPAC recommendations for nomenclature on organic and inorganic chemistry can be found in Hellwinkel (2001; The Blue Book) and Connelly *et al.* (2005; The Red Book), respectively.

If chemical species are part of the model, it is necessary to differentiate between protonated and non protonated molecules, which may be uncharged or ionic molecules depending on specific compounds, as well as ion activity and concentration. In accordance with established chemical notation, we propose to use:

- Brackets for the ion concentration (e.g. $S_{[NH4]}$ or $S_{[Ac]}$). Standard units should be [kmol·m^{-3}]
- Parentheses for ion activity (e.g. $S_{(NH4)}$) in [kmol·m^{-3}]
- Use of "H" for undissociated acids (e.g. acetic acid concentration would be $S_{[HAc]}$)
- No brackets or parentheses for total compounds (e.g. S_{NHx} [g N·m^{-3}] for total ammonia consisting of $NH_3 + NH_4$ (the x is used to lump both) or S_{Ac} [g COD·m^{-3}] for the sum of acetate and acetic acid)

For example, the description of total ammonia in the system is frequently required (e.g. in ASM1, S_{NHx}, as substrate for autotrophic nitrifying organisms, ANOs). Other times the model needs to consider one of the ionic species (e.g. inhibition by ammonia, $S_{[NH3]}$).

Specifications: In certain cases it is necessary to include extra information in the name of the variable (fourth and next levels). The following cases are considered:

- Structured biomass compounds will appear in the symbol next to the name of the organism, separated by a comma. With cell internal storage products, different levels of detail can be considered. For example, $X_{PAO,PHA}$ would be preferred when glycogen is included in the model as another state variable (i.e. modelling more than one storage polymer), while $X_{PAO,Stor}$ would be fine in cases where glycogen is not considered (i.e. only one organic storage polymer is modeled).
- The origin of the products can be specified to indicate whether the compound is originating from endogenous processes (E) or from the influent (Inf) (e.g. $X_{U,E}$ or $X_{U,Inf}$ to describe the ASM1 state variables X_P and X_I, respectively).
- For some models it is important to specify the compartment. For instance, in the case of biofilm or anaerobic digestion models, different compounds are in equilibrium between different compartments/phases. The symbols considered for the compartments are the following (Morgenroth, 2008): L for liquid, G for gas, F for the inner biofilm, LF for the biofilm surface (e.g. $S_{CO2,L}$ or $S_{CO2,G}$). If all variables of the model belong to the same compartment, there may be no need to specify the compartment.
- If required, the valence of an ion, for example. in the case that $S_{Fe,2}$ and $S_{Fe,3}$ are considered in the same model.

- If required, the units can be defined as an additional subscript. They should be written as shown in Grau *et al.* (1987), indicating the power (can be negative or positive) in the superscript (e.g. g COD·m^{-3}).

D1.3.3.1 Naming lumped variables

A lumped variable is the single variable obtained after grouping several variables. The first two levels proposed in the new framework allow the grouping of variables according to the degradability and the organic-inorganic properties (e.g. see in Figure D1.2, $X_{U,Org}$ and $X_{B,Org}$). It is also possible within this framework to lump variables according to their particle size. In this case, the main symbol will contain the different particle size letters, following the sequence X→C→S (for example, X_S in ASM1 is XC_B according to the new notation). For some of the lumped variables, the specific name is normally provided (e.g. "Stor" for storage products or "Bio" for total biomass).

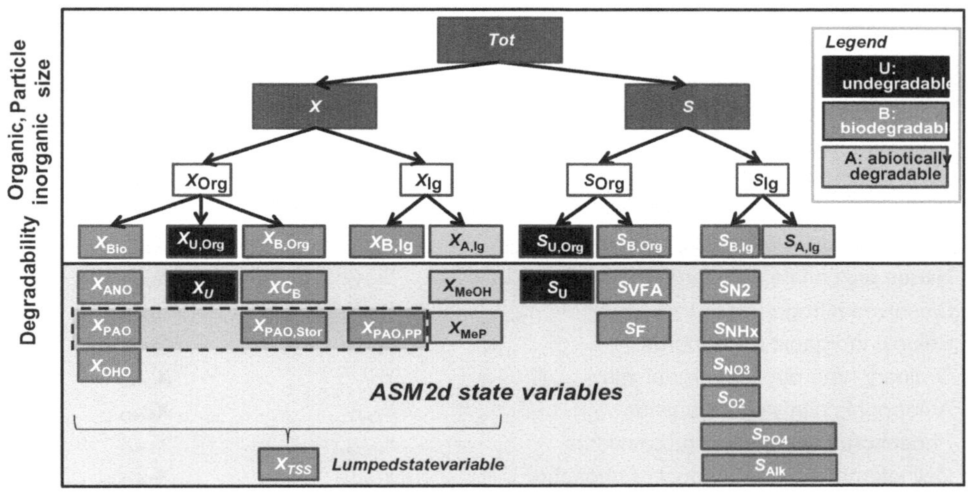

Figure D1.2 ASM2d state variables using the proposed notational framework.

Composite variables (calculated from multiple state variables, facilitating the comparison of model results with experimental measurements) are not discussed in this paper.

Example

ASM2d using the new framework

Figure D1.2 shows an example for the use of the new state variable notational framework for ASM2d (Henze *et al.* 2000). The variables are organized according to particle size, organic/inorganic properties and degradability. Table D1.2 presents a comparison of the symbols used for ASM2d according to the former and new notational systems.

It can be seen that the main symbols are kept identical (except for the former X_S, which becomes XC_B) in the proposed framework and that some modifications are introduced in the subscripts. For simple variables describing specific molecules, the chemical formula is used in both notational systems (e.g. S_{N2}, S_{O2}, S_{PO4}). For total ammonia an "x" is added at the end of the subscript with the new notation (the "x" combines NH_4^+ and NH_3); the same applies for S_{NOx}, where the "x" combines NO_2^- and NO_3^-. Regarding volatile fatty acids, the subscript "VFA" is used in the new notation instead of the abbreviation "A" used previously. For variables that do not have a specific name or formula, the degradability is specified in the subscript (e.g. X_U, XC_B). Organism variable symbols have the main symbol "X" and the subscript finishes with an "O" (e.g. X_{ANO} for ammonium nitrifying organisms, X_{OHO} for ordinary heterotrophic organisms). Internal cell compound symbols are linked to the organism ($X_{PAO,Stor}$).

Table D1.2 Comparison of the former and new notational systems for ASM2d. Bold signifies a proposed change to the state variable name.

Description	Former notation	New notation
Fermentable organic matter	S_F	S_F
Fermentation products, considered to be acetate	S_A	$\boldsymbol{S_{VFA}}$
Soluble undegradable organics	S_I	$\boldsymbol{S_U}$
Dissolved oxygen	S_{O2}	S_{O2}
Slowly biodegradable substrates(1)	X_S	$\boldsymbol{XC_B}$
Particulate undegradable organics (2)	X_I	$\boldsymbol{X_U}$
Ammonium plus ammonia nitrogen	S_{NH4}	$\boldsymbol{S_{NHx}}$
Nitrate plus nitrite nitrogen(3)	S_{NO3}	$\boldsymbol{S_{NOx}}$
Dissolved nitrogen gas	S_{N2}	S_{N2}
Soluble inorganic phosphorus	S_{PO4}	S_{PO4}
Ordinary heterotrophic organisms	X_H	$\boldsymbol{X_{OHO}}$
Autotrophic nitrifying organisms	X_{AUT}	$\boldsymbol{X_{ANO}}$
Phosphorus accumulating organisms	X_{PAO}	X_{PAO}
Cell internal storage product of phosphorus-accumulating organisms(4)	X_{PHA}	$\boldsymbol{X_{PAO,Stor}}$
Stored polyphosphates in PAOs	X_{PP}	$\boldsymbol{X_{PAO,PP}}$
Metal-hydroxides	X_{MeOH}	X_{MeOH}
Metal-phosphates	X_{MeP}	X_{MeP}
Alkalinity	S_{ALK}	$\boldsymbol{S_{Alk}}$
Total suspended solids	X_{TSS}	X_{TSS}

(1)In the ASM2d definitions X_S includes colloidal substrates and (2)X_I does not include colloidal substrates; (3)normally NO_3 stands for nitrate only; (4)X_{PHA} is not directly the measured PHA.

D1.4. MODEL PARAMETERS

It is an insurmountable task to define a framework that covers the naming of every parameter used in all present and future biokinetic models. Therefore, the authors' goal was to provide a framework for standard, frequently used parameters or for cases where problems were encountered in current practice. The comparison of the parameter symbols used in different models (see Table D1.3) reveals some challenges that the new notation faces (e.g. avoiding the use of different main symbols and subscripts for the same parameter).

Table D1.3 Parameter symbols used in different models.

Description		ASM1 [1]	ASM2d [1]	ASM3 [1]	GenASDM [2]	UCTPHO [3]	UCTPHO+ [4]	TU Delft-P [5]	ADM1 [6]
Yield for autotrophic biomass	$g\ X_{ANO}.(g\ N)^{-1}$	Y_A	Y_A	Y_A		Y_{ZA}	Y_{NIT}	Y_A	
Correction factor for max growth rate under anoxic conditions	-	η_g	η_{NO3}	$\eta_{NO,H}$	η_G	η_G	η_H	η_{NO}	
Saturation/inhibition coeff. for oxygen	$g\ O_2.m^{-3}$	$K_{O,H}$	K_{O2}	$K_{O,H}$	K_{OH}	K_{OH}	K_{OH}	K_O	
Rate constant for storage of X_{PHA} (base X_{PP})	$g\ X_{PAO,PHA}.(g\ X_{PAO})^{-1}.d^{-1}$		q_{PHA}		K_{SCFA}	K_P	q_{PHA}	q_{Ac}	
Saturation coeff. for phosphate (nutrient)	$g\ P.m^{-3}$		K_P		K_P	K_{PS}	$K_{PO4\text{-}gro}$	K_P	
Saturation coeff. for ammonium (substrate)	$g\ N.m^{-3}$	K_{NH}	K_{NH4}	$K_{NH,A}$	K_{NH}	K_{SA}	K_{NH4}	K_{NH}	
Max. rate for fermentation	$g\ S_F\ (g\ X_{OHO})^{-1}.d^{-1}$		q_{fe}		$\mu_{ZBH,ANA}$	K_C	K_{FE}	q_{fe}	
Max. specific hydrolysis rate	d^{-1}	k_h	k_h	k_h	k_h	K_{MP}		k_h	$k_{hyd,i}$

[1] IWA Activated Sludge Models, Henze *et al.* (2000).
[2] "General Activated Sludge Digestion Model" proprietary model in simulation software BioWin 3 (ESA, 2007).
[3] University of CapeTown model, Wentzel *et al.* (1992).
[4] latest version of UCT model, Hu *et al.* (2007).
[5] TU Delft phosphorus removal model, Meijer (2004).
[6] IWA Anaerobic Digestion Model No.1, Batstone *et al.* (2002).

This section describes the stoichiometric and kinetic parameters separately, in accordance with the setup of the Gujer Matrix.

D1.4.1 Stoichiometric parameters

Yield

In the proposed notation a "yield" represents a stoichiometric parameter describing the amount of a specified product that is obtained from specified amounts of reactants.

Specific problems encountered

- For the biomass growth yield coefficients, there is no standardization to specify the substrate source (not considered in the evaluated models) and the environmental conditions (e.g. for aerobic conditions, "O" is used in TU Delft-P and the subindex 1 in UCTPHO+).
- Naming yields, such as for cell-internal storage (e.g. Y_{PO_4} in ASM2, which represents the requirement of X_{PP} per X_{PHA} storage during P-release), is not straightforward and does not allow a clear understanding of the parameter on the basis of the symbol only.

Framework

The main symbol for yield is Y (upper case letter and italics). Subscripts start with the reactant (or substrate source) and, through an underscore, describe the product (e.g. the cell-internally stored compounds). They continue with the name of the organism followed by the environmental conditions, which allows differentiating yields depending on the availability of oxygen and nitrate/nitrite (Ox: oxic; Ax: anoxic; Ax2: anoxilic, nitrite present; Ax3: anoxalic, nitrate present; and An: anaerobic). The "reactant_product" subscript with the underscore between the two compounds for the yield is used in other fields as well. For instance in Roels (1983), Y_{SX} represents the yield of biomass on substrate and Y_{SP} the yield of product on substrate. Figure D1.3 shows the proposed framework and some examples that illustrate the notational procedure. In the cases when only one substrate is consumed for direct growth the "reactant_product" pair is not required (e.g. Y_{OHO}).

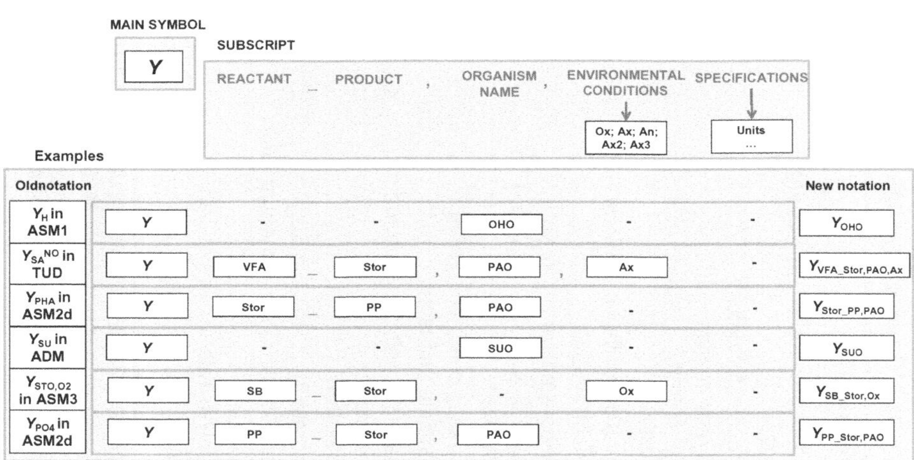

Figure D1.3 Notational framework and examples for the naming of yield coefficients.

D1.4.1.1 Composition and fractionation coefficients

In the proposed framework, composition coefficients refer to the conversion factors used in the continuity equations. Within this context they are defined as a part of a larger entity to explain the composition of a compound. For instance, composition factors are used to specify the content of an element (N, P), charge or any other part (e.g. COD, TSS) of a compound or organism (e.g. nitrogen content of ordinary heterotrophic organisms).

Fractionation coefficients are used to indicate the portion of a state variable that is transformed via a specific process (e.g. f_P in ASM1 describes the fraction of biomass leading to unbiodegradable particulate decay products).

Specific problems encountered

- Need for clarification of the different use of fractions (composition vs fractionation).
- Order of fraction and organism/main compound in the symbol is not defined.

Framework

The main symbol defines the meaning of the stoichiometric coefficient used. The letter "*i*" is used for composition coefficients and "*f*" for fractionation coefficients. When using "*i*", the first subscript represents the smaller portion (e.g. nitrogen content) and the second subscript represents the main compound or organism (larger entity). When using "*f*", the same order of subscripts is used ("smaller"_"larger") and the process type can be added in the specifications level. Figure D1.4 presents the proposed framework and some examples that illustrate the notational procedure. *f* can eventually be used to express ratios (e.g. PP/PAO in ASM2d would become f_{PP_PAO}).

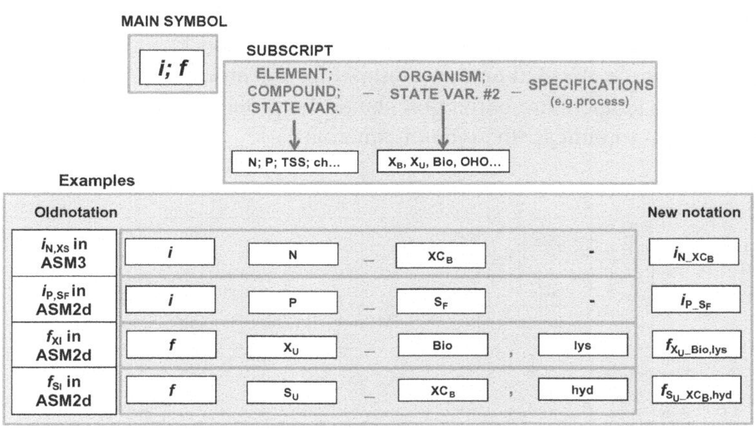

Figure D1.4 Notational framework and examples for the naming of fractions.

As a general rule simplification is recommended for state variables specified in one of the subscript levels of the parameters. The main letter (X,C,S) is used only if the subscript is not meaningful by itself. Normally, the organism names and the chemical compounds can be written without main letter (e.g. Bio). Lumped variables will need the main letter (e.g. XC_B). The comas separating the subscripts of a state variable name will not be used (e.g. i_{P_XUE}). This applies to the rest of the parameters as well.

D1.5.1 Kinetic parameters

D1.5.1.1 Rate Coefficients and reduction factors

Reaction rates characterize the kinetics of a process. In ASM-type models, process rate equations (ρj) normally include the maximum rate and several saturation terms (e.g. Monod term, Michaelis-Menten, …). Reduction factors account for a reduced rate under specific environmental conditions (e.g. anoxic conditions). This framework focuses on the rate coefficients and reduction factors used in these equations.

Specific problems encountered

The letter "k" was used for both rates (lower case "*k*") and saturation coefficients (upper case "*K*"), which could lead to confusion (e.g. for hydrolysis rate k_H and saturation coefficient K_H).

- Not all rate constants were defined in all models (e.g. maintenance was missing in most notational systems)
- A framework to include different substrate sources was missing (e.g. the growth of OHOs on acetate and propionate is usually modelled as two processes).

Framework

The main symbol (lower case letter, italics) specifies growth (μ), decay or endogenous respiration (b) and maintenance (m) (van Loosdrecht & Henze, 1999) and "q" is used for all other rates. As additional information, correction factors are specified in the framework since they can also be used for these parameters (the main letter is "η") and temperature correction (θ) as well.

The first subscript is used for the correction factors to specify the main symbol. The second subscript includes the organism in upper case and in the third level the substrate source or the "reactant_product" pair is specified. Other specifications may be given in the fourth level. Figure D1.5 provides an overview of the framework and some examples, including one for a reduction factor under anoxic conditions for the heterotrophic growth rate in ASM2d and an example for a temperature correction factor. In the latter case, the equation used for temperature correction should be properly documented ("pow" or "exp" can be used in the specifications to indicate the type of equation).

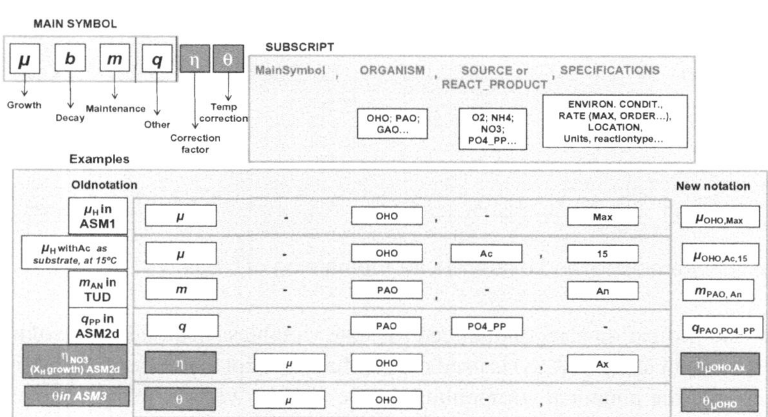

Figure D1.5 Notational framework and examples for the naming of kinetic rates and reduction factors.

Additional explanations and examples

Some common abbreviations for processes (e.g. "hyd" for hydrolysis, "ab" for acid-base reactions) can be found in Table D1.4. In the last examples of Figure D1.5, "η" and "θ" have been used as main symbols and the parameter symbol to which they refer is found in the subscript.

D1.5.1.2 Saturation or inhibition coefficients

These coefficients are used in reduction terms (e.g. Monod, inhibition Monod, Haldane, etc.) to reduce the maximum process rate according to the existence or limitation of another compound.

Specific problems encountered

- Non-unique names for some coefficients (e.g. K_{PP} and K_{IPP} in ASM2d or K_{O_2} without reference to specific biomass or a ratio).
- Additional information is sometimes required to understand the meaning of a parameter.

Framework

The main symbol is an upper case *K* in italics. The first subscript level describes the type of the reduction term (saturation or inhibition). The second level relates to the main compound. The name of the related organism can be given in the third level. For a surface saturation term an underscore is used to highlight the ratio between main compound and organism or second compound (e.g. K_{fPHA_PAO} in $\frac{X_{PAO,PHA}/X_{PAO}}{K_{fPHA\_PAO}+X_{PAO,PHA}/X_{PAO}}$). Figure D1.6 presents the proposed framework and some examples that illustrate the notational procedure.

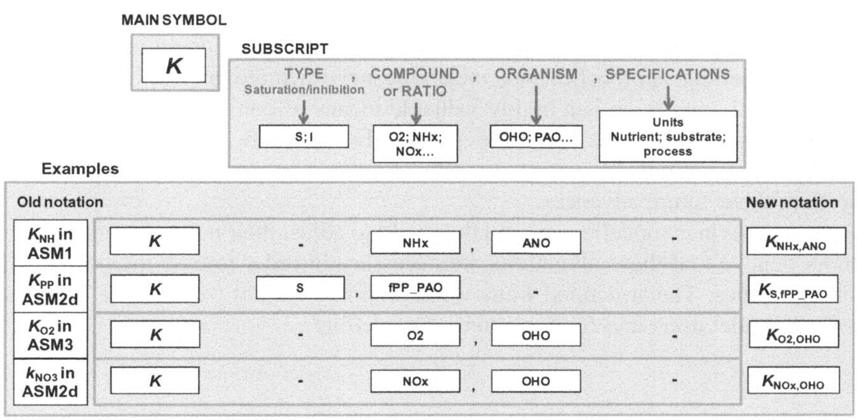

Figure D1.6 Notational framework and examples for the naming of saturation/inhibition constants

D1.6 CONTRIBUTIONS OF THE NEW FRAMEWORK

The new framework provides a structured system to specify the symbols for state variables and parameters used in wastewater treatment modelling. Different symbol levels, providing physical, biological and chemical information, are introduced in a systematic and intuitive way with the intention to provide a straightforward, simple and easy to understand framework. Necessarily, there must be compromise in order to keep symbols simple, yet meaningful. This is achieved by providing only those subscript levels that are required to make the symbol unique within the model context. The characters chosen for the framework originated from previously proposed notational examples and the symbols that result are often similar or identical to the ones that were most commonly used in practice. A list of abbreviations is provided as an attempt to standardize selected words and symbols (see Table D1.4). The structured framework allows the development of new symbols for new models (in different subfields of WWT modelling) and the extension of the same framework for future developments. The authors have converted state variables and parameters of several accessible models in order to check the applicability of the new framework. The results of this exercise show that conversions are most often straightforward, while naming gaps or other problems were not encountered. In an effort to assist with the implementation of the proposed framework by model users, a comprehensive list of the former and new notational symbols for the most commonly used models is provided on the GMP WaterWiki website. A macro is available along with this paper for modellers to update their notation in MS-Word 2003 + 2007 documents.

CONCLUSIONS

It is the hope of the authors that the proposed framework combines the advantages of different notational systems, such as the UCT and IWA systems, resulting in a standardised methodology for expressing nomenclature that is useful for the WWT modelling community. Using common notation should facilitate communication amongst modellers and other experts. It should help to achieve better "readability" of new models and help prevent misinterpretation and implementation errors. Since coding is an essential and error-prone part of model implementation, the new notation also provides naming rules for programming.

In view of emerging fields in WWT modelling, like the fate of micropollutants and the inclusion of water chemistry, or new modelling approaches like metabolic or structured biomass models, a standardised framework for notational expression is a highly valuable means of conveying modelling advances to the entire WWT modelling community. With the proposed framework, it should be possible to give meaningful, distinct and commonly accepted names to the new variables and parameters that will inevitably arise from these future advances.

The next step is to convince modellers around the world to adjust their notation and use the new naming rules. The authors believe that these alterations are necessary in order to ease the transfer of knowledge between modelling studies. The structured framework proposed should be directive, yet flexible enough for the benefit of all model users and for the future of modelling.

Acknowledgements

The authors thank the following persons for critically reviewing the paper and providing valuable comments to improve the proposed notation: Damien Batstone (AWMC, Australia), Adriano Joss (Eawag, Switzerland), Ulf Jeppsson (IEA of Lund, Sweden), Thomas Hug (UBC, Vancouver) and Eberhard Morgenroth (University of Illinois, USA). The authors also thank the Good Modelling Practice (GMP) IWA Task Group and the GMP members that supported this initiative, the Government of Canada for the "Canada Research Chair on Water Quality Modelling" and the Government of Catalonia for the postdoctoral fellowship "Beatriu de Pinós".

REFERENCES

Busby J. B. and Andrews J. F. (1975). Dynamic modeling and control strategies for the activated sludge process. *J. Water Pollut. Control Fed.*, **47**, 1055–1080.

Barker P. S. and Dold P. L. (1997). General model for biological nutrient removal activated-sludge systems: Model presentation. *Water Environ. Res.*, **69**(5), 969–984.

Batstone D. J., Keller J., Angelidaki R. I., Kalyuzhnyi S. V., Pavlostathis S. G., Rozzi A., Sanders W. T. M., Siegrist H. and Vavilin V. A. (2002). Anaerobic Digestion Model No. 1. Scientific and Technical Report No 13. IWA Publishing, London, UK.

Connelly N. G., Damhus T., Hartshorn R. M. and Hutton A. T. (eds)(2005). Nomenclature of Inorganic Chemistry: IUPAC Recommendations 2005. *The Royal Society of Chemistry*, [ISBN 978-0-85404-438-2].

Dold P. L., Ekama G. A. and Marais G. V. R. (1980). A general model for the activated sludge process. *Prog. Water Technol.*, **12**, 47–77.

Ekama G. A. and Marais G. V. R. (1977). The Activated Sludge process (Part II) – Dynamic behaviour. *Water SA*, **3**(1), 18–50.

ESA (2007). General activated sludge-digestion model (General ASDM). BioWin3 software, EnviroSim Associates, Flamborough, Ontario, Canada.

Gernaey K. V., Rosen C., Batstone D. J. and Alex J. (2006). Efficient modelling necessitates standards for model documentation and exchange. *Water Sci. Technol.*, **53**(1), 277–285.

GMP TG (2008). Website of the IWA Task Group on 'Good Modelling Practice – Guidelines for Use of Activated Sludge Models': https://iwa-gmp-tg.irstea.fr/.

Grau P., Copp J., Vanrolleghem P. A., Takács I. and Ayesa E. (2009). A comparative analysis of different approaches for integrated WWTP modeling. *Water Sci. Technol.*, **59**(1), 141–147.

Grau P., Sutton P. M., Henze M., Elmaleh S., Grady C. P. L., Gujer W. and Koller J. (1982a). Editorial: A common system of notation for use in biological wastewater treatment. *Water Res.*, **16**(11), 1499–1500.

Grau P., Sutton P. M., Henze M., Elmaleh S., Grady C. P. L., Gujer W. and Koller J. (1982b). Report: Recommended notation for use in the description of biological wastewater treatment processes. *Water Res.*, **16**(11), 1501–1505.

Grau P., Sutton P. M., Henze M., Elmaleh S., Grady C. P. L., Gujer W. and Koller J. (1987). Notation for the use in the description of wastewater treatement processes. *Water Res.*, **21**(2), 135–139.

Gujer W. (2006). Activated sludge modelling: past, present and future. *Water Sci. Technol.*, **53**(3), 111–119.

Hellwinkel D. (2001). Systematic nomenclature in organic chemistry: a directory to comprehension and application of its basic principles. *Springer*, Berlin; New York.

Henze M., Sutton P. M., Gujer W., Koller J., Grau P., Elmaleh S. and Grady C. P. L. (1982). The use and abuse of notation in biological wastewater treatment. *Water Res.*, **16**(6), 755–757.

Henze M., Gujer W., Mino T. and van Loosdrecht M. C. M. (2000). Activated Sludge Models: ASM1, ASM2, ASM2d and ASM3. Scientific and Technical Report n°9. *IWA Publishing*, London, UK.

Henze M., van Loosdrecht M. C. M., Ekama G. A. and Brdjanovic D. (Eds.) (2008). Biological Wastewater Treatment – Principles, Modelling and Design. *IWA Publishing*, London, UK. ISBN: 9781843391883.

Horn H., Reiff H. and Morgenroth E. (2003). Simulation of growth and detachment in biofilm systems under defined hydrodynamic conditions. *Biotechnol. Bioeng.*, **81**(5), 607–617.

Howard P. H., Boethling R. S., Jarvis W. F, Meylan W. M. and Michalenko E. M. (1991). Handbook of environmental degradation rates. Lewis Publishers, Chelsea, Michigan.

Hu Z., Wentzel M. C. and Ekama G. A. (2007). A general kinetic model for biological nutrient removal activated sludge systems: Model development. *Biotechnol. Bioeng.*, **98**(6), 1242–1258.

Jiang T., Myngheer S., De Pauw D. J. W., Spanjers H., Nopens I., Kennedy M. D., Amy G. and Vanrolleghem P. A. (2008). Modelling the production and degradation of soluble microbial products (SMP) in membrane bioreactors (MBR). *Water Res.*, **42**, 4955–4964.

Joss A., Zabczynski S., Göbel A., Hoffmann B., Löffler D., McArdell C. S., Ternes T. A., Thomsen A. and Siegrist H. (2006). Biological degradation of pharmaceuticals in municipal wastewater treatment: proposing a classification scheme. *Water Res.*, **40**(8), 1686–1696.

Lavallée B., Frigon D., Lessard P., Vanrolleghem P. A., Yuan Z. and van Loosdrecht M. C. M (2008). Modelling using rRNA-structured biomass models. In: *Proc. 1st IWA/WEF Wastewater Treatment Modelling Seminar*, Mont-Sainte–Anne, Québec, Canada, June 1–3, 2008.

Lee B. J., Wentzel M. C. and Ekama G. A. (2006). Measurement and modelling of ordinary heterotrophic organism active biomass concentrations in anoxic/aerobic activated sludge mixed liquor. *Water Sci. Technol.*, **54**(1), 1–10.

Lopez-Vazquez C. M., Oehmen A., Hooijmans C. M., Brdjanovic D., Gijzen H. J., Yuan Z. G. and van Loosdrecht M. C. M. (2009). Modeling the PAO-GAO competition: effects of carbon source, pH and temperature. *Water Res.*, **43**(2), 450–462.

Lu S. G., Imai T., Ukita M., Sekine M., Higuchi T. and Fukagawa M. (2001). A model for membrane bioreactor process based on the concept of formation and degradation of soluble microbial products. *Water Res.*, **35**(8), 2038–2048.

Melcer H., Dold P. L., Jones R. M., Bye C. M., Takacs I., Stensel H. D., Wilson A. W., Sun P. and Bury S. (2003). Methods for Wastewater Characterization in Activated Sludge Modeling. *Water Environment Research Foundation*, (WERF), Alexandria, VA, USA.

Meijer S. C. F. (2004). Theoretical and Practical Aspects of Modelling Activated Sludge Processes. PhD Thesis. Delft University of Technology, The Netherlands.

Morgenroth E. (2008). Modelling Biofilm Systems, in Biological Wastewater Treatment – Principles, Modelling, and Design, edited by Henze M., van Loosdrecht M. C. M., Ekama G. A. and Brdjanovic D., *IWA Publishing*, London, 2008.

Murnleitner E., Kuba T., van Loosdrecht M. C. M. and Heijnen J. J. (1997). An integrated metabolic model for the aerobic and denitrifying biological phosphorus removal. *Biotechnol. Bioeng.*, **54**(5) 434–450.

Rittmann B. E. and McCarty P. L. (1980). Model of steady-state-biofilm kinetics. *Biotechnol. Bioeng.*, **22**(11), 2343–2357.

Roeleveld P. J. and Van Loosdrecht M. C. M. (2002). Experience with guidelines for wastewater characterisation in The Netherlands. *Water Sci. Technol.*, **45**(6), 77–87.

Roels J. A. (1983). Energetics and kinetics in biotechnology. *Elsevier Biomedical Press*, Amsterdam.

Schönerklee M., Peev M., De Wever H., Reemtsma T. and Weiss S. (2009). Modelling the degradation of micropollutants in wastewater – Parameter estimation and application to pilot (laboratory-scale) MBR data in the case of 2,6-NDSA and BTSA. *Water Sci. Technol.*, **59**(1), 149–157.

Sin G., Kaelin D., Kampschreur M. J., Takács I., Wett B., Gernaey K. V., Rieger L., Siegrist H. and van Loosdrecht M. C. M. (2008). Modelling nitrite in wastewater treatment systems: A discussion of different modelling concepts. *Water Sci. Technol.*, **58**(6), 1155–1171.

Smolders G. J. F., Van der Meij J., van Loosdrecht M. C. M. and Heijnen J. J. (1995). A structured metabolic model for anaerobic and aerobic stoichiometry and kinetics of the biological phosphorus removal process. *Biotechnol. Bioeng.*, **47**(3), 277–287.

van Loosdrecht M. C. M. and Henze M. (1999). Maintenance, endogeneous respiration, lysis, decay and predation. *Water Sci. Technol.*, **39**(1), 107–117.

Wanner O. and Gujer W. (1986). A multispecies biofilm model. *Biotechnol. Bioeng.*, **28**(3), 314–328.

Wentzel M. C., Ekama G. A. and Marais G. V. R. (1992). Process and modelling of nitrification denitrification biological excess phosphorus removal systems—A review. *Water Sci. Technol.*, **25**(6), 59–82.

Wild D., von Schulthess R. and Gujer W. (1995). Structured modeling of denitrification intermediates. *Water Sci. Technol.*, **31**(2), 45–54.

ADDITIONAL MATERIAL

Table D1.4 Abbreviations proposed for the new notational framework.

Lumped variables	
A	Abiotically convertible compound
B	Biodegradable compound (substrate)
Bio	Organisms (biomass)
Ig	Inorganic compound
Inh	Inhibitory compound
ISS	Inorganic suspended solids
MP	Micropollutant
Org	Organic compound
Stor	Cell-internal storage compound
Tot	Total
Tox	Toxic compound
TSS	Total suspended solids
U	Undegradable compound
VSS	Volatile suspended solids
Abbreviations	
AA	Amino acid
Ac	Acetate
AcCoA	Acetyl-CoA
Ads	Adsorbed compound
Alk	Alkalinity
BAP	Biomass-associated products
Bu	Butyrate
Ca	Calcium
CHO	Carbohydrates
F	Fermentable organic matter[(1)]
Gly	Glycogen
HAc	Acetic acid
HAP	Hydroxyapatite
HBu	Butyric acid
HDP	Hydroxy dicalcium phosphate
Hi	High Molecular Weight
HPr	Propionic acid
Hva	Valeric acid
LCFA	Long chain fatty acid

Organisms	
OHO	Ordinary heterotrophic organisms
AAO	Amino acid degrading organisms
ACO	Acetoclastic methanogenic organisms
AMO	Anaerobic ammonia oxidizing organisms
ANO	Autotrophic nitrifying organisms (NH_4^+ to NO_3^-)
AOO	Ammonia oxidizing organisms
FAO	Fatty acid degrading organisms
FeOO	Ferrous oxidizing organisms
FeRO	Ferric reducing organisms
GO	G-Bacteria
GAO	Glycogen accumulating organisms
GAO_GB	GAO *Competibacter*
GAO_DEF	GAO *Defluviicoccus*
HMO	Hydrogenotrophic methanogenic organisms
LOO	Lipid oxidizing organisms
MEOLO	Methylotrophic organisms
NOO	Nitrite oxidizing organisms
PAO	Phosphorus accumulating organisms
PRO	Propionic acetogenic organisms
SOO	Sulfide oxidizing organisms
SRO	Sulfate reducing organisms
SUO	Sugar utilizing organisms
VBO	Valerate and butyrate degrading organisms
Chemical formula	
CH_3OH	Methanol
CH_4	Methane
CO_2	Carbon dioxide
CO_3	Carbonate
H_2	Hydrogen
H_2O	Water
HCO_3	Bicarbonate
HNO_2	Nitrous acid
HNO_3	Nitric acid
N	Nitrogen

(Continued)

Table D1.4 Abbreviations proposed for the new notational framework (*Continued*).

Li	Lipids	N_2O	Nitrous oxide
Lo	Low Molecular Weight	NH_3	Ammonia
MAP	Struvite (magnesium ammonium phosphate)	NH4	Ammonium
		NHx	Sum of ammonium and ammonia
Me	Metals	NO_2	Nitrite
MeOH	Metal hydroxide precipitate	NO_3	Nitrate
MeP	Metal phosphate compound	NOx	Sum of nitrite and nitrate
Mg	Magnesium	O_2	Oxygen
MW	Molecular Weight	P	Phosphorus
PH2MV	Poly-β-hydroxy-2-methylvalerates	PO_4	Phosphate
PHA	Poly-β-hydroxyalkanoates	S	Sulfur
PHB	Poly-β-hydroxybutyrates	SO_3	Sulfite
PHV	Poly-β-hydroxyvalerates	SO_4	Sulfate
PP	Polyphosphate	***Micropollutants***	
Pr	Propionate	BpA	Bisphenol A
PrCoA	Propionyl-CoA	Dcf	Diclofenac
Prot	Protein	DEHP	Di(2-ethylhexyl) phthalate
Su	Sugar	Ibp	Ibuprofen
UAP	Utilization-associated product	LAS	Linear alkyl benzene sulfonate (anionic detergent)
Va	Valerate		
VFA	Volatile fatty acid	***Processes*** [(2)]	
		ab	Acid-base reaction
Main symbols		ads	Adsorption
Parameters		am	Ammonification
b	Decay rate	dis	Dissociation
f	Fraction	fe	Fermentation
μ	Growth rate	gro	Growth
i	Composition coefficient	hyd	Hydrolysis
K	Saturation coefficient	lys	Lysis
m	Maintenance rate	pre	Precipitation
η	Reduction factor	red	Redisolution
r	Reaction rate	stor	Storage of cell-internal compounds
q	Other rates than μ, *b* and *m*	***Environmental conditions***	
Y	Yield	An	Anaerobic
State variables		Ax	Anoxic (nitrite and nitrate present)
C	Colloidal	Ax2	Anoxilic (nitrite present)
S	Soluble	Ax3	Anoxalic (nitrate present)
X	Particulate	Ox	Oxic or aerobic

(*Continued*)

Table D1.4 Abbreviations proposed for the new notational framework (*Continued*).

Origin		***Compartments***	
E	Endogenous product	F	Inner biofilm[(1)]
Inf	Compound originating from the influent	G	Gas
Other		L	Liquid
Max	Maximum	LF	Biofilm surface
Plim	Rate limited by phosphorus		

(1) The same abbreviation (F) is used for fermentable organic matter and for the inner biofilm compartment. However, the compartment is specified in the last subscript and the variable name in the first subscript, avoiding confusion.
(2) All letters are lower case for process abbreviations to minimize confusion (e.g. Stor and stor).

Appendix D2

Example Fractionation according to New Notation

Notes:

note 1: biomass is considered completely biodegradable. Only after decay, death, maintenance does it get transformed in unbiodegradable and biodegradable organics.

note 2: the "*" represents the state variable part of the symbol

measurements vs model

state var = modeling

BOD_U is about 10% less than the corresponding COD_b components

BOD components are carbonaceous BOD

The typical COD composition of a primary municipal effluent is shown (adapted from ASM2d; Henze et al., 1999)

Ranges of fractions of COD_t are shown for raw and primary effluent wastewaters (EnviroSim, 2007)

Activated sludge concentrations are shown for the aerobic zone of a UCT process treating raw wastewater at a 5 d SRT, 12C XSRO: SRB

Relationships between various components and combinations of these, and their ThOD, BOD, solids, N and P content
Applicable to any influent, reactor content and effluent stream
Developed by Yves Comeau according to the notation by Corominas et al., 2010

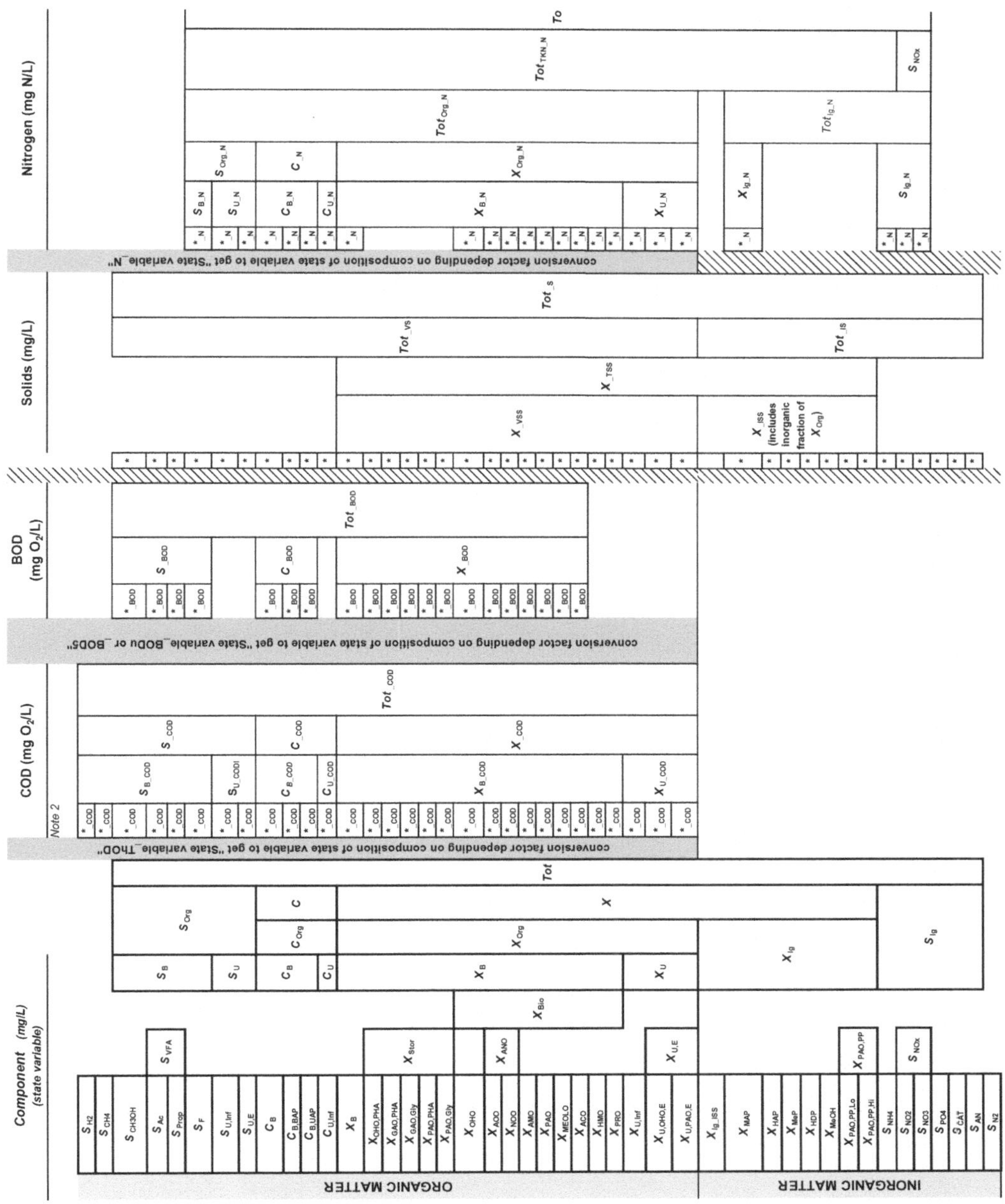

Appendix E

A Systematic approach for model verification – application on seven published activated sludge models

H. Hauduc[1,2], L. Rieger[2,3], I. Takács[4], A. Héduit[1], P. A. Vanrolleghem[2] and S. Gillot[1]

[1]Cemagref, UR HBAN, Parc de Tourvoie, BP 44, F-92163 Antony Cedex, France. (E-mail: sylvie.gillot@irstea.fr).
*[2]model*EAU*, Université Laval, Département de génie civil, Pavillon Adrien-Pouliot, 1065 av. de la Médecine, Quebec, G1V 0A6, Canada.*
[3]EnviroSim Associates Ltd., 7 Innovation Drive, Suite 205, Flamborough (ON) L9H 7H9, Canada
[4]EnviroSim Europe, 15 Impasse Fauré, 33000 Bordeaux, France

ABSTRACT

The quality of simulation results can be significantly affected by errors in the published model (typing, inconsistencies, gaps or conceptual errors) and/or in the underlying numerical model description. Seven of the most commonly used activated sludge models have been investigated to point out the typing errors, inconsistencies and gaps in the model publications: (1) ASM1 (Henze *et al.* 1987; republished in Henze *et al.* 2000a); (2) ASM2d (Henze *et al.* 1999; republished in Henze *et al.* 2000b); (3) ASM3 (Gujer *et al.* 1999; corrected version published in Gujer *et al.* 2000); (4) ASM3 + BioP (Rieger *et al.* 2001); (5) ASM2d + TUD (Meijer, 2004); (6) New General (Barker & Dold, 1997); (7) UCTPHO+ (Hu *et al.* 2007).

A systematic approach to verify models by tracking typing errors and inconsistencies in model development and software implementation is proposed. Then, stoichiometry and kinetic rate expressions are checked for each model and the errors found are reported in detail. An attached spreadsheet provides corrected matrices with the calculations of all stoichiometric coefficients for the discussed biokinetic models and gives an example of proper continuity checks.

Keywords: ASM, continuity, composition matrix, errors, model implementation, model verification

Previously published as: Hauduc, H., Rieger, L., Takács, I., Héduit, A., Vanrolleghem, P.A. & Gillot, S. (2010). A systematic approach for model verification – Application on seven published Activated Sludge Models. *Water Science and Technology,* **61**(4), 825–839.

E.1 INTRODUCTION

The quality of simulation results can be affected by several error sources (Refsgaard *et al.* 2007): (i) context and framing; (ii) input uncertainty; (iii) model structure uncertainty; (iv) parameter uncertainty and (v) model technical uncertainty, for example model implementation errors. Gernaey *et al.* (2006) detailed the error sources when models are implemented into a simulation software: (i) simplification of the original model; (ii) typing errors; (iii) incomplete model description in the paper; (iv) scattered description of the model in the paper (v) misinterpretation of the model description; (vi) errors when coding model; (vii) general programming bugs.

Surprisingly no error report has been published, except for ASM2d and ADM1 in Gernaey *et al.* (2006). Tracking those errors is indeed difficult and time consuming for model users, and the potential publication formats are not adapted to publish such information. Furthermore, some typing errors seem to appear or disappear following the version of the papers describing a given model (e.g. ASM2d where typing errors appeared in the paper Henze *et al.* 2000b compared to previous publications: Henze *et al.* 1998 and Henze *et al.* 1999). This work aims thus to provide (i) a systematic approach to track typing errors and inconsistencies in models, (ii) a thorough list of errors in the commonly used activated sludge model publications and (iii) the corrected Gujer Matrices and matrices in a new standardised notation format (Corominas *et al.* 2010) in a spreadsheet (available as additional material in electronic format on the GMP WaterWiki website). This work does not intend to address model structure problems linked either to modelling concepts or to simplifications used in the model.

Seven of the most commonly used activated sludge models have been investigated: (1) ASM1 (Henze *et al.* 1987; republished in Henze *et al.* 2000a); (2) ASM2d (Henze *et al.* 1999; republished in Henze *et al.* 2000b); (3) ASM3 (Gujer *et al.* 1999; corrected version published in Gujer *et al.* 2000); (4) ASM3 + BioP (Rieger *et al.* 2001); (5) ASM2d + TUD (Meijer, 2004); (6) New General (Barker & Dold, 1997); (7) UCTPHO+ (Hu *et al.* 2007). To keep the article readable, those references will not be repeated each time.

E.2 HOW TO TRACK TYPING ERRORS AND INCONSISTENCIES IN MODEL DEVELOPMENT AND SOFTWARE IMPLEMENTATION

Before using or implementing a published model or when developing a new model one should first verify the model by checking the continuity of the stoichiometry and the consistency of the kinetic rate expressions. Because typing errors could stem from the original model publication or could occur during software implementation, this step should be done directly in the simulation software (simulator). However, not all simulators provide adequate tools to track such errors.

One way of verifying model implementations would consist in performing a ring test between several simulators with independent implementation (by several modellers). The simulation results for the same modelling project are compared to validate the model implementations. This method was chosen by the BSM task group to validate their implementation of ASM1 (Copp *et al.* 2002; Jeppsson *et al.* 2007) and revealed errors in the model codes, in the simulator codes and in the aeration models of the evaluated simulators. However, this task necessitates considerable effort and different simulators, which is not usually available to ASM users.

The following paragraphs propose functionalities of model editors to allow model verification. Some alternative ways to track errors are also suggested.

E.2.1 How to track stoichiometric discontinuities

As state variables are typically expressed in terms of COD, elements (e.g. N, P) or charge, a composition matrix (Gujer & Larsen, 1995) was developed complementary to the Gujer Matrix (Henze *et al.* 1987). It contains the required conversion coefficients for all state variables (in rows) to check the continuity for conservatives (e.g. COD, elements and charge) and observables (e.g. TSS) (in columns) for each process. The continuity check is carried out by multiplying (analytically or numerically) the stoichiometric matrix with the composition matrix as shown in Figure E.1. The resulting matrix should contain only zeros, or near zeros in case of rounding problems.

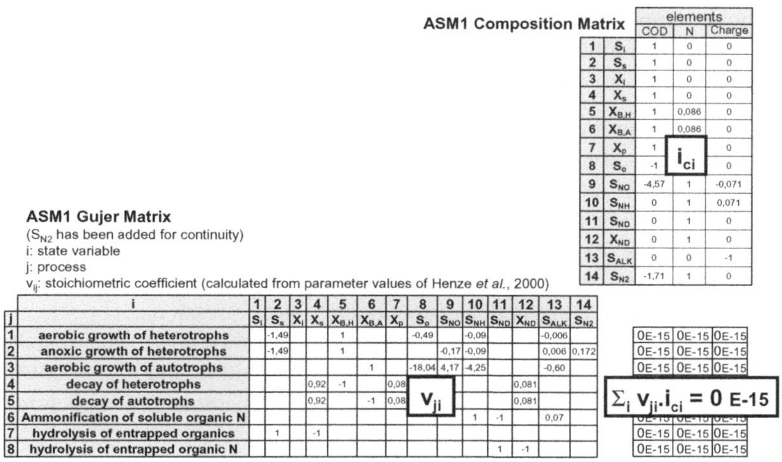

Figure E.1 How to check continuity (numerically).

The common way to check continuity is a numerical analysis starting with default parameter values. For this study the tolerance is set to 10^{-15}. The numerical analysis is an option available in most simulators, or can be performed using spreadsheets (see additional material).

However, when some parameters are fixed to zero (e.g. f_{SU}), a stoichiometric coefficient could be forgotten without any impact on continuity (see hydrolysis process in ASM3 + BioP and processes 5 to 12 in UCTPHO+). Furthermore, errors could be compensated by other parameter values (e.g. when using rounded values everywhere in the model). Thus, another check has to be done in changing parameter values one after the other to track any discontinuity. To change parameter values of parameters calculated from elemental molecular weights (see Table E.14), the molecular weight of the element has to be changed for all the concerned parameters at once.

A better way to track discontinuity that prevent from numerical problems is to use a symbolic analysis. Symbolic analysis allows to re-calculate stoichiometric coefficients from the basic stoichiometric coefficients (e.g. yields) and the composition matrix. The symbolic analysis could be carried out by appropriate tools such as Maple (Maplesoft).

E.2.2 How to track kinetic inconsistencies

Some simulators provide the kinetic rates in symbolic form, what allows an easy check of the proper implementation (mainly parentheses errors). However, it is not possible to track kinetic inconsistencies in model editors so far. A tool to check kinetic rate expressions that could be implemented in simulators is proposed. This tool is based on four questions that modellers should answer for every process:

- Which are the consumed components (every state variable with a negative stoichiometric coefficient)? For every consumed component the kinetic rate expression should include a limitation function (e.g. Monod term). Concerning alkalinity, see the discussion in the chapter "Common published errors".

- Which biomass is involved in the process as biocatalyst? The kinetic rate expression is typically proportional to this biomass concentration.
- Are other components required for the process (e.g. an electron acceptor that is not consumed: Oxygen in ASM2d aerobic hydrolysis)? The kinetic rate expression should include a limitation function for those components (e.g. Monod term).
- Are other components inhibitory (e.g. oxygen in an anoxic process)? The kinetic rate expression should include an inhibitory function for those components (e.g. inhibitory Monod term).

In the attached spreadsheet it is proposed to perform this analysis by colouring Gujer Matrix cells with different colours for each question. When this exercise is done, modellers have to carefully check for each process whether the kinetic rate expression includes a term for each coloured component.

The two first questions could be easily automated in a model editor through the stoichiometric values of the Gujer Matrix. Nevertheless, the two last questions should involve the modeller to indicate electron acceptor conditions of the process and inhibitors. With those pieces of information, model editors should be able to automatically check the consistency of the kinetic rate expressions.

In the presented work, the kinetic rate expressions were checked carefully to ensure that i) every reactant of the process is limiting (to stop a reaction when a reactant is limiting and to prevent the calculation of negative concentrations); ii) every switching function or kinetic parameter is coherent; and iii) kinetic rate expressions are consistent from one model to another.

E.3 COMMON PUBLISHED ERRORS

E.3.1 Rounding parameters

An error that occurs systematically and that may hinder the continuity of a model is to round parameters to 2 significant figures or even to use rounded and "exact values" of parameters (i.e. fractions in calculated parameters, see Table E.14 in appendix) in the very same model. To avoid an accumulation of rounding problems, it is recommended to keep "exact values" everywhere in the model.

The "exact values" of conversion coefficients can be calculated from theoretical (conceptual) COD of elements as defined by Gujer and Larsen (1995) (see Table E.13) and from molecular weights (periodic table of elements). Table E.14 summarizes the main conversion coefficients, their calculation explanation and their exact values to be used in ASM-type models.

E.3.2 Temperature adjustment of kinetic parameters

Kinetic parameter values depend on temperature. Three different ways have been proposed to provide temperature adjustment of kinetic parameters (with θ the temperature adjustment coefficient, $k_{20°C}$ the kinetic coefficient at 20°C and k_T the kinetic coefficient at temperature T):

- In ASM1 and ASM2d (Henze *et al.* 2000a, b), kinetic parameters are given at 10 and 20°C
- In ASM3 (Gujer *et al.* 2000), ASM3 + BioP (Rieger *et al.* 2001) and ASM2d + TUD (Meijer, 2004), θ values are provided using the following equation: $k_T = k_{20°C} * e^{\theta*(T-20)}$
- In ASM2 (Henze *et al.* 2000c), New General (Barker & Dold, 1997) and UCTPHO+ (Hu *et al*,. 2007) provide θ values using: $k_T = k_{20°C} * \theta^{T-20}$

The two last equations are similar: the temperature adjustment e^{θ} in equation $k_T = k_{20°C} * e^{\theta*(T-20)}$ is equivalent to θ in the equation $k_T = k_{20°C} * \theta^{T-20}$. It is thus easy to convert temperature coefficient from one equation to the other one. Unfortunately the same symbol (θ) is given to these two different parameters. As suggested in Corominas *et al.* (2010), an extended notation should be used. The first parameter could be noted θ_{exp} and the second one θ_{pow}. Then $\theta_{pow} = \exp(\theta_{exp})$. However, it should be easier for model comparison to use a single temperature adjustment equation among the modellers community. The second equation ($k_T = k_{20°C}{*}\theta^{T-20}$) is chosen in this work as it is the simplest one and the most commonly used (Vavilin, 1982).

E.3.3 Impact of alkalinity on kinetic rates

Three ways to deal with alkalinity have been proposed in the models:

- Alkalinity is not taken into account in the model at all (New General and UCTPHO+)
- Alkalinity is taken into account in the stoichiometry but is not limiting in the kinetic rates (ASM1)
- Alkalinity is taken into account in both stoichiometry and kinetic rates (ASM2d, ASM3, ASM3 + BioP and ASM2d + TUD).

For the latter models, the stoichiometric coefficients for alkalinity were compared using the parameter sets from the original publications (see attached spreadsheet). Those calculations reveal that:

- The use of alkalinity as a limiting factor in kinetic rates is not consistent, as shown in Table E.1. Indeed, alkalinity is a limiting factor for all processes of those models where alkalinity is consumed, except in process 11, 21 and 15 of ASM2d + TUD (see Paragraph concerning ASM2d + TUD below). However, alkalinity is sometimes considered as a limiting factor in the processes where it is produced.
- The stoichiometric coefficients for alkalinity highly depend on parameter values (e.g. yield values or conversion coefficients that change the proportions of consumed or released nutrients). Table E.2 shows the consequences of different parameter values on the stoichiometric coefficient of alkalinity for ASM2d + TUD.

Table E.1 The use of alkalinity as a limiting factor. Stoichiometric coefficients have been calculated with published parameter values (In parenthesis, number of processes with no alkalinity production or consumption).

Number of processes	Total	Consumption limiting factor	Production limiting factor	Consumption not limiting	Production not limiting
ASM2d	21	8	7	0	6
ASM3	12	2	1	0	8 (+1)
ASM3 + BioP	23	6	4	0	11 (+2)
ASM2d + TUD	22	3	4	3	12

Table E.2 Examples of changes in stoichiometric coefficient values of ASM2d + TUD depending on parameter values.

Parameter	Parameter values		Processes	Alkalinity
i_{NBM}	default	0.07	15	Consumed
	tested	0.08	15	Produced
i_{NSF}	default	0.03	1, 2, 3, 4	Produced Consumed
	tested	0.045	1, 2, 3, 4	Consumed Produced

For a consistent use of alkalinity as a limiting factor, three solutions are proposed:

- Alkalinity should be limiting for all processes involving alkalinity. However, in case of low alkalinity, the process rate is slowed down, even if the process produces alkalinity.
- Alkalinity should be limiting only for processes where it is consumed. Thus, from one modelling project to another, the processes where alkalinity is consumed could be different, and the model user should appropriately correct the kinetic rate expression. However, this solution would introduce additional implementation effort and could require adaptations of kinetic rate expressions after changes of stoichiometric parameters.
- Alkalinity should be limiting only for processes where it is significantly consumed (depending both from stoichiometry and kinetic coefficient).

The authors suggest to use the first solution, because it is the most rigorous one.

E.4 TYPING ERRORS, INCONSISTENCIES AND GAPS IN PUBLISHED MODELS

During the checks performed on the stoichiometric continuity and the evaluation of the kinetic rate expressions, several implementation errors and inconsistencies were identified. They are presented below for each model and separated into 3 different error types: (i) typing errors; (ii) inconsistencies when it is not clearly an error but a potentially risky simplification; and (iii) gaps in stoichiometry and kinetics due to oversight or purposeful omission to keep the model simple.

E.4.1 ASM1 (Henze *et al.* 2000)

Inconsistencies. There is no term in the kinetic rate expression to model nutrient (ammonia) limitation in the heterotrophic growth process, what could induce negative ammonia concentration values (Table E.3).

Table E.3 Inconsistencies in kinetic rate expressions in ASM1 model publication.

Process	Description	Missing Monod term	Correct Monod term
1, 2	Heterotrophic growth	Ammonia limitation	$\frac{S_{NH}}{K_{NH}+S_{NH}}$

Gaps. In order to close mass balances, N_2 should be included in the Gujer Matrix in process 2 (anoxic growth of heterotrophic biomass). This variable is only useful to verify the model continuity but has no impact on model results.

To perform a full nitrogen balance, variables S_{NI} (soluble non-biodegrabable organic nitrogen) and X_{NI} (particulate non-biodegrabable organic nitrogen) should be estimated in the influent. As non-biodegradable compounds, they do not appear in the Gujer Matrix. S_{NI} should be added to total soluble nitrogen in the effluent and X_{NI} should be added to total nitrogen in activated sludge.

E.4.2 ASM2d (Henze *et al.* 2000b)

Typing errors. Table E.4 summarises ASM2d typing errors. Those typing errors have previously been pointed out by Gernaey *et al.* (2006).

Table E.4 Typing errors in ASM2d model publication.

Process	Description	Kinetic or stoichiometry	Wrong	Correct
6, 7	Anoxic growth of heterotrophs on S_F and S_A	kinetic rate	$\frac{K_{NO_3}}{K_{NO_3}+S_{NO_3}}$	$\frac{S_{NO_3}}{K_{NO_3}+S_{NO_3}}$
7	Anoxic growth of heterotrophs on S_A	Stoichiometry of S_{N2}	$-\frac{1-Y_H}{40/14.Y_H}$	$\frac{1-Y_H}{40/14.Y_H}$
8	Fermentation	kinetic rate	K_F	K_{fe}
11	Aerobic storage of X_{PP}	kinetic rate	K_{PP}	K_{IPP}
13,14	Aerobic and anoxic growth of X_{PAO}	Stoichiometry of X_{PHA}	$-1/Y_H$	$-1/Y_{PAO}$

Stoichiometric coefficients for S_{O_2}, S_{NH_4}, S_{N_2}, S_{NO_3}, S_{PO_4}, S_{ALK} and X_{TSS} are not given in detail, so that users have to apply continuity equations to implement them. The corrected matrix provided in the attached spreadsheet details those coefficients.

Inconsistencies. The same parameter name, η_{NO_3}, is given to the correction factor for hydrolysis under anoxic conditions (process 2) and to the reduction factor for anoxic heterotrophic processes (anoxic OHO growth (6 and 7) and anoxic PAO processes (12 and 14)). However, those parameters have different values in the parameter set provided in the publication (η_{NO_3}(hydrolysis) = 0.6 and η_{NO_3}(heterotrophs) = 0.8). This problem is fixed when defining extended symbols using the standardised notation (Corominas *et al.* 2010).

E.4.3 ASM3 (Gujer *et al.* 2000)

Typing errors. The original publication (Gujer *et al.* 1999) had several typing errors. The corrected version (Gujer *et al.* 2000) should be used.

Stoichiometric coefficients for S_{O_2}, S_{NH_4}, S_{N_2}, S_{NO_x}, S_{PO_4}, S_{ALK} and X_{TSS} are not given in detail, so that users have to apply continuity equations to implement them. The corrected matrix in spreadsheet format is provided on the GMP WaterWiki website and details those coefficients.

E.4.4 ASM3 + BioP (Rieger *et al.* 2001)

Typing errors. Table E.5 summarises ASM3 + BioP typing errors.

Table E.5 Typing errors in ASM3 + BioP model publication.

Process	Description	Kinetic or stoichiometry	Wrong	Correct
1	Hydrolysis	Stoichiometry of S_I	No coefficient	f_{SI}
8, 9	Aerobic and anoxic respiration of internal storage	kinetic rate	b_H	b_{Sto}
11	Aerobic endogenous respiration	kinetic rate	$K_{O,H}$	$K_{O,A}$
12	Anoxic endogenous respiration	kinetic rate	$K_{O,H}$	$K_{O,A}$
P9	Anoxic lysis of X_{PP}	kinetic rate	$\frac{S_{NO}}{K_{NO,PAO}}$	$\frac{S_{NO}}{K_{NO,PAO}+S_{NO}}$
P11	Anoxic respiration of X_{PHA}	kinetic rate	$\frac{S_{NO}}{K_{NO,PAO}}$	$\frac{S_{NO}}{K_{NO,PAO}+S_{NO}}$

Inconsistencies. The kinetic parameter $K_{NO,A}$ is missing: the kinetic rate in process 12 (Anoxic endogenous respiration) uses $K_{NO,H}$.

E.4.5 ASM2d + TUD (Meijer, 2004)

Typing errors. Table E.6 summarises ASM2d + TUD typing errors.

Table E.6 Typing errors in ASM2d + TUD model publication.

Process	Description	Kinetic or stoichiometry	Wrong	Correct
21	Autotrophic growth	kinetic rate	K_{PO}	S_{PO}

Inconsistencies. The kinetic check reveals missing Monod terms to insure consistency with the process. Table E.7 summarises ASM2d + TUD inconsistencies in kinetic rate expressions.

Table E.7 Inconsistencies in kinetic rate expressions in ASM2d + TUD model publication.

Process	Description	Missing Monod term	Correct Monod Term
1	Aerobic hydrolysis	Oxygen limitation	$\frac{S_O}{K_O+S_O}$
11, 15, 21	PAO anaerobic maintenance, Anoxic glycogen formation, Autotrophic growth	Alkalinity limitation	$\frac{S_{HCO}}{K_{HCO}+S_{HCO}}$

E.4.6 New general (Barker & Dold, 1997)

Inconsistencies. In the kinetic rate expressions of processes 1 to 8 (growth on S_{BSC} or S_{BSA}) there is no substrate preference switch function (as e.g. in ASM2d) such as $S_{BSC}/(S_{BSC}+S_{BSA})$. This substrate preference switch function avoids that the heterotrophic specific growth rate increases above a maximum value if both substrates are present in high concentration (Henze *et al.* 2000b). Even if Barker and Dold (1997) specify that S_{BSA} concentration entering the anoxic and aerobic zones is usually very low, this substrate preference switch function could be added to enhance the robustness of the model. The preference switch function $S_{BSC}/(S_{BSC}+S_{BSA})$ is proposed (as e.g. used in ASM2d), other function types are described in Dudley *et al.* (2002).

The kinetic check reveals other missing Monod terms to insure consistency with the stoichiometry of the process. Table E.8 summarises New General inconsistencies in kinetic rate expressions.

Table E.8 Inconsistencies in kinetic rate expressions in New General model publication.

Process	Description	Missing Monod term	Correct Monod Term
1 to 4	Heterotrophic growth on S_{BSC}	Substrate preference switch function	$\frac{S_{BSC}}{S_{BSC}+S_{BSA}}$
5 to 8	Heterotrophic growth on S_{BSA}	Substrate preference switch function	$\frac{S_{BSA}}{S_{BSC}+S_{BSA}}$

(*Continued*)

Table E.8 Inconsistencies in kinetic rate expressions in New General model publication (*Continued*).

Process	Description	Missing Monod term	Correct Monod Term
15	Fermentation of S_{BSC} to S_{BSA} (Anaerobic growth)	Phosphate limitation	$\frac{P_{O_4}}{K_{LP,GRO}-P_{O_4}}$
15	Fermentation of S_{BSC} to S_{BSA} (Anaerobic growth)	Ammonia limitation	$\frac{N_{H_3}}{K_{NA}+N_{H_3}}$
16	Autotrophic growth	Phosphate limitation	$\frac{P_{O_4}}{K_{LP,GRO}-P_{O_4}}$
20 and 21	Aerobic growth of PAO, PO_4 limited	$P_{PP\text{-}LO}$ limitation (phosphorous source in case of PO_4 depletion)	$\frac{P_{PP-LO}}{K_{XP}+P_{PP-LO}}$

Gaps. In order to keep the continuity, N_2 (processes 2, 4, 6, 8, 22 and 27) should be included in the Gujer Matrix as a state variable. As in ASM1, this variable is only useful to verify the model continuity but has no impact on model results.

The "COD losses" mentioned in Barker and Dold (1997) (processes 11, 12, 15 and 36) have been detected based on experimental data. It is modelled through the introduction of an efficiency parameter in hydrolysis processes (11, 12) and a yield parameter in fermentation and sequestration processes (15, 36). However the fate of the resulting COD is not described by the model and leads to a lack of continuity. In the model ASDM (BioWin, EnviroSim), the "COD loss" is considered to be due to H_2 gas formation (Kraemer *et al.* 2008). A state variable S_H is then to be added to the model (Table E.9).

Table E.9 Gaps in stoichiometry in new general model publication.

Process	Description	Gap in stoichiometry	Corrected stoichiometry*
2, 4, 6, 8	Anoxic growth of heterotrophs	S_{N_2} variable	$(1\text{-}Y_{H\cdot ANOX})/(i_{SNO3\_SN2}{*}Y_{H\cdot ANOX})$
22	Anoxic growth of PolyP organisms	S_{N_2} variable	$(1\text{-}Y_P)/(i_{SNO3\_SN2}{*}Y_P)$
27	Anoxic decay of PolyP organisms	S_{N_2} variable	$(1\text{-}f_{EP.P}\text{-}f_{ES\cdot P})/i_{SNO3\_SN2}$
11	Anoxic hydrolysis of stored/enmeshed COD	S_H variable	$(1\text{-}E_{ANOX})/i_{COD\_SH}$
12	Anaerobic hydrolysis of stored/enmeshed COD	S_H variable	$(1\text{-}E_{ANA})/i_{COD\_SH}$
15	Fermentation of S_{BSC} to S_{BSA}	S_H variable	$(1\text{-}(1\text{-}Y_{H,ANA}){*}Y_{AC}\text{-}Y_{H,ANA})/i_{COD\_SH}$
36	Sequestration of S_{CFA} by PolyP organisms	S_H variable	$(1\text{-}Y_{PHB})/i_{COD\_SH}$
3, 7	Aerobic growth of heterotrophs on S_{BSC}/S_{BSA} with NO_3	Oxygen from consumed NO_3 not included in S_O	$-(1\text{-}Y_{H\cdot AER})/Y_{H\cdot AER}$ $-i_{COD\_SNO_3}{*}f_{N\cdot ZH}$

(*Continued*)

Table E.9 Gaps in stoichiometry in new general model publication (*Continued*).

Process	Description	Gap in stoichiometry	Corrected stoichiometry*
19, 21	Aerobic growth of PolyP organisms on S_{PHB} with N_{O_3} without and with P_{O_4} limited	Oxygen from consumed NO_3 not included in S_O	$-(1-Y_P)/Y_P$ $-i_{COD\_SNO3}*f_{N\cdot ZP}$
4, 8	Anoxic growth of heterotrophs on S_{BSC} / S_{BSA} with N_{O_3}	Different yield for S_{BSC} or S_{BSA} consumption with NO_3	$-1/Y_{H.ANOX} +$ $i_{COD\_SNO3}*f_{N\cdot ZH}$

*The parameters and variables newly introduced in this study are named according to the standardised notation rules (Corominas *et al.* 2010) and thus may not be consistent with the original model notation. The conversion factors are described in Table E.14.

In processes 3, 4, 7, 8, 19 and 21, there is another discontinuity for COD that is not mentioned in the paper, due to the potential use of NO_3^- as a nitrogen source by heterotrophs. Indeed, when NO_3^- is used as a nitrogen source, the fate of the oxygen content of NO_3^- is not considered. The O_2 stoichiometric coefficient should be lower for growth with NO_3^- as nitrogen source than O_2 consumption for growth with NH_3 (Grady *et al.* 1999).

To match the continuity of aerobic processes (3, 7, 19 and 21), the authors suggest to decrease the oxygen stoichiometric coefficient by subtracting the COD content in the consumed nitrates (Table E.9). This correction is not possible for anoxic processes (4 and 8). The proposed solution is to consider that more substrate is needed for the same growth: the stoichiometric coefficient of the substrate (S_{BSC} or S_{BSA}) is increased by the COD consumed in using nitrates as nitrogen source (Table E.9).

Polyphosphate Accumulating Organisms (PAOs) (Z_P in the model's notation) have not the same nitrogen content as autotrophs (Z_A) and heterotrophs (Z_H) ($f_{N,ZP} = 0.07$ and $f_{N,ZA}$ and $f_{N,ZH} = 0.068$). In the decay process, all organisms turn into endogenous mass (Z_E) that has the same nitrogen content as the biomass it comes from ($f_{N,ZEP} = 0.07$ and $f_{N,ZEA}$ and $f_{N,ZEH} = 0.068$). Thus, the model structure allows different nitrogen fractions for the endogenous masses, however all the biomasses are turned into a single Z_E, which only has a single nitrogen fraction. Consequently, there is a lack in nitrogen continuity of -5×10^{-15} gN for the processes 23, 27 and 31 (aerobic, anoxic and anaerobic decay of PAOs). All biomass nitrogen fractions $f_{N,ZEP}$, $f_{N,ZEA}$ and $f_{N,ZEH}$ should be corrected with the same value. A value of 0.07 $gN{\cdot}gCOD^{-1}$ is proposed.

E.4.7 UCTPHO+ (Hu *et al.* 2007)

Typing errors. Table E.10 summarises UCTPHO+ typing errors.

Table E.10 Typing errors in UCTPHO+ model publication.

Process	Description	Kinetic or stoichiometry	Wrong	Correct
14, 17	Heterotrophic and autotrophic decay	Stoichiometry of X_{ENM}	$f_{XI,H}$	$f_{XE,H}$
14, 17	Heterotrophic and autotrophic decay	Stoichiometry of S_{NH_4}	No coefficient	$i_{NBM}-(1-f_{XE\text{-}H})*i_{NENM}-f_{XE\text{-}H}*i_{NXE}$ or

(*Continued*)

Table E.10 Typing errors in UCTPHO+ model publication (*Continued*).

Process	Description	Kinetic or stoichiometry	Wrong	Correct
				i_{NBM} -(1-$f_{XE\text{-}NIT}$)* i_{NENM} -$f_{XE\text{-}NIT}$*i_{NXE}
5 to 8	Heterotrophic growth on S_F	Stoichiometry of S_{PO_4}	-i_{PBM} No P contained in S_F	-i_{PBM} + i_{PSF}/Y_{H1} or -i_{PBM} + i_{PSF}/Y_{H2}
9 to 12	Heterotrophic growth on X_{ads}	Stoichiometry of S_{PO_4}	-i_{PBM} No P contained in X_{ads}	-i_{PBM} + i_{PENM}/Y_{H1} or -i_{PBM} + i_{PENM}/Y_{H2}
18	Aerobic growth of X_{PAO} on X_{PHA} with S_{NH_4}	Stoichiometry of S_{PO_4}	No coefficient	-i_{PBM}-Y_{PP1}/Y_{PAO1}
24, 27, 30	Decay of X_{PAO}	Stoichiometry of S_{NH4}	In coefficients A, B and C, nitrogen fraction of X_E is i_{NBM} instead of i_{NXE}	A: i_{NBM}-$f_{XE\cdot PAO}$*i_{NXE}-$f_{SI\cdot PAO}$*i_{NSI} B: i_{NBM}-$f_{XE\cdot PAO}$*i_{NXE}-$f_{SI\cdot PAO}$*i_{NSI}-i_{NENM}*(1-η_{PAO})*(1-$f_{XE\cdot PAO}$-$f_{SI\cdot PAO}$) C: i_{NBM}-$f_{XE\cdot PAO}$*i_{NXE}-$f_{SI\cdot PAO}$*i_{NSI}-i_{NENM}*(1-$f_{XE\cdot PAO}$-$f_{SI\cdot PAO}$)
14, 17, 24, 27, 30	OHO, ANO and PAO decay	Stoichiometry of S_{PO_4}	i_{PBM}*(1-f_{XE}) P fraction of X_E is i_{PBM} instead of i_{PXE}	i_{PBM}-$f_{XE\cdot H}$*i_{PXE} or i_{PBM}-$f_{XE\cdot NIT}$*i_{PXE} or i_{PBM}-$f_{XE\cdot PAO}$*i_{FXE}
24, 27, 30	PAO decay	Stoichiometry of S_{PO4}	i_{PBM}-$f_{XE.PAO}$*i_{PXE} No P contained in S_I	i_{PBM}-$f_{XE\cdot PAO}$*i_{PXE}-$f_{SI\cdot PAO}$*i_{PSI}
14, 17, 26, 27, 29, 30, 32	OHO and ANO decay, anoxic and anaerobic PAO decay, X_{PHA} lysis	Stoichiometry of S_{PO4}	No P contained in X_{ENM}	Depends on X_{ENM} stoichiometry: $\nu_{ij,XENM}$*i_{PENM} should be added (see attached spreadsheet)

Inconcistencies. The kinetic check reveals some missing Monod terms to insure consistency with the stoichiometry of the processes. Table E.11 summarises UCTPHO+ inconsistencies in the kinetic rate expressions.

Table E.11 Inconsistencies in kinetic rate expressions in UCTPHO+ model publication.

Process	Description	Missing Monod term	Correct Monod Term
1 to 4	Heterotrophic growth on S_A	Substrate preference switch function	$\frac{S_A}{S_F+S_A+X_{Ads}}$
5 to 8	Heterotrophic growth on S_F	Substrate preference switch function	$\frac{S_F}{S_F+S_A+X_{Ads}}$
9 to12	Heterotrophic growth on X_{Ads}	Substrate preference switch function	$\frac{X_{Ads}}{S_F+S_A+X_{Ads}}$

(*Continued*)

Table E.11 Inconsistencies in kinetic rate expressions in UCTPHO+ model publication (*Continued*).

Process	Description	Missing Monod term	Correct Monod Term
3, 7, 11	Anoxic growth with S_{NH_4}	Nitrate limitation	$\frac{S_{NO3}}{K_{NO3}+S_{NO3}}$
26, 29, 32	X_{PHA} lysis during (aerobic, anoxic, anaerobic) PAO decay	Ammonia limitation (X_{PHA} is turned into X_{ENM}, which contains nitrogen. Ammonia has thus to be consumed)	$\frac{S_{NH4}}{K_{NH4}+S_{NH4}}$
20 and 21	Aerobic growth of PAO, PO_4 limited	X_{PP} limitation (phosphorous source in case of PO_4 depletion)	$\frac{X_{PP}}{K_{PP}+X_{PP}}$

Gaps. In order to keep the continuity, N_2 as a state variable should be included in the Gujer Matrix for the processes 3, 4, 7, 8, 11, 12, 22, 23 and 27. As in ASM1 and New General, this variable is only useful to verify the model continuity but has no impact on model results.

As previously seen in the New General, a discontinuity for COD in processes 2, 4, 6, 8, 10, 12, 19, 21 and 23 is due to the use of NO_3^- as nitrogen source.

In the same way as in the New General, authors suggest to lower the oxygen stoichiometric coefficient in aerobic processes (2, 6, 10, 19 and 21) and to increase the stoichiometric coefficient of substrate in anoxic processes (4, 8, 12 and 23). In contrast to the New General, some of the substrates contain a fraction of nitrogen and phosphorous (S_F for process 8 and X_{Ads} for process 12). The stoichiometric coefficients of S_{NH4} and S_{PO4} should be corrected to match the continuity.

Table E.12 Gaps in stoichiometry of UCTPHO+ model publication.

Process	Description	Gap	Corrected stoichiometry
2, 4, 6, 8, 11, 12	Anoxic growth of heterotrophs	S_{N_2} variable missing	$(1-Y_{H_2})/(i_{SNO3\_SN2}*Y_{H2})$
22, 23	Anoxic growth of PolyP organisms	S_{N_2} variable missing	$(1-Y_{PAO2})/(i_{SNO3\_SN2}*Y_{PAO2})$
27	Anoxic decay of PolyP organisms	S_{N_2} variable missing	$\eta_{PAO}*(1-f_{XE\cdot PAO}-f_{SI\cdot PAO})/i_{SNO3\_SN2}$
2, 6, 10	Aerobic growth of heterotrophs on $S_A/S_F/X_{ads}$ with S_{NO_3}	Oxygen from consumed S_{NO_3} not include in S_{O_2}	$-(1-Y_{H1})/Y_{H1}$ $-i_{COD\_SNO3}*i_{NBM}$
19, 21	Aerobic growth of PolyP organisms on X_{PHA} with S_{NO_3} without and with S_{PO_4} limited	Oxygen from consumed S_{NO_3} not include in S_{O_2}	$-(1-Y_{PAO1})/Y_{PAO1}$ $-i_{COD\_SNO3}*i_{NBM}$
4	Anoxic growth of heterotrophs on S_A with S_{NO_3}	Different yield of S_A consumption with S_{NO_3}	$-1/Y_{H2}+i_{COD\_SNO3}*i_{NBM}$
8	Anoxic growth of heterotrophs on S_F with S_{NO_3}	Different yield of S_F consumption with S_{NO_3} S_{NH_4} coefficient correction S_{PO_4} coefficient correction	$-1/Y_{H2}+i_{COD\_SNO3}*i_{NBM}$ $i_{NSF}*(1/Y_{H2}-i_{COD\_SNO3}*i_{NBM})$ $-i_{PBM}+i_{PSF}*(1/Y_{H2}-i_{COD\_SNO3}*i_{NBM})$

(*Continued*)

Table E.12 Gaps in stoichiometry of UCTPHO+ model publication (*Continued*).

Process	Description	Gap	Corrected stoichiometry
12	Anoxic growth of heterotrophs on X_{Ads} with S_{NO_3}	Different yield of X_{Ads} consumption with NO_3 S_{NH4} coefficient correction S_{PO4} coefficient correction	$-1/Y_{H2} + i_{COD\_SNO3}*i_{NBM}$ $i_{NENM}*(1/Y_{H2}-i_{COD\_SNO3}*i_{NBM})$ $-i_{PBM} + i_{PENM}*(1/Y_{H2}-i_{COD\_SNO3}*i_{NBM})$
23	Anoxic growth of PolyP organisms on S_{PHA} with S_{NO_3}	Different yield of X_{PHA} consumption with S_{NO_3}	$-1/Y_{PAO2} + i_{COD\_SNO3}*i_{NBM}$

CONCLUSION

Several error sources can impact model quality. This paper points out typing errors, inconsistencies and gaps in the publications of seven selected models. Some of the errors corrected in this paper are mainly theoretical errors and will only have a minor impact on model results in typical conditions, but may have a significant impact in case of peculiar treatment conditions (e.g. near or outside model limits).

It is necessary to verify both a published model before using it and the model implementation in simulators to avoid typing errors and inconsistencies. A simple spreadsheet, as presented on the GMP WaterWiki website, could be used for continuity checks. The evaluation of the kinetic rate expressions is only possible based on a detailed check of the individual expressions but should be carried out with great care. The spreadsheet provides corrected matrices with all stoichiometric coefficients for the discussed biokinetic models and gives an example on a proper continuity and kinetic rate expressions check.

Model verification is a time-consuming task that could be facilitated and automated by appropriate model editor tools as part of a simulator. Albeit model verification is facilitated with these tools, it remains that model users have to redo this work each time they implement a new model.

Acknowledgments

This work has been carried out in close collaboration with the IWA Good Modelling Practice Task Group. The authors would like to especially thank Andrew Shaw and Imre Takács for their valuable contributions. The GMP Task Group is sponsored by IWA and its home institutions (Black&Veatch, Cemagref, EnviroSim Associates Ltd., Kurita Water Industries Ltd., TU Vienna, Canada Research Chair in Water Quality modelling, Univ. Laval, École Polytechnique de Montreal, EAWAG). The GMP task group would like to especially thank its sponsors Hydromantis Inc. and MOSTforWATER. N.V.

The authors would also like to thank Sebastiaan Meijer and Jeremy Dudley for their fruitful comments about this article.

Hélène Hauduc is a PhD student at Cemagref (France) and model*EAU* (Université Laval, Québec, Canada).

Peter A. Vanrolleghem holds the Canada Research Chair in Water Quality Modelling.

REFERENCES

Barker P. S. and Dold P. L. (1997). General model for biological nutrient removal activated-sludge systems: model presentation. *Water Environment Research*, **69**(5), 969–984.

Copp J. B., Jeppsson U. and Vanrolleghem P. A. (2008). The benchmark simulation models – a valuable collection of modelling tools. *Proceedings of the iEMSs Fourth Biennial Meeting: International Congress on Environmental Modelling and Software (iEMSs 2008)*, Barcelona, Catalonia, July 2008.

Corominas L., Rieger L., Takács I., Ekama G., Hauduc H., Vanrolleghem P. A., Oehmen A., Gernaey K. V. and Comeau Y. (2010). New framework for standardized notation in wastewater treatment modelling. *Water Science and Technology*, **61**(4), 841–857.

Dudley J., Buck G., Ashley R. and Jack A. (2002). Experience and extensions to the ASM2 family of models. *Water Science and Technology*, **45**(6), 177–186

Gernaey K. V., Rosen C., Batstone D. J. and Alex J. (2006). Efficient modelling necessitates standards for model documentation and exchange. *Water Science and Technology*, **53**(1), 277–285.

Grady C. P. L., Jr, Daigger G. T. and Lim H. C. (1999). Biological wastewater treatment: Principles and practices. M. Dekker, CRC Press, New York.

Gujer W. and Larsen T. A. (1995). The implementation of biokinetics and conservation principles in ASIM. *Water Science and Technology*, **31**(2), 257–266.

Gujer W., Henze M., Mino T. and van Loosdrecht M. C. M. (1999). Activated Sludge model No. 3. *Water Science and Technology*, **39**(1), 183–193.

Gujer W., Henze M., Mino T. and van Loosdrecht M. C. M. (2000). Activated Sludge Model No. 3. In M. Henze, W. Gujer, T. Mino and M. C. M. van Loosdrecht (2000). Activated Sludge Models ASM1, ASM2, ASM2d and ASM3. Scientific and Technical Report No. 9, IWA Publishing, London, UK.

Henze M., Grady C. P. L., Gujer W., Marais G. v. R. and Matsuo T. (1987). Activated Sludge Model No. 1. Scientific and Technical Report No. 1, IAWPRC, London, UK.

Henze M., Gujer W., Mino T., Matsuo T., Wentzel M. C., Marais G. v. R. and van Loosdrecht M. C. M. (1998). Outline activated sludge model No. 2d. In *Preprints of the 4th Kollekolle Seminar on Activated Sludge Modelling*. Modelling and Microbiology of Activated Sludge Processes, 16–18 March, Copenhagen, Denmark.

Henze M., Gujer W., Mino T., Matsuo T., Wentzel M.C., Marais G. v. R. and van Loosdrecht M. C. M. (1999). Activated sludge model No. 2d, ASM2d. *Water Science and Technology*, **39**(1), 165–182.

Henze M., Grady C. P. L., Jr, Gujer W., Marais G. v. R. and Matsuo T. (2000a). Activated sludge model No. 1. In M. Henze, W. Gujer, T. Mino and M. C. M. van Loosdrecht (2000). Activated Sludge Models ASM1, ASM2, ASM2d and ASM3. Scientific and Technical Report No. 9, IWA Publishing, London, UK.

Henze M., Gujer W., Mino T., Matsuo T., Wentzel M.C., Marais G. v. R. and van Loosdrecht M. C. M. (2000b). Activated Sludge Model No. 2d. In M. Henze, W. Gujer, T. Mino and M. C. M. van Loosdrecht. (2000). Activated Sludge Models ASM1, ASM2, ASM2d and ASM3. Scientific and Technical Report No. 9, IWA Publishing, London, UK.

Henze M., Gujer W., Mino T., Matsuo T., Wentzel M.C. and Marais G. v. R. (2000c). Activated Sludge Model No. 2. In M. Henze, W. Gujer, T. Mino and M. C. M. van Loosdrecht (2000). Activated Sludge Models ASM1, ASM2, ASM2d and ASM3. Scientific and Technical Report No.9, IWA Publishing, London, UK.

Hu Z. R., Wentzel M. C. and Ekama G. A. (2007). A general kinetic model for biological nutrient removal activated sludge systems: model development. *Biotechnology and Bioengineering*, **98**(6), 1242–1258.

Jeppsson U., Pons M. N., Nopens I., Alex J., Copp J. B., Gernaey K. V., Rosen C., Steyer J. P. and Vanrolleghem P. A. (2007). Benchmark simulation model no 2: General protocol and exploratory case studies. *Water Science and Technology*, **56**(8), 67–78.

Kraemer J., Constantine T., Crawford G., Erdal Z., Katehis D., Johnson B. and Daigger G. (2008). Process design implications due to differences in anaerobic COD stabilisation models. *Proc. 81st Annual Water Environment Federation Technical Exhibition and Conference*, Chicago, Illinois, USA, October 18–22, 2008.

Meijer S. C. F. (2004). Theoretical and practical aspects of modelling activated sludge processes. PhD thesis, Department of Biotechnological Engineering. *Delft University of Technology*, The Netherlands.

Refsgaard J. C., van der Sluijs J. P., Højberg A. L. and Vanrolleghem P. A. (2007). Uncertainty in the environmental modelling process – A framework and guidance. *Environmental Modelling and Software*, **22**(11), 1543–1556.
Rieger L., Koch G., Kühni M., Gujer W. and Siegrist H. (2001). The eawag bio-P module for activated sludge model No.3. *Water Research*, **35**(16), 3887–3903.
Vavilin V. A. (1982). The effects of temperature, inlet pollutant concentration, and microorganism concentration on the rate of aerobic biological treatment. *Biotechnology and Bioengineering*, **24**(12), 2609–2625.

ADDITIONAL MATERIAL

Table E.13 Theoretical COD of electrical charge and main elements (from Gujer & Larsen, 1995).

Element description	Symbol	Theoritical COD ($gCOD \cdot mol^{-1}$)	Molecular weight ($g \cdot mol^{-1}$)
Negative charge	(–)	+8	–
Positive charge	(+)	-8	–
Carbon	C	+32	12
Nitrogen	N	-24	14
Hydrogen	H	+8	1
Oxygen	O	-16	16
Sulphur	S	+48	32
Phosphorus	P	+40	31
Iron	Fe	+24	55.8

Table E.14 Explanation and exact values of the main coefficients used in ASM-type models.

Description	Symbol	Calculation	Exact value	Unit
Conversion factor for NO_3^- in COD	i_{COD_SNO3}	$(-24 + 3*(-16) + 8)$ $gCOD \cdot mol^{-1}/$ $14\ gN \cdot mol^{-1}$	-64/14	$gCOD \cdot gN^{-1}$
Conversion factor for N_2 in COD	i_{COD_SN2}	$(-24*2)$ $gCOD \cdot mol^{-1}/$ $(14*2)\ gN \cdot mol^{-1}$	-24/14	$gCOD \cdot gN^{-1}$
Stoichiometric factor for NO_3^- reduction to N_2 (amount of COD provided by reduction)	i_{SNO3_SN2}	$(64–24)\ gCOD \cdot mol^{-1}/14\ gN \cdot mol^{-1}$	40/14	$gCOD \cdot gN^{-1}$
Conversion factor for NH_4^+ in charge	i_{Charge_SNH4}	$1\ Charge \cdot mol^{-1}/14\ gN \cdot mol^{-1}$	1/14	$Charge \cdot gN^{-1}$
Conversion factor for NO_3^- in charge	i_{Charge_SNO3}	$-1\ Charge \cdot mol^{-1}/14\ gN \cdot mol^{-1}$	-1/14	$Charge \cdot gN^{-1}$
Conversion factor for Ac (CH_3COO^-) in charge	i_{Charge_SAc}	$-1\ Charge \cdot mol^{-1}/$ $(2*32 + 3*8–2*16 + 8) gCOD \cdot mol^{-1}$	-1/64	$Charge \cdot gCOD^{-1}$
Conversion factor for PolyP in charge $(K_{0,33}Mg_{0,33}PO_3)_n$	i_{Charge_XPP}	K+ and Mg2+ not considered: $(PO_3)_n^-$ $-1\ Charge \cdot mol^{-1}/31\ gP \cdot mol^{-1}$	-1/31	$Charge \cdot gP^{-1}$
Conversion factor for PO_4^{3-} in charge	i_{Charge_SPO4}	PO_4^{3-}: 50% $H_2PO_4^-$ + 50% HPO_4^{2-} $(-1-2)\ Charge \cdot mol^{-1}/(2*31)\ gP. mol^{-1}$	-1.5/31	$Charge \cdot gP^{-1}$

(Continued)

Table E.14 Explanation and exact values of the main coefficients used in ASM-type models (*Continued*).

Description	Symbol	Calculation	Exact value	Unit
Conversion factor for MeP ($FePO_4$) in P	i_{P_XMeP}	$FePO_4$: $55.8 + 31 + 4*16 = 150.8$ $g \cdot mol^{-1}$ 31 $gP.mol^{-1}$/ 150.8 $gTSS.mol^{-1}$	31/150.8	$gP \cdot gTSS^{-1}$
Stoichiometric factors for precipitation and redissolution of PO_4^{3-} (ASM2d)	v_{XMeOH}	$Fe(OH)_3 + PO_4^{3-} \leftrightharpoons FePO_4 + 3HCO_3^-$ $Fe(OH)_3$: $55.8 + 3*16 + 3 = 106.8$ $g \cdot mol^{-1}$ $FePO_4$: $55.8 + 31 + 4*16 = 150.8$ $g \cdot mol^{-1}$ Coefficient are normalised on PO_4^{3-} (=31$gP \cdot mol^{-1}$) TSS decrease : (150.8–106.8)/31	-106.8 /31	$gTSS \cdot L^{-1}$

Appendix F

Activated sludge modelling: Development and potential use of a practical applications database

H. Hauduc[a,b], L. Rieger[c,b], T. Ohtsuki[d], A. Shaw[e,f], I. Takács[g], S. Winkler[h], A. Héduit[a], P.A. Vanrolleghem[b] and S. Gillot,[a]

[a]Cemagref, UR HBAN, Parc de Tourvoie, BP 44, F-92163 Antony Cedex, France (Email: sylvie.gillot@irstea.fr).
[b]modelEAU, Département de génie civil, Université Laval, Pavillon Adrien-Pouliot, 1065 av. de la Médecine, Quebec (QC) G1V 0A6, Canada.
[c]EnviroSim Associates Ltd., McMaster Innovation Park, 175 Longwood Rd S, Suite 114A, Hamilton, ON L8P 0A1, Canada.
[d]Kurita Water Industries Ltd, 4-7, Nishi-Shinjuku 3-Chome, Shinjuku-Ku, Tokyo 160-8383, Japan.
[e]Black and Veatch, 8400 Ward Parkway, Kansas City, MO, 64114 USA.
[f]Department of Civil, Architectural, and Environmental Engineering, Illinois Institute of Technology, Alumni Memorial Hall, Room 228, 3201 S. Dearborn St., Chicago, IL 60616, USA.
[g]EnviroSim Europe., 15 Impasse Fauré, Bordeaux 33000, France.
[h]Institute for Water Quality and Waste Management, Vienna Technical University, Karlsplatz 13/E 226, 1040 Vienna, Austria).

ABSTRACT

This study aims at synthesizing experiences in the practical application of ASM type models. The information is made easily accessible to model users by creating a database of modelling projects. This database includes answers to a questionnaire that was sent out to model users in 2008 to provide inputs for a Scientific and Technical Report of the IWA Task Group on Good Modelling Practice – Guidelines for use of activated sludge models, and a literature review on published modelling projects.

The database is analysed to determine which biokinetic model parameters are usually changed by modellers, in which ranges, and what values are typically used for seven selected activated sludge models. These results should help model users in the calibration step, by providing typical parameter values as a starting point and ranges as a guide. However, the proposed values should be used with great

Previously published as: Hauduc, H., Rieger, L., Ohtsuki, T., Shaw, A., Takács, I., Winkler, S., Héduit, A., Vanrolleghem, P.A. and Gillot, S. (2011). Activated sludge modelling: Development and potential use of a practical applications database. *Water Science and Technology*, **63**(10), 2164–2182.

care since they are the result of averaging practical experience and not taking into account specific parameter correlations.

Keywords: Good modelling practice; ASM; Database; Parameter sets; Parameter ranges; Survey

F.1 INTRODUCTION

The International Water Association (IWA) Task Group on *Good Modelling Practice – Guidelines for use of activated sludge models* (GMP-TG, https://iwa-gmp-tg.irstea.fr/) is collecting knowledge and experience on how to use activated sludge (AS) models in engineering practice. The group developed and sent out a first questionnaire to current and potential users of activated sludge models to better define the profile of ASM users and to identify the tools and procedures used. Ninety-six answers were received that provided useful insights into the use of activated sludge models and highlighted the main limitations of modelling and the expectations of users for improvements (Hauduc *et al.* 2009). The calibration step was pointed out especially as one of the most time-consuming steps and is considered as an obstacle for widespread model use. Respondents also asked for better knowledge transfer.

A second, more detailed, questionnaire was sent out in 2008 to provide inputs for the GMP-TG report regarding typical parameter values and case studies from several countries and for different wastewater treatment conditions. In addition and as a second source of information, a literature review was carried out on published modelling projects. The objective of this work was to collect available experiences of practical applications using AS models. A database was constructed to synthesise the answers from the second questionnaire and literature data.

The database includes parameters for seven published activated sludge models: (1) ASM1 (Henze *et al.* 2000a); (2) ASM2d (Henze *et al.* 2000b); (3) ASM3 (Gujer *et al.* 2000); (4) ASM3 + BioP (Rieger *et al.* 2001); (5) ASM2d + TUD (Meijer, 2004); (6) Barker and Dold model (Barker & Dold, 1997); (7) UCTPHO+ (Hu *et al.* 2007). In order to keep this paper readable, these references will not be repeated each time. Prior to this parameter study, all models were analysed for typos and errors (Hauduc *et al.* 2010).

F.1.1 Method

F.1.1.1 Source of data

Questionnaire

In order to completely describe each modelling study, the questionnaire asked for the objectives of the project, the description of the wastewater treatment plant (WWTP) and the parameter set used for the biokinetic model. The questionnaire was sent out in 2008 to the respondents of the first survey, to the attendees of the WWTmod2008 seminar in Mont-St-Anne, QC, Canada, and could be downloaded from GMP-TG sponsor websites.

Probably due to the higher complexity of this questionnaire, only 28 answers were received, among which 17 were usable for this study (i.e. at least one model parameter set provided).

Literature review

In order to have a homogeneous database, only published modelling projects applied to full-scale WWTPs or pilot plants with a major domestic wastewater influent were selected. The review includes 50 articles containing 59 parameter sets.

F.1.1.2 Database description

Structure

In order to store all the information in an efficient way, a database composed of three main tables was constructed:

(1) *Parameter sets*: model, country, temperature, parameter values
(2) *WWTP description*: information on influent, wastewater characteristics, processes and environmental conditions
(3) *Model users*: user information

To facilitate parameter comparison, a new standardised notation (Corominas *et al.* 2010) was used.

Classification of parameter sets

Two classes of model parameter sets were distinguished:

- *Optimised parameter sets* obtained for a specific modelling project. These parameter sets were provided with the description of the WWTP under study. Parameter values may have different sources (see below).
- *Proposed new default parameter sets* based on personal expertise. These parameters were used as starting points for the calibration step during the project and given with an approximate number of WWTPs on which this experience was gained.

Sources of parameter values:

- *Original*: value given in the original publication of the model;
- *New Default*: value given in a proposed new default parameter set;
- *Measured*: value is obtained using a dedicated experimental protocol;
- *Calibrated*: value is changed either using a manual or an automatic procedure to fit simulation results to the data collected on the WWTP.

Temperature adjustment

For comparison purposes, the parameter values were standardised at 20°C. The correction factor was either provided with the dataset or extracted from the original publication. For instance, in the original ASM1 and ASM2d publications, the kinetic parameters only are given at 10°C and 20°C. The correction factor ϑ_{pow} has thus been recalculated following the equation $k_{10°C} = k_{20°C} * \theta_{pow}^{10-20}$.

F.1.1.3 Database analysis

The database was analysed for the three topics:

- Original vs. proposed new default parameter sets: The parameter sets were compared and differences identified and discussed.
- Parameters changed in modelling projects: Most often changed parameters (in more than 50% of the projects) are highlighted.
- Parameter ranges and statistics: For each model, the following variables have been calculated:
  - The median values, which should not be misinterpreted as new default parameters, because the median values are not from a single parameter set and some parameters may be correlated (e.g. growth and decay rate),

- The 25th and 75th percentiles. These percentiles have been chosen to exclude extreme values and to obtain a representative range of the typical parameter values,
- The variability (V), the difference between the two percentiles divided by the median.

These results are discussed and confronted to the knowledge on parameter values and parameter ranges of other published overview studies.

F.2 RESULTS

F.2.1 Modelling project characteristics

The database contains 76 parameter sets, which can be differentiated into 57 optimised parameter sets and 19 proposed new default parameter sets, and distributed as shown in Figure F.1. ASM1 and ASM2d are the most represented models in the database.

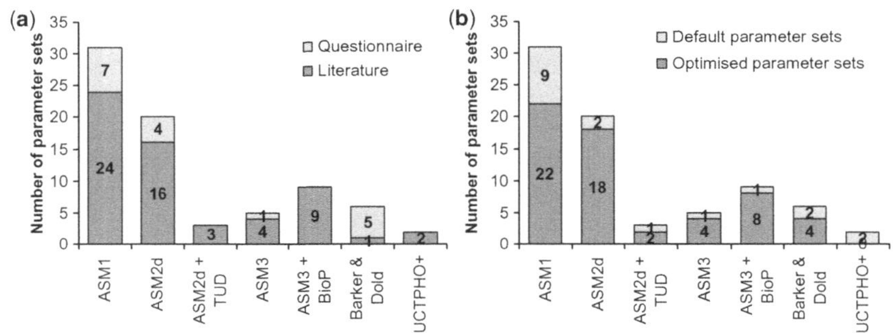

Figure F.1 Distribution of parameter sets per model: a) per source; b) per class of parameter set.

The following paragraphs describe the main information extracted from the current database for ASM1 and ASM2d. An insufficient number of modelling studies is available for the other models, thus no comments are given on the results for these models. However, synthesis tables are presented in the additional material section at the end of this appendix for ASM3, ASM3 + BioP and Barker and Dold model. Concerning ASM2d + TUD and UCTPHO +, no additional modelling projects than those in the original publication were found, the tables are thus not presented.

F.2.2 ASM1

Data description

The database contains 31 parameter sets for ASM1, of which 9 are "proposed new default parameter" sets and 22 are "optimised parameter sets" from specific modelling projects. The modelling studies were mainly carried out at full scale WWTPs (19) mostly in Europe (18), with only 1 application in North-America and 3 in Asia. The sludge ages of the specific modelling projects are between 4 and 40 days.

Table F.1 presents the main results extracted from the database including the original and the proposed new default parameter sets. Basic statistics for optimised parameter sets comprise the number of values for each parameter (n), if the parameter has been modified in more than 50% of the cases (Modif. >50%), the median value (Med.), the 25th and 75th percentiles and the variability (V). The proposed new default parameter sets are by definition based on several simulation studies and therefore present more experience than single studies. Consequently, they are presented on the same level as the median of all the optimised parameter sets.

Table F.1 Synthesis of database results for ASM1, only modified parameters are mentioned. Parameter values are standardised at a temperature of 20°C.

Parameter*	Unit	Description	Original parameter set		Optimised parameter sets						Proposed new default parameters
			Notation	Value (a)	n	Modif >50%	Med.	Perc. 25%	Perc. 75%	V (%)	Parameter sets: b/c/d/e/f/g/h/i
Stoichiometric parameters											
Y_{OHO}	$g\ X_{OHO} \cdot g\ XC_B^{-1}$	Yield for X_{OHO} growth	Y_H	0.67	26	X	0.67	0.62	0.67	7	0.6*(c;i)*$Y_{OHO.Ox}$: 0.67 and $Y_{OHO.Ax}$: 0.54 (b;f;h)
Conversion coefficient											
i_{N_XBio}	$g\ N \cdot g\ X_{Bio}^{-1}$	N content of biomass (X_{OHO}, X_{ANO})	$i_{X,B}$	0.086	31		0.086	0.079	0.086	8	0.08*(c;g)*
Kinetic parameters											
Hydrolysis											
$q_{XCB\_SB,hyd}$	$g\ XC_B \cdot g\ X_{OHO}^{-1} \cdot d^{-1}$	Maximum specific hydrolysis rate	k_h	3	31		3	2.2	3	26	2*(c)*/2.21*(i)*/5.2*(g)*
$\theta_{qXCB\_SB,hyd}$	–	Temperature correction factor for $q_{XCB\_SB,hyd}$	θ_{kh}	1.116	11	X	1.116	1.072	1.12	4	1.072*(f)*
$K_{XCB,hyd}$	$g\ XC_B \cdot g\ X_{OHO}^{-1}$	Half-saturation coefficient for XC_B/X_{OHO}	K_X	0.03	30		0.03	0.03	0.03	0	0.02*(c)*/0.17*(g)*/ 0.15*(i)*
$\theta_{KXCB,hyd}$	–	Temperature correction factor for $K_{XCB,hyd}$	θ_{KX}	1.116	10	X	1.116	1.116	1.12	0	1*(f)*
$\eta_{qhyd,Ax}$	–	Correction factor for hyd. under anoxic cond.	η_h	0.4	31		0.4	0.4	0.5	25	0.5*(g)*/ 0.6*(d)* $n_{qhyd.An}$: 0.75*(d)*

(*Continued*)

Table F.1 Synthesis of database results for ASM1, only modified parameters are mentioned. Parameter values are standardised at a temperature of 20°C (*Continued*).

Parameter*	Unit	Description	Original parameter set		Optimised parameter sets						Proposed new default parameters
			Notation	Value (a)	n	Modif >50%	Med.	Perc. 25%	Perc. 75%	V (%)	Parameter sets: b/c/d/e/f/g/h/i
Ordinary Heterotrophic Organisms											
$\mu_{OHO,Max}$	d^{-1}	Maximum growth rate of X_{OHO}	μ_H	6	31		6	5.7	6	6	4*(d)*/5.7*(g)*
$\theta_{\mu OHO,Max}$	–	Temperature correction factor for $\mu_{OHO,Max}$	$\theta_{\mu H}$	1.072	11	X	1.072	1.071	1.09	2	
$\eta_{\mu OHO,Ax}$	–	Reduction factor for anoxic growth of X_{OHO}	η_g	0.8	31		0.8	0.8	0.8	0	0.6*(c)*
$K_{SB,OHO}$	$g\ S_B \cdot m^{-3}$	Half-saturation coefficient for S_B	K_S	20	31		20	10	20	50	5*(d)*/10*(g)*
b_{OHO}	d^{-1}	Decay rate for X_{OHO}	b_H	0.62	31		0.62	0.61	0.62	2	0.4*(d)*/0.41*(i)*/0.5*(c)*/0.53*(g)*
θ_{bOHO}	–	Temperature correction factor for b_{OHO}	θ_{bH}	1.12	11	X	1.1	1.029	1.12	8	1.029*(f)*/1.071*(c;d)*
$K_{O2,OHO}$	$g\ S_{O2} \cdot m^{-3}$	Half-saturation coefficient for S_{O2}	K_{OH}	0.2	31		0.2	0.2	0.2	0	0.05*(f)*/0.1*(i)*
$K_{NOx,OHO}$	$g\ S_{NOx} \cdot m^{-3}$	Half-saturation coefficient for S_{NOx}	K_{NO}	0.5	31		0.5	0.1	0.5	80	0.1*(f)*/ 0.2*(i)*

Autotrophic Nitrifying Organisms

$\mu_{ANO,Max}$	d^{-1}	Maximum growth rate of X_{ANO}	μ_A	0.8	30	X	0.8	0.66	0.9	30	0.77*(i)*/ 0.82*(g)*/ 0.85*(c)*/0.9*(b; d)*
$\theta_{\mu ANO,Max}$	–	Temperature correction factor for $\mu_{ANO,Max}$	$\theta_{\mu A}$	1.103	14	X	1.103	1.059	1.11	5	1.059*(f; h)*/1.072*(b)*
b_{ANO}	d^{-1}	Decay rate for X_{ANO}	b_A	0.5-0.15	30	X	0.1	0.08	0.15	70	0.07*(g)*/0.096*(i)*/ 0.17*(b; f; h)*
θ_{bANO}	–	Temperature correction factor for b_{ANO}	θ_{bA}	1.072	12	X	1.07	1.029	1.072	4	1.027*(f; h)*/ 1.083 *(d)*/1.103*(c)*
q_{am}	$m^3 \cdot g\ X_{CB,N}^{-1} \cdot d^{-1}$	Rate constant for ammonification	k_a	0.08	29		0.08	0.07	0.08	12	0.05*(g)*/0.16*(i)*
θ_{qam}	–	Temperature correction factor for q_{am}	θ_{ka}	1.072	11		1.07	1.07	1.07	0	1.071*(d; c)*
$K_{O2,ANO}$	$g\ S_{O2} \cdot m^{-3}$	Half-saturation coefficient for S_{O2}	K_{OA}	0.4	31		0.4	0.4	0.4	0	0.2*(f)*/0.5*(c)*/0.75*(i)*
$K_{NHx,ANO}$	$g\ S_{NHx} \cdot m^{-3}$	Half-saturation coefficient for S_{NHx}	K_{NH}	1	31		1	0.75	1	25	0.1*(f)*/ 0.5*(d)*

a: Henzee*t al.* (2000a); b, c: 2 answers from the questionnaire; d: Bornemann *et al.* (1998); e: Hulsbeek *et al.* (2002); f: Marquot (2006); g: Spanjers *et al.* (1998); h: Choubert *et al.* (2009b); i: Grady *et al.* (1999).

*Standardised notation from Corominas *et al.*(2010) is used. n: number of parameter values in the database.Please refer to the additional material section at the end of the appendix for the parameter definitions.

Original vs. proposed new default parameter set: Only 3 parameters (out of 26) have not been changed compared to the original value: the autotrophic growth yield (Y_{ANO}), the fraction of particulate unbiodegradable organics generated in biomass decay ($f_{XU\_Bio,lys}$) and the nitrogen content of unbiodegradable organics generated in biomass decay (i_{N_XUE}).

A change in ASM1 model structure for the ordinary heterotrophic yield (Y_{OHO}) value by introducing an ordinary heterotrophic yield under anoxic conditions is suggested in three of the proposed new default parameter sets.

Parameters changed in modelling projects (compared to original values): for each parameter set, a majority of parameters are kept at their default values. Only the autotrophic growth yield (Y_{ANO}) is always kept at its original value. Nine parameters were changed in more than half of the modelling studies: 6 temperature correction factors, the heterotrophic yield (Y_{OHO}) and the autotrophic growth and decay rate pair ($\mu_{ANO,Max}$, b_{ANO}).

Only a few parameter sets contain measured parameters (Makinia and Wells, 2000; Nuhoglu *et al.* 2005; Stamou *et al.* 1999 and Petersen *et al.* 2002). Most of the measured values are close to the values used in other modelling projects, except for Stamou *et al.* (1999) who determined very low values for the heterotrophic and autotrophic growth related parameters.

Parameter ranges and statistics: All median values are the same as in the original parameter set. The variability is quite narrow ($<33\%$), except for the half-saturation coefficients for substrate ($K_{SB,OHO}$) and nitrate ($K_{NOx,OHO}$) and the autotrophic decay rate (b_{ANO}).

F.2.3 Discussion

Original vs. proposed new default parameter sets: The need to change the ASM1 model structure by introducing a heterotrophic yield under anoxic conditions ($Y_{OHO,Ax}$) to properly model the nitrate and COD consumption was experimentally proven by Orhon *et al.* (1996). A new default value of 0.54 g X_{OHO}.g XC_B^{-1} is proposed by Choubert *et al.* (2009a), based on full-scale modelling studies.

The change of the maximum autotrophic growth rate ($\mu_{ANO,Max}$) and decay rate (b_{ANO}) is discussed in Dold *et al.* (2005). The authors showed that it was no longer necessary to modify $\mu_{ANO,Max}$ when the sludge retention time (SRT) varies if a higher b_{ANO} value is used (experimentally measured to 0.19 ± 0.4 d^{-1}). Choubert *et al.* (2009b) proposed the values of $\mu_{ANO,Max} = 0.8$ d^{-1} and $b_{ANO} = 0.17$ d^{-1} at 20°C as new default values validated on 13 full-scale WWTPs in France.

Parameters changed in modelling projects (compared to original values): Similar to the proposed new default parameter sets a reduced heterotrophic growth rate (Y_{OHO}) is often associated with plants with anoxic and/or anaerobic zones. This confirms the need to differentiate aerobic and anoxic growth yields.

The couple ($\mu_{ANO,Max}$, b_{ANO}) is modified in most studies. However, in the analysed modelling projects a high maximum growth rate was not always compensated by a high decay rate.

In addition, the temperature correction factor values are sometimes re-evaluated in the course of a project. They are deduced from the parameter determination at a different temperature and therefore include measurement uncertainties.

Parameter ranges and statistics: The ranges provided by the 25th and 75th percentiles of the database are generally in agreement with other overview studies, which ranges were not included in the database (Weijers and Vanrolleghem, 1997; Bornemann *et al.* 1998; Hulsbeek *et al.* 2002; Cox, 2004 and Sin *et al.* 2009). However, the ranges from the database differ from these studies for the following parameters:

- $\mu_{OHO,Max}$ and b_{OHO} ranges proposed by Weijers and Vanrolleghem (1997) are wider (respectively 2–10 d^{-1}; 0.1–1.5 d^{-1});

- $K_{SB,OHO}$, b_{OHO} and $K_{NHx,ANO}$ in Bornemann *et al.* (1998) have different and not overlapping ranges (respectively 1–5 g $S_B.m^{-3}$; 0.3-0.5 d^{-1}; 0.1–0.7 g $S_{NHx}.m^{-3}$);
- The median values provided by Cox (2004) are quite different from the database ones (up to 100% of relative difference); whereas the 25th and 75th percentiles are in agreement. An exception is for the heterotrophic growth and decay rates ($\mu_{OHO,Max}$, b_{OHO}) and the half-saturation coefficient for substrate ($K_{SB,OHO}$), for which the ranges provided by Cox (2004) are not overlapping the database ones (respectively 2.06–4.69 d^{-1}; 0.2–0.6 d^{-1}; 2.54–7.06 g $S_B.m^{-3}$);
- Sin *et al.* (2009) provided "uncertainties" (or better variabilities) based on expert knowledge. Two parameter variabilities ($\mu_{ANO,Max}$, b_{ANO}) are narrower than the observed variability in this study (respectively 5% and 25%) and 8 much wider (50% of variability for i_{N_XBio}, $K_{O2,OHO}$, q_{am}, $K_{NHx,ANO}$; 25% of variability for $K_{XCB,hyd}$, $\mu_{OHO,Max}$, $\eta_{\mu OHO,Ax}$, $K_{O2,ANO}$).

It is noticeable that the above mentioned parameters correspond to the ones with the greatest variability in Table F.1 and/or to those modified in more than 50% of the cases, although the observed variations of these parameters are often lower than those provided in these studies.

Finally, all of the overview studies present a parameter range or an "uncertainty" for the autotrophic yield (Y_{ANO}), whereas its value was modified in none of the 22 modelling projects.

Conclusion

Regarding ASM1, six parameters have been pointed out as subject to changes: Y_{OHO}, $K_{SB,OHO}$, $K_{NOx,OHO}$, $\mu_{ANO,Max}$, b_{ANO}, and $K_{NHx,ANO}$. In addition to the variability of Y_{OHO}, $\mu_{ANO,Max}$, and b_{ANO} already discussed, the three other parameters are half-saturation coefficients, suspected to depend on environmental conditions. These results are supported by the literature data although the chosen 25th and 75th percentiles provide a narrower range for some of the parameters than specified in literature.

F.2.3.1 ASM2d

Data description

The database contains 20 parameter sets for ASM2d, of which 2 are "proposed new default parameter sets" and 18 are "optimised parameter sets" from specific modelling projects. The modelling studies were mainly carried out in Europe (16), with only two applications in Asia; and mainly on full scale WWTPs (12). Table F.2 synthesises the main results for ASM2d. The sludge ages of the specific modelling projects are between 7 and 22 days.

Original vs. proposed new default parameter sets: Only the original parameter set is presented. A new default parameter set was proposed by Cinar *et al.* (1998) but it concerns in fact ASM2 and not ASM2d.

Parameters changed in modelling projects (compared to original values): The majority of the parameters are kept at their original values, from which 33 (of the 83 parameters) have never been changed:

- 4 of the 11 stoichiometric parameters: the inert fractions generated in hydrolysis and biomass decay processes ($f_{SU\_XCB,hyd}$, $f_{XU\_Bio,lys}$); the yield of polyphosphate storage per organic stored compound used (Y_{PHA_PP}) and the autotrophic growth yield (Y_{ANO}).
- 7 of the 15 conversion coefficients: i_{N_SF}, i_{N_XBio}, i_{P_SF}, i_{P_SU}, i_{TSS_XCB}, $i_{TSS\_XPAO,PHA}$ and $i_{TSS\_XPAO,PP}$.
- 22 of the 57 kinetic parameters: the alkalinity half-saturation parameters ($K_{Alk,OHO}$, $K_{Alk,PAO}$, $K_{Alk,ANO}$); heterotrophic half-saturation parameters for nutrients ($K_{NHx,OHO}$, $K_{PO4,OHO}$); autotrophic half-saturation parameters for nutrients ($K_{PO4,ANO}$); 5 phosphorus accumulating organism half-saturation parameters (K_{S,fPP_PAO}, $K_{O2,PAO}$, $K_{NOx,PAO}$, $K_{NHx,PAO}$, $K_{PO4,PAO,upt}$); the

Table F.2 Synthesis of database results for ASM2d model, only modified parameters are mentioned. Parameter values are standardised at a temperature of 20°C.

Parameter*	Unit	Description	Original notation	Original parameter set	Optimised parameter sets					
				j	n	Modif. >50%	Median	Perc. 25%	Perc. 75%	V (%)
Kinetic parameters										
Hydrolysis										
$\eta_{qhyd,Ax}$	–	Correction factor for hydrolysis under anoxic conditions	η_{NO3}	0.6	20		0.6	0.6	0.8	33
$\eta_{qhyd,An}$	–	Correction factor for hydrolysis under anaerobic conditions	η_{fe}	0.4	20		0.4	0.2	0.4	50
Ordinary Heterotrophic Organisms										
$\mu_{OHO,Max}$	d^{-1}	Maximum growth rate of X_{OHO}	μ_H	6	20	X	6	4	6	33
$\eta_{\mu OHO,Ax}$	–	Reduction factor for anoxic growth of X_{OHO}	η_{NO3}	0.8	20		0.8	0.8	0.8	0
Phosphorus Accumulating Organisms										
q_{PAO,VFA_Stor}	$g\ X_{Stor} \cdot g\ X_{PAO}^{-1} \cdot d^{-1}$	Rate constant for S_{VFA} uptake rate ($X_{PAO,Stor}$ storage)	q_{PHA}	3	20	X	3.4	3	6	90
$q_{PAO,PO4\_PP}$	$g\ X_{PP} \cdot g\ X_{PAO}^{-1} \cdot d^{-1}$	Rate constant for storage of $X_{PAO,PP}$	q_{PP}	1.5	20	X	1.5	1.5	3.3	120
$\mu_{PAO,Max}$	d^{-1}	Maximum growth rate of X_{PAO}	μ_{PAO}	1	20		1	1	1.04	4
$\theta_{\mu PAO,Max}$	–	Temperature correction factor for $\mu_{PAO,Max}$	$\theta_{\mu PAO}$	1.041	3		1.041	1.041	1.058	2
m_{PAO}	d^{-1}	Endogenous respiration rate of X_{PAO}	b_{PAO}	0.2	20	X	0.2	0.15	0.2	25
b_{PP_PO4}	d^{-1}	Rate constant for lysis of $X_{PAO,PP}$	b_{PP}	0.2	20	X	0.2	0.15	0.2	25
b_{Stor_VFA}	d^{-1}	Rate constant for respiration of $X_{PAO,Stor}$	b_{PHA}	0.2	20	X	0.2	0.15	0.2	25
Autotrophic Nitrifying Organisms										
$\mu_{ANO,Max}$	d^{-1}	Maximum growth rate of X_{ANO}	μ_{AUT}	1	20	X	1	1	1.15	15
b_{ANO}	d^{-1}	Decay rate for X_{ANO}	b_{AUT}	0.15	20		0.15	0.15	0.16	7
$K_{NHx,ANO}$	$g\ S_{NHx} \cdot m^{-3}$	Half-saturation coefficient for S_{NHx}	K_{NH4}	1	20	X	1	0.5	1	50

j: Henze *et al.* (2000b). Please refer to the additional material section for the parameter definitions.
*Standardised notation from Corominas *et al.*(2010) is used. n: number of parameter values in the database

half-saturation parameters for dissolved oxygen and nitrates in the hydrolysis process ($K_{O2,hyd}$, $K_{NOx,hyd}$); 6 of the 12 temperature correction factors ($\theta_{q\_XCB\_SB,hyd}$, $\theta_{\mu\_OHO,Max}$, $\theta_{q\_SF\_Ac,Max}$, θ_{b_OHO}, $\theta_{\mu\_ANO,Max}$, θ_{b_ANO}); and the chemical phosphorus precipitation parameters ($q_{P,pre}$, $q_{P,red}$, $K_{Alk,pre}$).

Two types of modelling studies could be distinguished:

- Studies with a calibrated parameter subset (12 studies). These are mainly composed of kinetic parameters;
- Studies with measured parameters (6 studies), among which 4 studies use the calibration protocol of Penya-Roja *et al.* (2002) (Penya-Roja *et al.* 2002; 2 by Ferrer *et al.* 2004; Garcia-Usach *et al.* 2006). This protocol is based on batch tests that allow the measurement of many stoichiometric and kinetic coefficients for autotrophs, ordinary heterotrophs and phosphorus accumulating organisms.

Among the 18 modelling studies, 8 parameters - all of which are kinetic parameters - were changed in more than half of the cases: the heterotrophic and autotrophic maximum growth rates ($\mu_{OHO,Max}$, $\mu_{ANO,Max}$), the autotrophic half-saturation coefficient for ammonia ($K_{NHx,ANO}$), the rate constants for volatile fatty acids (VFA) uptake (q_{PAO,VFA_Stor}) and for polyphosphate storage ($q_{PAO,PO4\_PP}$) of the PAO and their storage pools' decay (m_{PAO}, b_{PP_PO4}, b_{Stor_VFA}).

Parameter ranges and statistics: The median values are the same as the original publication values except for the rate constant for VFA uptake (qPAO,VFA_Stor). The ranges of kinetic parameter values between 25th and 75th percentiles are quite narrow (<33%), except for the reduction factor for hydrolysis under anaerobic conditions (ηqhyd,Aη), for the rate constants for VFA uptake (qPAO,VFA_Stor) and polyphosphate storage (qPAO,PO4_PP) and the half-saturation coefficient for ammonia (KNHx,ANO).

F.2.3.2 Discussion

Parameters changed in modelling projects (compared to original values): Among the eight parameters that were changed most, two have a particularly wide range of values: the rate constants for VFA uptake (qPAO, VFA_Stor) and polyphosphate storage (qPAO,PO4_PP). Furthermore, the users of the Penya-Roja *et al.* (2002) protocol observed large parameter ranges for PAO growth and polyphosphates storage yields (YPAO, YPP_Stor,PAO). This could indicate a problem in the ASM2d model structure, such as the simplification of not taking into account glycogen storage and glycogen accumulating organisms in ASM2d (Penya-Roja *et al.* 2002).

Another explanation could be that the ASM2d model describes polyphosphate uptake and the growth of PAOs as two independent kinetic processes. However, experimental results show that oxidation of organic stored compounds provides energy for both PAO growth and polyphosphate storage (Wentzel *et al.* 1989). Consequently, PAO growth and polyphosphate storage yield are linked and depend on the oxidation of organic stored compounds. Therefore some models link both yields to energy production (in metabolic models, e.g.: Meijer, 2004) or model PAO growth and polyphosphate storage as a single process (Barker & Dold model, UCTPHO+). Fixing the ratio between growth and phosphate storage would then assist ASM2d calibration.

Parameter ranges and statistics: Based on expert knowledge, Brun *et al.* (2002) assigned an uncertainty to each parameter. The database is in agreement with the low uncertainties (between 5% and 20%) attributed to stoichiometric parameters and conversion coefficients parameters by Brun *et al.* (2002). In Brun *et al.* (2002), high uncertainty is attributed to kinetic parameters (between 20% and 50% of uncertainty), which are overestimated based on the database results for all the parameters, except for the reduction factors for hydrolysis under anoxic and anaerobic conditions (ηqhyd,Ax, ηqhyd,An), the rate constants for VFA uptake (qPAO,VFA_Stor) and polyphosphate storage (qPAO,PO4_PP).

Conclusion

The main potential pitfalls in calibrating ASM2d seem to come from the determination of the rate constants for VFA uptake (qPAO,VFA_Stor) and polyphosphate storage (qPAO,PO4_PP). These two parameters are used in organic compound storage and consumption processes, and their high variability could indicate a problem in the model structure leading to difficulties in the calibration process.

F.3 GENERAL DISCUSSION

F.3.1 Inter-model comparison

In both ASM1 and ASM2d, few parameters have been changed in more than half the cases considered. This shows that either model users are in most cases relying on the original values, or that the model outputs are not sensitive to these parameters. In an inter-model comparison taking into account the results for other models presented in the additional material section at the end of this appendix (ASM3, ASM3 + BioP, ASM2d + TUD, Barker & Dold), the following parameters are most often modified:

- growth and decay rates of autotrophs,
- PAO storage processes rates,
- Heterotrophic half-saturation coefficients for substrate and oxygen
- Autotrophic half-saturation coefficient ammonia.

The half-saturation coefficients are thought to be dependent on specific environmental conditions.

Several modelling protocols suggest measuring some kinetic and stoichiometric parameters: WERF (Melcer *et al.* 2003), BIOMATH (Vanrolleghem *et al.* 2003) and HSG (Langergraber *et al.* 2004). However, in current practice few, if any, biokinetic parameters are measured.

F.3.2 Limitations of modelling project articles

The large literature review on modelling projects revealed that important information is often missing from these articles to enable them to be fully used. Lacking information included:

- information on plant: tank configuration, tank dimensions, aeration time;
- information on environmental conditions: temperature, rain events, diurnal variations;
- information on measurement campaign: duration, number of samples, measurement methods;
- information on influent characteristics and characterisation method used;
- method used for data validation and reconciliation;
- method used to optimise the parameter set: protocol, parameters set to original value;
- temperature for which the optimised parameter set is provided.

This lack of information prevents further analysis of the database, such as an investigation of correlations. It also makes it difficult to evaluate the quality of the modelling projects. Thus, the modelling projects included in the database had to be considered to be of equal quality. The differences within parameter values therefore are supposed to be linked to the WWTP conditions and not to a wrong calibration or poor data quality.

It should also be noted that there seems to be a lack of (published) experimental data with respect to parameter values. If parameters were measured it is often difficult to evaluate the results due to missing information on the measurement method.

F.3.3 Potential use of the database

A number of correlations were searched for in the database including: correlations between parameters; between changed values; between parameters and WWTP conditions (Food/Microorganism ratio, nitrogen loads, Sludge Retention Time). No significant correlations were found which is probably due to the limited number of datasets.

The database has been designed to allow future extensions with new data sets. A larger database could allow further analysis to:

- determine model parameter ranges and typical values to define current practice and help model users in the calibration step for the commonly used models;
- provide a synthesis of practical modelling experiences that could help model users to find appropriate case studies similar to their simulation project;
- examine correlation between parameters and analyse its impact on calibration practice (e.g. b_{ANO} and μ_{ANO});
- examine correlations between changes in parameter values and WWTP conditions;
- determine practical model limits from various modelling experiences;
- identify research needs.

CONCLUSION

This study synthesises the practical knowledge of activated sludge models through a database that combines experience from modelling projects and expert knowledge. For now, this database provides parameter ranges for ASM1 and ASM2d. These values should help model users in the calibration step, by showing typical practice in model calibration. However, these parameters should be used with great care since they are the result of averaging practical experience without taking into account parameter correlations or specific environmental conditions.

These results contribute to the knowledge transfer on activated sludge modelling that was requested by respondents of the first survey. This database can be expanded with more modelling projects which would enable further analysis to be carried out. The authors would like to make this database accessible to the AS modelling community and several solutions are currently under study.

The questionnaire provides further information that has not been presented in this study, but will be included in the Scientific and Technical Report of the Good Modelling Practice – Guidelines for use of activated sludge models (GMP-TG), such as typical ratios and key numbers currently measured at WWTPs.

Acknowledgement

The GMP Task Group kindly thanks all the respondents of this questionnaire for their valuable contribution.

Hélène Hauduc is a PhD student at Cemagref (France) and modelEAU (Université Laval, Québec, Canada).

Peter A. Vanrolleghem holds the Canada Research Chair in Water Quality Modelling.

The GMP Task Group is sponsored by IWA and its home institutions (Black & Veatch, Cemagref, EnviroSim Associates Ltd., Kurita Water Industries Ltd., TU Vienna, Université Laval, École Polytechnique de Montreal, EAWAG). The GMP task group would like to especially thank its sponsors Hydromantis Inc. and MOSTforWATER N.V.

REFERENCES

Barker P.S. and Dold P.L. (1997). General model for biological nutrient removal activated-sludge systems: Model presentation. *Water Environment Research*, **69**(5), 969–984.

Bornemann C., Londong J., Freund M., Nowak O., Otterpohl R. and Rolfs T. (1998). Hinweise zur dynamischen Simulation von Belebungsanlagen mit dem Belebtschlammodell Nr. 1 der IAWQ. Korrespondenz Abwasser, **45** (3), 455–462, [in German].

Brun R., Kuehni M., Siegrist H., Gujer W. and Reichert P. (2002). Practical identifiability of ASM2d parameters - Systematic selection and tuning of parameter subsets. Water Research, **36**(16), 4113–4127.

Choubert J.-M., Marquot A., Stricker A.-E., Racault Y., Gillot S. and Héduit A. (2009a). Anoxic and aerobic values for the yield coefficient of the heterotrophic biomass: Determination at full-scale plants and consequences on simulations. *Water SA*, **35**(1), 103–110.

Choubert J.-M., Stricker A.-E., Marquot A., Racault Y., Gillot S. and Héduit A. (2009b). Updated Activated Sludge Model n1 parameter values for improved prediction of nitrogen removal in activated sludge processes: Validation at 13 full-scale plants. *Water Environment Research*, **81**, 858–865.

Cinar O., Daigger G.T. and Graef S.P. (1998). Evaluation of IAWQ Activated Sludge Model No. 2 using steady-state data from four full-scale wastewater treatment plants. *Water Environment Research*, **70**(6), 1216–1224.

Corominas L., Rieger L., Takács I., Ekama G., Hauduc H., Vanrolleghem P. A., Oehmen A., Gernaey K.V. and Comeau Y. (2010). New framework for standardized notation in wastewater treatment modelling. *Water Science and Technology*, **61**(4), 841–857.

Cox C.D. (2004). Statistical distributions of uncertainty and variability in activated sludge model parameters. *Water Environment Research*, **76**(7), 2672–2685.

Dold P.L., Jones R.M. and Bye C.M. (2005). Importance and measurement of decay rate when assessing nitrification kinetics. *Water Science and Technology*, **52**(10–11), 469–477.

Ferrer J., Morenilla J.J., Bouzas A. and Garcia-Usach F. (2004). Calibration and simulation of two large wastewater treatment plants operated for nutrient removal. *Water Science and Technology*, **50**(6), 87–94.

Garcia-Usach F., Ferrer J., Bouzas A. and Seco A. (2006). Calibration and simulation of ASM2d at different temperatures in a phosphorus removal pilot plant. *Water Science and Technology*, **53**(12), 199–206.

Grady C.P.L., Jr, Daigger G.T. and Lim H.C. (1999). Biological Wastewater Treatment - Second Edition, revised and expanded, 1076 pp., Marcel Dekker, New York.

Gujer W., Henze M., Mino T. and van Loosdrecht M.C.M. (2000). Activated Sludge Model No.3, in Activated Sludge Models ASM1, ASM2, ASM2d and ASM3. edited byM. Henze et al., IWA Publishing, London, UK.

Hauduc H., Gillot S., Rieger L., Ohtsuki T., Shaw A., Takács I. and Winkler S. (2009). Activated Sludge Modelling in Practice - An International Survey. *Water Science and Technology*, **60**(8), 1943–1951.

Hauduc H., Rieger L., Takács I., Héduit A., Vanrolleghem P.A. and Gillot S. (2010). Systematic approach for model verification – Application on seven published Activated Sludge Models. *Water Science and Technology*, **61**(4), 825–839.

Henze M., Grady C.P.L., Jr, Gujer W., Marais G.v.R. and Matsuo T. (2000a). Activated Sludge Model No.1, in Activated Sludge Models ASM1, ASM2, ASM2d and ASM3. edited by M. Henze et al., IWA Publishing, London, UK.

Henze M., Gujer W., Mino T., Matsuo T., Wentzel M.C., Marais G.v.R. and van Loosdrecht M.C.M. (2000b). Activated Sludge Model No.2d, in Activated Sludge Models ASM1, ASM2, ASM2d and ASM3. edited by M. Henze et al., IWA Publishing, London, UK.

Hu Z.R., Wentzel M.C. and Ekama G.A. (2007). A general kinetic model for biological nutrient removal activated sludge systems: Model development. *Biotechnology and Bioengineering*, **98**(6), 1242–1258.

Hulsbeek J.J.W., Kruit J., Roeleveld P.J. and van Loosdrecht M.C.M. (2002). A practical protocol for dynamic modelling of activated sludge systems. *Water Science and Technology*, **45**(6), 127–136.

Koch G., Kühni M., Gujer W. and Siegrist H. (2000). Calibration and validation of Activated Sludge Model no. 3 for Swiss municipal wastewater. *Water Research*, **34**(14), 3580–3590.

Langergraber G., Rieger L., Winkler S., Alex J., Wiese J., Owerdieck C., Ahnert M., Simon J. and Maurer M. (2004). A guideline for simulation studies of wastewater treatment plants. *Water Science and Technology*, **50**(7), 131–138.

Makinia J. and Wells S.A. (2000). A general model of the activated sludge reactor with dispersive flow - I. Model development and parameter estimation. *Water Research*, **34**(16), 3987–3996.

Marquot A. (2006). Modelling nitrogen removal by activated sludge on full-scale plants: Calibration and evaluation of ASM1. PhD thesis, Ecole doctorale des Sciences Exactes et de leurs Applications, Université de Pau et des Pays de l'Adour (UPPA), Bordeaux, France.

Meijer S.C.F. (2004). Theoretical and practical aspects of modelling activated sludge processes. PhD thesis, Department of Biotechnological Engineering, Delft University of Technology, Delft, The Netherlands.

Melcer H., Dold P.L., Jones R.M., Bye C.M., Takacs I., Stensel H.D., Wilson A.W., Sun P. and Bury S. , (2003). Methods for wastewater characterisation in activated sludge modeling, Water Environment Research Foundation (WERF), Alexandria, VA, US.

Nuhoglu A., Keskinler B. and Yildiz E. (2005). Mathematical modelling of the activated sludge process - The Erzincan case. *Process Biochemistry*, **40**(7), 2467–2473.

Orhon D., Sozen S. and Artan N. (1996). The effect of heterotrophic yield on the assessment of the correction factor for anoxic growth. *Water Science and Technology*, **34**(5-6), 67–74.

Penya-Roja J.M., Seco A., Ferrer J. and Serralta J. (2002). Calibration and validation of Activated Sludge Model No.2d for Spanish municipal wastewater. *Environmental Technology*, **23**(8), 849–862.

Petersen B., Gernaey K., Henze M. and Vanrolleghem P.A. (2002). Evaluation of an ASM1 model calibration procedure on a municipal-industrial wastewater treatment plant. *J. Hydroinformatics*, **4**(1), 15–38.

Rieger L., Koch G., Kuhni M., Gujer W. and Siegrist H. (2001). The EAWAG Bio-P module for Activated Sludge Model No. 3. *Water Research*, **35**(16), 3887–3903.

Sin G., Gernaey K.V., Neumann M.B., van Loosdrecht M.C.M. and Gujer W. (2009). Uncertainty analysis in WWTP model applications: A critical discussion using an example from design. *Water Research*, **43**(11), 2894–2906.

Spanjers H., Vanrolleghem P. A., Nguyen K., Vanhooren H. and Patry G.G. (1998). Towards a simulation-benchmark for evaluating respirometry-based control strategies. *Water Sci. Technol.*, **37**(12), 219–226.

Stamou A., Katsiri A., Mantziaras I., Boshnakov K., Koumanova B. and Stoyanov S. (1999). Modelling of an alternating oxidation ditch system. *Water Science and Technology*, **39**(4), 169–176.

Vanrolleghem P.A., Insel G., Petersen B., Sin G., Pauw D.D., Nopens I., Dovermann H., Weijers S. and Gernaey K. (2003). A comprehensive model calibration procedure for activated sludge models. In: Proceedings 76th Annual WEF Conference and Exposition. Los Angeles, USA, October 11-15, 2003. (on CD-ROM).

Weijers S.R. and Vanrolleghem P.A. (1997). A procedure for selecting best identifiable parameters in calibrating Activated Sludge Model No. 1 to full-scale plant data. *Water Science and Technology*, **36**(5), 69–79.

Wentzel M.C., Ekama G.A., Loewenthal R.E., Dold P.L. and Marais G.v.R. (1989). Enhanced polyphosphate organism cultures in activated sludge systems. Part II: Experimental behaviour. *Water S.A.*, **15**(2), 71–88.

ADDITIONAL MATERIAL
Parameter definitions

ASM3

Data description

The database contains 5 parameter sets for ASM3, of which 1 is a "proposed new default parameter sets", and 3 are "optimised parameter sets". The modelling studies were exclusively carried out in the North of Europe (Belgium, Finland, Germany) on full scale WWTPs. Table F.4 synthesises the main results for ASM3.

ASM3 + BioP

Data description

The database contains 9 parameter sets for ASM3 + BioP, 1 original parameter set and 8 optimised parameter sets. The modelling studies were exclusively carried out in Germany. Half of them were carried out on full scale WWTPs.

Table F.3 Parameter definitions and original notation of the studied models.

Description	Parameter*	Unit	ASM1	ASM2d	ASM3	ASM3 +BioP	Barker and Dold
State Variables							
COD soluble							
Soluble biodegradable organics	S_B	g COD · m^{-3}	S_S		S_S	S_S	
Fermentable organic matter	S_F	g COD · m^{-3}		S_F			S_{BSC}
Fermentation product (volatile fatty acids)	S_{VFA}	g COD · m^{-3}		S_A			S_{BSA}
Soluble undegradable organics	S_U	g COD · m^{-3}	S_I	S_I	S_I	S_I	S_{US}
Dissolved oxygen	S_{O2}	- g COD · m^{-3}	S_O	S_{O2}	S_O	S_O	S_O
COD particulate and colloidal							
Particulate biodegradable organics	X_{CB}	g COD · m^{-3}	X_S	X_S	X_S	X_S	S_{ENM}
Adsorbed slowly biodegradable substrate	X_{Ads}	g COD · m^{-3}					
Particulate undegradable organics	X_U	g COD · m^{-3}		X_I	X_I	X_I	
Particulate undegradable organics from the influent	$X_{U,Inf}$	g COD · m^{-3}	X_I				S_{UP}
Particulate undegradable endogenous products	$X_{U,E}$	g COD · m^{-3}	X_P				Z_E
Nitrogen and Phosphorus							
Ammonium and ammonia nitrogen (NH_4+NH_3)	S_{NHx}	g N.m^{-3}	S_{NH}	S_{NH4}	S_{NH}	S_{NH}	N_{H3}
Nitrate and nitrite (NO_3+NO_2) (considered to be NO_3 only for stoichiometry)	S_{NOx}	g N.m^{-3}	S_{NO}	S_{NO3}	S_{NO}	S_{NO}	N_{O3}
Particulate biodegradable organic N	$X_{CB,N}$	g N.m^{-3}	X_{ND}				N_{BP}
Soluble biodegradable organic N	$S_{B,N}$	g N.m^{-3}	S_{ND}				N_{BS}
Soluble inert organic N	$S_{U,N}$	g N.m^{-3}					N_{US}
Soluble inorganic phosphorus	S_{PO4}	g P.m^{-3}		S_{PO4}		S_{PO4}	P_{O4}
Biomasses							
Ordinary heterotrophic organisms	X_{OHO}	g COD · m^{-3}	X_{BH}	X_H	X_H	X_H	Z_H

Autotrophic nitrifying organisms (NH_4 to NO_3)	X_{ANO}	$g\ COD \cdot m^{-3}$	X_{BA}	X_{AUT}	X_A	X_A	Z_A
Phosphorus accumulating organisms	X_{PAO}	$g\ COD \cdot m^{-3}$		X_{PAO}		X_{PAO}	Z_P
Organisms (biomass)	X_{Bio}	$g\ COD \cdot m^{-3}$					
Storage compound in OHOs	$X_{OHO,Stor}$	$g\ COD \cdot m^{-3}$			X_{STO}	X_{STO}	
Storage compound in PAOs	$X_{PAO,Stor}$	$g\ COD \cdot m^{-3}$		X_{PHA}		X_{PHA}	S_{PHB}
Stored glycogen in PAOs	$X_{PAO,Gly}$	$g\ COD \cdot m^{-3}$					
Stored polyphosphates in PAOs	$X_{PAO,PP}$	$g\ P.m^{-3}$		X_{PP}		X_{PP}	
Stoichiometric parameters							
Yield for X_{OHO} growth	Y_{OHO}	$g\ X_{OHO}.g\ XC_B^{-1}$	Y_H				
Yield for X_{OHO} growth per $X_{OHO,Stor}$ (Aerobic)	$Y_{Stor\_OHO,Ox}$	$g\ X_{OHO}.g\ X_{Stor}^{-1}$			$Y_{H,O2}$		
Yield for X_{OHO} growth per $X_{OHO,Stor}$ (Anoxic)	$Y_{Stor\_OHO,Ax}$	$g\ X_{OHO}.g\ X_{Stor}^{-1}$			$Y_{H,NOX}$		
Yield for $X_{OHO,Stor}$ formation per S_B (Aerobic)	$Y_{SB\_Stor,Ox}$	$g\ X_{Stor}.g\ S_B^{-1}$			$Y_{STO,O2}$		
Yield for $X_{OHO,Stor}$ formation per S_B (Anoxic)	$Y_{SB\_Stor,Ax}$	$g\ X_{Stor}.g\ S_B^{-1}$			$Y_{STO,NOX}$		
Conversion coefficient							
N content of S_U	i_{N_SU}	$g\ N \cdot g\ S_U^{-1}$					$f_{N,SEP}$
N content of X_U	i_{N_XU}	$g\ N \cdot g\ X_U^{-1}$			$i_{N,XI}$		
N content of XC_B	i_{N_XCB}	$g\ N \cdot g\ XC_B^{-1}$			$i_{N,XS}$		
N content of X_{OHO}	i_{N_OHO}	$g\ N \cdot g\ X_{OHO}^{-1}$					$f_{N,ZH}$
N content of biomass (X_{OHO}, X_{PAO}, X_{ANO})	i_{N_XBio}	$g\ N \cdot g\ X_{Bio}^{-1}$	$i_{X,B}$				
N content of products from X_{OHO}	$i_{N\_XUE,OHO}$	$g\ N \cdot g\ X_{UE}^{-1}$					$f_{N,ZEH}$
N content of products from X_{PAO}	$i_{N\_XUE,PAO}$	$g\ N \cdot g\ X_{UE}^{-1}$					$f_{N,ZEP}$
N content of products from X_{ANO}	$i_{N\_XUE,ANO}$	$g\ N \cdot g\ X_{UE}^{-1}$					$f_{N,ZEA}$
Kinetic parameters							
Hydrolysis							
Maximum specific hydrolysis rate	$q_{XCB\_SB,hyd}$	$g\ XC_B.g\ X_{OHO}^{-1}.d^{-1}$	k_h		k_H		
	$\theta_{qXCB\_SB,hyd}$	–	θ_{kh}				

(*Continued*)

Table F.3 Parameter definitions and original notation of the studied models (*Continued*).

Description	Parameter*	Unit	ASM1	ASM2d	ASM3	ASM3 +BioP	Barker and Dold
Temperature correction factor for $q_{XCB\_SB,hyd}$							
Half-saturation coefficient for XC_B/X_{OHO}	$K_{XCB,hyd}$	$g\ XC_B.g\ X_{OHO}^{-1}$	K_X				
Temperature correction factor for $K_{XCB,hyd}$	$\theta_{KXCB,hyd}$		θ_{KX}				
Correction factor for hydrolysis under anoxic conditions	$\eta_{qhyd,Ax}$	–	η_h	η_{NO3}			η_{gro}
Correction factor for hydrolysis under anaerobic conditions	$\eta_{qhyd,An}$	–		η_{fe}			
Ordinary Heterotrophic Organisms							
Rate constant for $X_{OHO,Stor}$ storage	q_{SB_Stor}	$g\ XC_B.g\ X_{OHO}^{-1}.d^{-1}$			k_{STO}		
Maximum growth rate of X_{OHO}	$\mu_{OHO,Max}$	d^{-1}	μ_H	μ_H	μ_H		
Temperature correction factor for $\mu_{OHO,Max}$	$\theta_{\mu OHO,Max}$	–	$\theta_{\mu H}$				
Reduction factor for anoxic growth of X_{OHO}	$\eta_{\mu OHO,Ax}$	–	η_g	η_{NO3}	η_{NOX}		
Half-saturation coefficient for S_B	$K_{SB,OHO}$	$g\ S_B.m^{-3}$	K_S		K_S		
Half-saturation coefficient $X_{OHO,Stor}/X_{OHO}$	K_{Stor_OHO}	$g\ X_{Stor}.g\ X_{OHO}^{-1}$			K_{STO}		
Decay rate for X_{OHO}	b_{OHO}	d^{-1}	b_H				
Temperature correction factor for b_{OHO}	θ_{bOHO}		θ_{bH}				
Endogenous respiration rate of X_{OHO} (Aerobic)	$m_{OHO,Ox}$	d^{-1}			$b_{H,O2}$		
Endogenous respiration rate of X_{OHO} (Anoxic)	$m_{OHO,Ax}$	d^{-1}			$b_{H,NOX}$		
Endogenous respiration rate of $X_{OHO,Stor}$ (Aerobic)	$m_{Stor,Ox}$	d^{-1}			$b_{STO,O2}$		
Endogenous respiration rate of $X_{OHO,Stor}$ (Anoxic)	$m_{Stor,Ax}$	d^{-1}			$b_{STO,NOX}$		
	$\eta_{mOHO,Ax}$	–				$\eta_{NO,end,H}$	

Reduction factor for anoxic endogenous respiration of X_{OHO}							
Half-saturation coefficient for S_{O2}	$K_{O2,OHO}$	$g\ S_{O2}.m^{-3}$	K_{OH}		K_{O2}	$K_{O,H}$	$K_{O,HET}$
Half-saturation coefficient for S_{NOx}	$K_{NOx,OHO}$	$g\ S_{NOx}.m^{-3}$	K_{NO}				
Phosphorus Accumulating Organisms							
Rate constant for S_{VFA} uptake rate ($X_{PAO,Stor}$ storage)	q_{PAO,VFA_Stor}	$g\ X_{Stor}.g\ X_{PAO}^{-1}.d^{-1}$		q_{PHA}			
Rate constant for storage of $X_{PAO,PP}$	$q_{PAO,PO4\_PP}$	$g\ X_{PP}.g\ X_{PAO}^{-1}.d^{-1}$		q_{PP}		q_{PP}	
Maximum ratio of $X_{PAO,PP}/X_{PAO}$	$f_{PP\_PAO,Max}$	$g\ X_{PP}.g\ X_{PAO}^{-1}$				$K_{max,PAO}$	
Maximum growth rate of X_{PAO}	$\mu_{PAO,Max}$	d^{-1}		μ_{PAO}			
Temperature correction factor for $\mu_{PAO,Max}$	$\theta_{\mu PAO,Max}$	–		$\theta_{\mu PAO}$			
Endogenous respiration rate of X_{PAO}	m_{PAO}	d^{-1}		b_{PAO}			
Rate constant for lysis of $X_{PAO,PP}$	b_{PP_PO4}	d^{-1}		b_{PP}			
Rate constant for respiration of $X_{PAO,Stor}$	b_{Stor_VFA}	d^{-1}		b_{PHA}			
Half-saturation coefficient for $X_{PAO,PP}$	$K_{PP,PAO}$	$g\ X_{PP}.m^{-3}$					K_{XP}
Autotrophic Nitrifying Organisms							
Maximum growth rate of X_{ANO}	$\mu_{ANO,Max}$	d^{-1}	μ_A	μ_{AUT}	μ_A	μ_A	μ_A
Temperature correction factor for $\mu_{ANO,Max}$	$\theta_{\mu ANO,Max}$	–	$\theta_{\mu A}$				
Decay rate for X_{ANO}	b_{ANO}	d^{-1}	b_A	b_{AUT}			b_A
Temperature correction factor for b_{ANO}	θ_{bANO}		θ_{bA}				
Endogenous respiration rate for X_{ANO} (Aerobic)	$m_{ANO,Ox}$	d^{-1}			$b_{A,O2}$		
Endogenous respiration rate for X_{ANO} (Anoxic)	$m_{ANO,Ax}$	d^{-1}			$b_{A,NOX}$		
Rate constant for ammonification	q_{am}	$M^3.g\ XC_{B,N}^{-1}.d^{-1}$	k_a				
Temperature correction factor for q_{am}	θ_{qam}	–	θ_{ka}				
Half-saturation coefficient for S_{O2}	$K_{O2,ANO}$	$g\ S_{O2}.m^{-3}$	K_{OA}			$K_{O,A}$	$K_{O,AUT}$
Half-saturation coefficient for S_{NHx}	$K_{NHx,ANO}$	$g\ S_{NHx}.m^{-3}$	K_{NH}	K_{NH4}	$K_{A,NH4}$		K_{NH}

*Standardised notation from Corominas *et al.* (2010) is used.

Table F.4 Synthesis of database results for ASM3 model, only modified parameters are given.

Parameter*	Unit	Original parameter set		Proposed new default parameter set	Optimised parameter sets					
Parameter sets		Notation	Value (k)	I	n	Modif >50%	Median	Perc. 25%	Perc. 75%	V (%)
Stoichiometric parameters										
$Y_{Stor\_OHO,Ox}$	$g\ X_{OHO} \cdot g\ X_{Stor}^{-1}$	$Y_{H,O2}$	0.85	0.8	5		0.80	0.80	0.80	0
$Y_{Stor\_OHO,Ax}$	$g\ X_{OHO} \cdot g\ X_{Stor}^{-1}$	$Y_{H,NOX}$	0.8	0.7	5		0.70	0.70	0.70	0
$Y_{SB\_Stor,Ox}$	$g\ X_{Stor} \cdot g\ S_B^{-1}$	$Y_{STO,O2}$	0.63	0.8	5		0.80	0.63	0.80	21
$Y_{SB\_Stor,Ax}$	$g\ X_{Stor} \cdot g\ S_B^{-1}$	$Y_{STO,NOX}$	0.54	0.65	5		0.65	0.54	0.65	17
Conversion coefficient										
i_{N_XU}	$g\ N \cdot g\ X_U^{-1}$	$i_{N,XI}$	0.02	0.04	5		0.040	0.035	0.040	13
i_{N_XCB}	$g\ N \cdot g\ X_{CB}^{-1}$	$i_{N,XS}$	0.04	0.03	5		0.030	0.030	0.030	0
Kinetic parameters										
Hydrolysis										
$q_{XCB\_SB,hyd}$	$g\ X_{CB}.g\ X_{OHO}^{-1}.d^{-1}$	k_H	3	9	5		9.0	3.0	9.0	67
q_{SB_Stor}	$g\ X_{CB}.g\ X_{OHO}^{-1}.d^{-1}$	k_{STO}	0.1		5		12.0	10.0	12.0	17
Ordinary Heterotrophic Organisms										
$\mu_{OHO,Max}$	d^{-1}	μ_H	2	3	5		3.0	2.0	3.0	33
$\eta_{\mu OHO,Ax}$	–	η_{NOX}	0.6	0.5	5		0.50	0.50	0.60	20
$K_{SB,OHO}$	$g\ S_B.m^{-3}$	K_S	2	10	5		10.0	2.0	10.0	80
K_{Stor_OHO}	$g\ X_{Stor} \cdot g\ X_{OHO}^{-1}$	K_{STO}	1	0.1	5		0.10	0.10	0.10	0
$m_{OHO,Ox}$	d^{-1}	$b_{H,O2}$	0.2	0.3	5		0.30	0.20	0.30	33
$m_{OHO,Ax}$	d^{-1}	$b_{H,,NOX}$	0.1	0.15	5		0.15	0.10	0.15	33
$m_{Stor,Ox}$	d^{-1}	$b_{STO,O2}$	0.2	0.3	5		0.30	0.20	0.30	33
$m_{Stor,Ax}$	d^{-1}	$b_{STO,NOX}$	0.1	0.15	5		0.15	0.10	0.15	33
$K_{O2,OHO}$	$g\ S_{O2}.m^{-3}$	K_{O2}	0.2		5		0.200	0.200	0.500	150
Autotrophic Nitrifying Organisms										
$\mu_{ANO,Max}$	d^{-1}	μ_A	1	1.3	5	X	1.00	1.00	1.30	30
$m_{ANO,Ox}$	d^{-1}	$b_{A,O2}$	0.15	0.2	5		0.20	0.15	0.20	25
$m_{ANO,Ax}$	d^{-1}	$b_{A,NOX}$	0.05	0.1	5		0.10	0.05	0.10	50
$K_{NHx,ANO}$	$g\ S_{NHx}.m^{-3}$	$K_{A,NH4}$	1	1.4	5		1.40	1.00	1.40	29

k: Gujer *et al.* (2000); I: Koch *et al.* (2000).

*Standardised notation from Corominas *et al.* (2010) is used. n: number of parameter values in the database.

Table F.5 Synthesis of database results for ASM3 + BioP model, only modified parameters are mentioned.

Parameter*	Unit	Original notation	Original parameter set	Optimised parameter sets					
Parameter sets			m	n	Modif > 50%	Median	Perc. 25%	Perc. 75%	V (%)
Kinetic parameters									
Ordinary Heterotrophic Organisms									
$\eta_{mOHO,Ax}$	-	$\eta_{NO,end,H}$	0.33	9		0.33	0.33	0.50	52
$K_{O2,OHO}$	$g\ S_{O2}\cdot m^{-3}$	$K_{O,H}$	0.2	9		0.200	0.200	0.500	150
Phosphorus Accumulating Organisms									
$q_{PAO,PO4\_PP}$	$g\ X_{PP}\cdot g\ X_{PAO}^{-1}\cdot d^{-1}$	q_{PP}	1.5	9	X	1.50	1.50	2.30	53
$f_{PP\_PAO,Max}$	$g\ X_{PP}\cdot g\ X_{PAO}^{-1}$	$K_{max,PAO}$	0.2	9	X	1.00	0.24	1.00	76
Autotrophic Nitrifying Organisms									
$\mu_{ANO,Max}$	d^{-1}	μ_A	0.9 - 1.8	9	X	1.20	1.10	1.60	42
$K_{O2,ANO}$	$g\ S_{O2}\cdot m^{-3}$	$K_{O,A}$	0.5	9	X	0.18	0.13	0.50	206

m: Rieger *et al.* (2001).
*Standardised notation from Corominas *et al.* (2010). n: number of parameter values in the database.

Barker and Dold model

The database contains 6 parameter sets for the Barker and Dold model, of which 1 is "proposed new default parameter sets" and 4 are "optimised parameter sets". Two of the modelling studies were carried out in North-America, one in Africa and one in Oceania. The modelling studies mainly concern full scale WWTPs with domestic influent (3).

Table F. 6 Synthesis of database results for Barker and Dold model, only modified parameters are mentioned.

Parameter*	Unit	Original parameter set		Proposed new default parameter set	n	Optimised parameter sets				
Parameter sets		Notation	Value (n)	o		Modif > 50%	Median	Perc. 25%	Perc. 75%	V (%)
Conversion coefficient										
i_{N_SU}	$g\ N\cdot g\ S_U^{-1}$	$f_{N,SEP}$	0.07	0.034	5		0.070	0.034	0.070	51
i_{N_OHO}	$g\ N\cdot g\ X_{OHO}^{-1}$	$f_{N,ZH}$	0.07		6	X	0.069	0.068	0.070	3
$i_{N\_XUE,OHO}$	$g\ N\cdot g\ X_{UE}^{-1}$	$f_{N,ZEH}$	0.07	0.034	6		0.069	0.034	0.070	52
$i_{N\_XUE,PAO}$	$g\ N\cdot g\ X_{UE}^{-1}$	$f_{N,ZEP}$	0.07	0.034	6		0.070	0.034	0.070	51
$i_{N\_XUE,ANO}$	$g\ N\cdot g\ X_{UE}^{-1}$	$f_{N,ZEA}$	0.07	0.034	6		0.068	0.034	0.068	50

(Continued)

Table F. 6 Synthesis of database results for Barker and Dold model, only modified parameters are mentioned (*Continued*).

Parameter*	**Unit**	**Original parameter set**		**Proposed new default parameter set**	**n**	**Optimised parameter sets**				
Parameter sets		**Notation**	**Value (n)**	**o**		**Modif > 50%**	**Median**	**Perc. 25%**	**Perc. 75%**	**V (%)**
Kinetic parameters										
$\eta_{\mu OHO,Ax}$	-	η_{gro}	0.37		6		0.37	0.37	0.50	35
$K_{O2,OHO}$	$g\ S_{O2}.m^{-3}$	$K_{O,HET}$	0.002	0.05	6		0.002	0.002	0.050	2400
Phosphorus Accumulating Organisms										
$K_{PP,PAO}$	$g\ X_{PP}.m^{-3}$	K_{XP}	0.05	0.01	6	X	0.010	0.010	0.010	0
Autotrophic Nitrifying Organisms										
$\mu_{ANO,Max}$	d^{-1}	μ_A	0.6	0.9	6	X	0.73	0.60	0.90	41
b_{ANO}	d^{-1}	b_A	0.04	0.17	6	X	0.08	0.04	0.17	163
$K_{O2,ANO}$	$g\ S_{O2}.m^{-3}$	$K_{O,AUT}$	0.5	0.25	6		0.50	0.25	0.50	50
$K_{NHx,ANO}$	$g\ S_{NHx}.m^{-3}$	K_{NH}	1	0.5	6		1.00	0.50	1.00	50

n: Barker and Dold (1997); o: Questionnaire (based on >100 modelling project studies)
*Standardised notation from Corominas *et al.* (2010) is used. n: number of parameter values in the database.

DATABASE REFERENCES

Abusam A., Keesman K.J., Van Straten G., Spanjers H. and Meinema K. (2001). Sensitivity analysis in oxidation ditch modelling: The effect of variations in stoichiometric, kinetic and operating parameters on the performance indices. *Journal of Chemical Technology & Biotechnology*, **76**(4), 430–438.

Baetens D. (2001). Enhanced biological phosphorus removal: Modelling and experimental design. *PhD thesis*, BIOMATH Faculteit Landbouwkundige en Toegepaste Biologische Wetenschappen, Ghent University, Gent, Belgium.

Barker P.S. and Dold P.L. (1997). General model for biological nutrient removal activated-sludge systems: Model presentation. *Water Environment Research*, **69**(5), 969–984.

Bornemann C., Londong J., Freund M., Nowak O., Otterpohl R. and Rolfs T. (1998). Hinweise zur dynamischen Simulation von Belebungsanlagen mit dem Belebtschlammodell Nr. 1 der IAWQ. *Korrespondenz Abwasser*, **45** (3), 455–462, [in German].

Brun R., Kuehni M., Siegrist H., Gujer W. and Reichert P. (2002). Practical identifiability of ASM2d parameters - Systematic selection and tuning of parameter subsets. *Water Research*, **36**(16), 4113–4127.

Chachuat B., Roche N. and Latifi M.A. (2005). Optimal aeration control of industrial alternating activated sludge plants. *Biochemical Engineering Journal*, **23**(3), 277–289.

Choubert J.-M., Stricker A.-E., Marquot A., Racault Y., Gillot S. and Héduit A. (2009). Updated Activated Sludge Model n1 parameter values for improved prediction of nitrogen removal in activated sludge processes: Validation at 13 full-scale plants. *Water Environment Research*, **81**, 858–865.

Cinar O., Daigger G.T. and Graef S. P. (1998). Evaluation of IAWQ Activated Sludge Model No. 2 using steady-state data from four full-scale wastewater treatment plants. *Water Environment Research*, **70(6)**, 1216–1224.

de la Sota A., Larrea L., Novak L., Grau P. and Henze M. (1994). Performance and model calibration of R-D-N processes in pilot plant. *Water Science and Technology*, **30**(6), 355–364.

Ferrer J., Morenilla J. J., Bouzas A. and Garcia-Usach F. (2004). Calibration and simulation of two large wastewater treatment plants operated for nutrient removal. *Water Science and Technology*, **50**(6), 87–94.

Funamizu N. and Takakuwa T. (1994). Simulation of the operating conditions of the municipal wastewater treatment plant at low temperatures using a model that includes the IAWPRC activated sludge model. *Water Science and Technology*, **30**(4), 105–113.

Garcia-Usach F., Ferrer J., Bouzas A. and Seco A. (2006). Calibration and simulation of ASM2d at different temperatures in a phosphorus removal pilot plant. *Water Science and Technology*, **53**(12), 199–206.

Gokcay C. F. and Sin G. (2004). Modeling of a large-scale wastewater treatment plant for efficient operation. *Water Science and Technology*, **50**(7), 123–130.

Grady C.P.L., Jr, Daigger G.T. and Lim H.C. (1999). Biological Wastewater Treatment - Second Edition, revised and expanded, 1076 pp., Marcel Dekker, New York.

Gujer W., Henze M., Mino T. and Van Loosdrecht M.C.M. (2000). Activated Sludge Model No.3, in Activated Sludge Models ASM1, ASM2, ASM2d and ASM3. edited byM. Henze et al. , IWA Publishing, London, UK.

Henze M., Grady C.P.L., Jr, Gujer W., Marais G.v.R. and Matsuo T. (2000a). Activated Sludge Model No.1, in Activated Sludge Models ASM1, ASM2, ASM2d and ASM3. edited byM. Henze et al. , IWA Publishing, London, UK.

Henze M., Gujer W., Mino T., Matsuo T., Wentzel M.C., Marais G.v.R. and van Loosdrecht M.C.M. (2000b). Activated Sludge Model No.2d, in Activated Sludge Models ASM1, ASM2, ASM2d and ASM3. edited byM. Henze et al. , IWA Publishing, London, UK.

Hu Z. R., Wentzel M.C. and Ekama G.A. (2007a). A general kinetic model for biological nutrient removal activated sludge systems: Model development. *Biotechnology and Bioengineering*, **98**(6), 1242–1258.

Hu Z. R., Wentzel M.C. and Ekama G.A. (2007b). A general kinetic model for biological nutrient removal activated sludge systems: Model evaluation. *Biotechnology and Bioengineering*, **98**(6), 1259–1275.

Hulsbeek J.J.W., Kruit J., Roeleveld P.J. and van Loosdrecht M.C.M. (2002). A practical protocol for dynamic modelling of activated sludge systems. *Water Science and Technology*, **45**(6), 127–136.

Koch G., Kühni M., Gujer W. and Siegrist H. (2000). Calibration and validation of activated sludge model no. 3 for Swiss municipal wastewater. *Water Research*, **34**(14), 3580–3590.

Lagarde F. (2003). Optimisation du traitement de l'azote et du carbone par boues activées en temps de pluie à basse température. *PhD thesis*, Sciences et Techniques de l'Environnement, Université Paris XII Val de Marne, Paris, France. [in French].

Lessard P., Tusseau-Vuillemin M.-H., Héduit A. and Lagarde F. (2007). Assessing chemical oxygen demand and nitrogen conversions in a multi-stage activated sludge plant with alternating aeration. *Journal of Chemical Technology & Biotechnology*, **82**, 367–375.

Lubken M., Wichern M., Rosenwinkel K.H. and Wilderer P.A. (2003). Efficiency of different mathematical models for simulating enhanced biological phosphorus removal in activated sludge systems. *Environmental Informatics Archives*, **1**, 339–347.

Makinia J., Rosenwinkel K.H. and Spering V. (2005a). Long-term simulation of the activated sludge process at the Hanover-Gummerwald pilot WWTP. *Water Research*, **39**(8), 1489–1502.

Makinia J., Rosenwinkel K.H. and Spering V. (2006a). Comparison of two model concepts for simulation of nitrogen removal at a full-scale biological nutrient removal pilot plant. *Journal of Environmental Engineering-ASCE*, **132** (4), 476–487.

Makinia J., Rosenwinkel K.H., Swinarski M. and Dobiegala E. (2006b). Experimental and model-based evaluation of the role of denitrifying polyphosphate accumulating organisms at two large scale WWTPs in northern Poland. *Water Science and Technology*, **54**(8), 73–81.

Makinia J., Swinarski M. and Rosenwinkel K.H. (2005b). Dynamic simulation as a tool for evaluation of the nitrogen removal capabilities at the "Gdansk-Wschod" WWTP. In: Proceedings IWA Specialized Conference - Nutrient Management in Wastewater Treatment Process and Recycle Streams, Krakow, Poland, 19-21 September 2005.

Makinia J. and Wells S.A. (2000). A general model of the activated sludge reactor with dispersive flow - I. Model development and parameter estimation. *Water Research*, **34**(16), 3987–3996.

Marquot A. (2006). Modelling nitrogen removal by activated sludge on full-scale plants: Calibration and evaluation of ASM1. *PhD thesis*, Ecole doctorale des Sciences Exactes et de leurs Applications, Université de Pau et des Pays de l'Adour (UPPA), Bordeaux, France.

Marquot A., Stricker A.-E. and Racault Y. (2006). ASM1 dynamic calibration and long-term validation for an intermittently aerated WWTP. *Water Science and Technology*, **53**(12), 247–256.

Meijer S.C.F. (2004). Theoretical and practical aspects of modelling activated sludge processes. *PhD thesis*, Department of Biotechnological Engineering, Delft University of Technology, Delft, The Netherlands.

Meijer S.C.F., van Loosdrecht M.C.M. and Heijnen J.J. (2001). Metabolic modelling of full-scale biological nitrogen and phosphorus removing WWTP's. *Water Research*, **35**(11), 2711–2723.

Nuhoglu A., Keskinler B. and Yildiz E. (2005). Mathematical modelling of the activated sludge process - The Erzincan case. *Process Biochemistry*, **40**(7), 2467–2473.

Obara T., Yamanaka O. and Yamamoto K. (2006). A sequential parameter estimation algorithm for activated sludge model no.2d based on mathematical optimization and a prior knowledge for parameters. In: Proceedings IWA World Water Congress, Beijing, CHN, 10-14 September 2006

Penya-Roja J.M., Seco A., Ferrer J. and Serralta J. (2002). Calibration and validation of Activated Sludge Model No.2d for Spanish municipal wastewater. *Environmental Technology*, **23**(8), 849–862.

Petersen B., Gernaey K., Henze M. and Vanrolleghem P.A. (2002). Evaluation of an ASM1 model calibration procedure on a municipal-industrial wastewater treatment plant. *J. Hydroinformatics*, **4**(1), 15–38.

Rieger L., Koch G., Kuhni M., Gujer W. and Siegrist H. (2001). The EAWAG Bio-P module for Activated Sludge Model No. 3. *Water Research*, **35**(16), 3887–3903.

Ronner-Holm S.G.E., Mennerich A. and Holm N.C. (2006). Specific SBR population behaviour as revealed by comparative dynamic simulation analysis of three full-scale municipal SBR wastewater treatment plants. *Water Science and Technology*, **54**(1), 71–80.

Sahlstedt K.E., Aurola A.M. and Fred T. (2003). Practical modelling of a large activated sludge DN-process with ASM3. In: Proceedings Ninth IWA Specialized Conference on Design, Operation and Economics of Large Wastewater Treatment Plants, Praha, Czech Republic, 1-4 September 2003.

Sin G. (2004). Systematic calibration of activated sludge models. *PhD thesis*, BIOMATH Faculteit Landbouwkundige en Toegepaste Biologische Wetenschappen, Ghent University, Gent, Belgium.

Sin G., De Pauw D.J.W., Weijers S. and Vanrolleghem P.A. (2008). An efficient approach to automate the manual trial and error calibration of activated sludge models. *Biotechnology and Bioengineering*, **100**(3), 516–528.

Spanjers H., Vanrolleghem P., Nguyen K., Vanhooren H. and Patry G.G. (1998). Towards a simulation-benchmark for evaluating respirometry-based control strategies. *Water Science and Technology*, **37**(12), 219–226.

Stamou A., Katsiri A., Mantziaras I., Boshnakov K., Koumanova B. and Stoyanov S. (1999). Modelling of an alternating oxidation ditch system. *Water Science and Technology*, **39**(4), 169–176.

Stricker A.-E. (2000). Application de la modélisation à l'étude du traitement de l'azote par boues activées en aération prolongée : comparaison des performances en temps sec et en temps de pluie. *PhD thesis*, Strasbourg I - Louis Pasteur, Strasbourg, France. [in French].

Sun P. (2006). Numerical modelling COD, N and P removal in a full-scale WWTP of China. *Journal of Applied Sciences*, **6**(15), 3155–3159.

Wentzel M.C., Ekama G.A. and Marais G.v.R. (1992). Processes and modelling of nitrification denitrification biological excess phosphorus removal systems - A review. *Water Science and Technology*, **25**(6), 59–82.

Wichern M., Lubken M., Blomer R. and Rosenwinkel K.H. (2003). Efficiency of the Activated Sludge Model no. 3 for German wastewater on six different WWTPs. *Water Science and Technology*, **47**(11), 211–218.

Wichern M., Obenaus F., Wulf P. and Rosenwinkel K.H. (2001). Modelling of full-scale wastewater treatment plants with different treatment processes using the Activated Sludge Model no. 3. *Water Science and Technology*, **44** (1), 49–56.

Xu S. and Hultman B. (1996). Experiences in wastewater characterization and model calibration for the activated sludge process. *Water Science and Technology*, **33**(12), 89–98.

Appendix G

Typical sources of measurement errors

Table G.1 Potential sources of errors for flow measurements.

Measurement principle	Potential sources of errors	To be checked
General		Easy and reliable check by increase or decrease of water level in AS or other tanks
Venturi or weir	Measurement of height	Calibration of ultra-sonic, echo-sonic, radar, bubbler At least zero and maximum check Check of temperature dependency
	Changing of cross-section	Cross-section cleanliness (algae or sediments) Leakages Cross-check dimension and installation with technical reference (physical check)
	Calculation of flow	Cross-check used formula with technical reference (for an independent check special experiments are required, for example tracer experiments or mass balances)
	Miscellaneous	Signal transmission or conversion
Velocity and height	Measurement of height	See above
	Changing of cross-section	See above

(*Continued*)

Previously published in: Rieger, L., Takács, I., Villez, K., Siegrist, H., Lessard, P., Vanrolleghem, P.A. and Comeau, Y. (2010). Data reconciliation for WWTP simulation studies – Planning for high quality data and typical sources of errors. *Water Environment Research*, **82**(5), 426–433.

Table G.1 Potential sources of errors for flow measurements (*Continued*).

Measurement principle	Potential sources of errors	To be checked
	Measurement of velocity	Depending on measuring principle (e.g. doppler effect, transit time, etc.) In general impact of particle conc., flow profile
	Calculation of flow	See above
Electromagnetic flow meter	Air in pipe	Filling of pipe
	Changing of cross-section	Fouling in pipe
	In and outflow distances	Check distances (not long enough or not straight)
	Miscellaneous	Calibration, signal transmission or conversion

Table G.2 Potential sources of errors during sampling.

Type of sampling	Potential sources of errors	To be checked
General	Incorrect location	Measured flow, process dynamics
	Insufficient homogeneity	Mixing/dead zones, location in the cross-section
	Time	Cross-check time stamp with SCADA system Start/stop intervals
Autosampler	Kind of sampling	Proportional to volume or time, fixed intervals
	Settings	Check settings of pump rate and time. Most rain peaks should be included to prevent overestimation of loads.
	Cooling	Always lower than 4°C (check for the smallest sampling interval due to the higher temperature emission of the distribution unit or the sampling pump)
	Volume and pumping speed per single sampling event	Unwanted sedimentation in pre-sampling vessel Proper flushing of the whole inlet hose. Undesired emptying of suction hose.
	Installation of inlet hose	Hose length, unwanted bends, proper siphon?
	Inlet hose	Biofilm, sediments

Table G.3 Potential sources of errors during analysis.

Steps	Potential sources of errors	To be checked
Sample storage	Biological degradation, precipitation	Best to measure directly after sampling Cooling/freezing Complete filling of bottle to prevent oxygen supply Depending on the composition prevent precipitation Sufficient addition of inhibitor to prevent biodegradation

(*Continued*)

Table G.3 Potential sources of errors during analysis (*Continued*).

Steps	Potential sources of errors	To be checked
Sample preparation	Insufficient homogeneity	Homogeneity
	Filtration	Filter pore size Content of unwanted compounds in the filter (COD, NO_3)
	Dilution	Correct volumes (dilution as little as possible, frequent source of errors)
	Digestion/fractionation	Suitability of digestion method suitable
Analysis	Micropipettes	Check of volume (with scale) Cut tip for sludge measurements to prevent filter effect
	Laboratory scales	Check with standard weights or Check by supplier
	Photometer	Maintenance → regular check by supplier recommended Correct calibration factors (updates) Cleanliness of optical system and cuvettes Disturbance due to air bubbles
	Analysis	Quality of reagents and standards (production, storage) Calibration (check with standard solutions) Matrix influence/cross sensitivities (check with standard addition/dilution experiments or reference measurements) Measuring range Cross-check with standard methods
	Data processing	Check unit, dilution rate, typing errors A well kept lab journal is strongly recommended

Table G.4 Potential sources of errors of on-line sensors.

Steps	Potential sources of errors	To be checked
Filtration unit for on-line sensors	Inlet hose/filtrate outlet	Biofilm or sediments. Check with measurements of influent/effluent of filtration unit.
	Installation of inlet hose/filtrate outlet	Hoses too long, sharp bend, siphon
	Coordination with sensor	Discontinuously working filtration devices should be triggered from the sensor.
	Membrane	Clogging or biofilms growth
	Pump rate	Pump rate not sufficient to reach low response time
Sensor/ analyzer	Installation	Location, gradients, flow velocity
	Condition	Maintenance, cleaning system
	Clogging/biofilm/fouling	Mainly a problem for in-situ sensors Check automatic cleaning system

(*Continued*)

Table G.4 Potential sources of errors of on-line sensors (*Continued*).

Steps	Potential sources of errors	To be checked
	Air in measuring cell	Check if incoming flow is high enough (particularly if several sensors are connected to one filtration unit) Check for leakage Check pump tubes
Analysis	Calibration standards	Check age/expiration date Check concentration of standard Settings of the analyzer
	Reagents	Check expiration date Even original reagents from the supplier can be wrong
	Calibration	Some methods require a calibration to the water matrix
	Matrix influence	Check regularly with reference measurements
	Measuring range	Check whether range is suitable for the meas. variations Check whether accuracy is sufficient over whole range
	Response time	Check response time of the whole measurement system (can be more than 30 min)
Data transmission and processing	Amplifier	Check filtering Check delay Check averaging Check for interferences (e.g. radio) Check noise profile
	Settings of sensor	Check settings for output
	SCADA system	Check incoming signal Check data aggregation and filtering (actual vs. processed value)

Appendix H
Sources of uncertainties

An overview regarding uncertainty and model prediction accuracy in the field of wastewater treatment modelling has been issued recently by Belia *et al.* (2009). Activities are ongoing by the International Water Association (IWA)/Water Environment Federation (WEF) Task Group on Design and Operations Uncertainty (DOUT, 2011), which will present their findings in a Scientific and Technical Report (Belia *et al.* in preparation). This section therefore only presents a list of sources of uncertainty introduced during the main steps of the GMP Unified Protocol.

H.1 DEFINITIONS

The terms required to understand the following discussion on uncertainties and model prediction quality are defined in Table H.1. Note that one of the tasks of the IWA Task Group DOUT is to come up with an accepted terminology in order to propose a common language to communicate on the subject. The definitions provided in Table H.1 may slightly differ from the ones that will soon be proposed by the DOUT Task Group.

H.2 SOURCES OF UNCERTAINTY

Sources of uncertainty can be classified in respect to their location in a generic model. Four main areas have been identified: model inputs (e.g. input data, physical data, operational settings, performance data), model structure (sub-models and their interfaces), model parameters and implementation in software packages.

The sources of uncertainty may also be identified according to *when* they are introduced in the course of the modelling project. The nature and level of uncertainty introduced during the 5 steps of the Unified Protocol have been described by Belia *et al.* (2009) and are reported in Figure H.1 and Table H.2.

Updated version, original version previously published in: Belia, E., Amerlinck, Y., Benedetti, L., Johnson, B., Sin, G., Vanrolleghem, P.A., Gernaey, K.V., Gillot, S., Neumann, M.B., Rieger, L., Shaw, A. and Villez, K. (2009). Wastewater treatment modelling: dealing with uncertainties. *Water Science and Technology*, **60**(8), 1929–1941.

Table H.1 Definitions referring to uncertainties and model prediction quality.

Category	Term	Definition
General	Model prediction accuracy	An estimate of how close a model predicted quantity is to the true or reference values of the described real system.
	Variability	A quantity is variable when subjected to random disturbances or fluctuations. For example, it is the degree to which repeated measurements show different or dissimilar results. The term variability is recommended to describe the concept qualitatively while standard deviation and variance are measures (numbers) used to quantify variability.
	Uncertainty	The degree of inability to determine or predict the exact behaviour of a system or process. Uncertainty relates to (1) the inability to determine truly and precisely what has happened in the past because several possibilities lead to similar observations and (2) the inability to predict truly and precisely what will happen in the future.
Nature of uncertainty	Reducible	Uncertainty that can be reduced with further research/efforts. (e.g. experimental determination of kinetic parameters).
	Irreducible	Uncertainty due to the inherent variability of a system that cannot be reduced with any further research/efforts (e.g. rainfall, toxic spills).
Level of uncertainty	Quantifiable uncertainty	Uncertainty that can be reduced with further research/efforts. (e.g. determination of kinetic parameters)
	Scenario uncertainty	Uncertainty due to the inherent variability of a system that cannot be reduced with any further research/efforts (e.g. rainfall, toxic spills)
	Recognized ignorance	Recognised ignorance is the state where fundamental uncertainty is acknowledged to exist and the scientific basis is insufficient to develop functional relationships, statistics, or scenarios.
	Total ignorance	Total ignorance is defined as the state where a deep level of uncertainty exists. It is unknown what is unknown.

Source: (Belia *et al.* 2009).

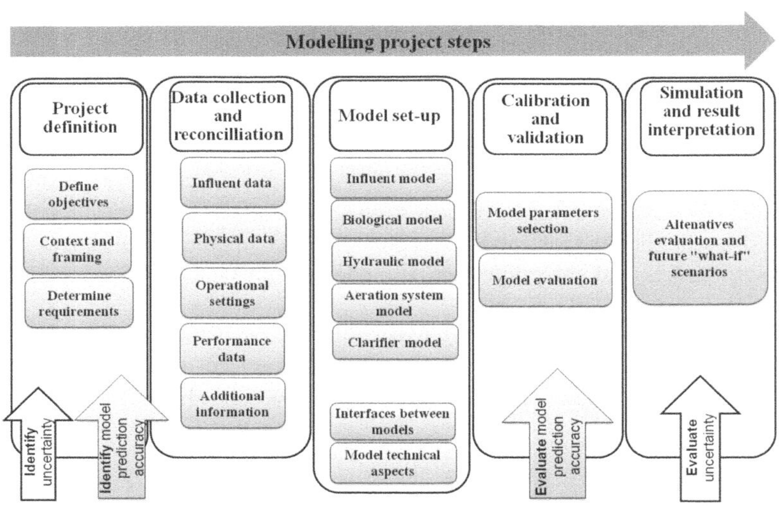

Figure H.1 Instances where model uncertainty and model prediction accuracy should be identified and evaluated (from Belia *et al.* 2009).

Table H-2 Nature and level of uncertainty introduced during each step of a typical activated sludge modelling project.

Typical modelling project steps		Details of each step	Nature and source of uncertainty	Level of uncertainty
Project definition	Objectives	Design, operation, training	The required prediction accuracy of the model is decided at this stage of the project. This will define which of the uncertainty items listed below will be taken into account	N/A
	Context and framing	The boundaries of the system to be modelled. Biological treatment only, whole plant or sewer and river		
	Requirements	Level of model prediction accuracy, what type of data		
Data collection and reconciliation	Influent data	Flow rate, concentrations, influent characterisation data, data from other models and other systems like sewers	Irreducible: due to the inherent variability of the real system like weather, unexpected demographic changes	Quantifiable, scenario, recognised ignorance
			Reducible: due to data collection e.g. sampling method, location, frequency, accuracy of sensors, accuracy of analytical techniques	Quantifiable
	Physical data	Process flow diagram, tank volumes, clarifier surface areas, flow splits	Irreducible: due to the unpredictable and dynamic behaviour of structures like splitters to flow changes	Scenario
			Reducible: due to e.g. unknown true volume constructed or operational depth of structures	Quantifiable
	Operational settings	Controller set-points, valve positions, pumped flows	Irreducible: due to the unpredictability of operator decisions	Quantifiable, scenario
			Reducible: due to actions different from planned or changes not logged, e.g. a change in set-points	Recognised ignorance
	Performance data	Effluent data and reactor concentrations such as MLSS (when not used as controller set-points)	Irreducible: due to the inherent variability of the real system e.g. response of microbial consortium	Quantifiable, scenario, recognised ignorance
			Reducible: due to data collection issues	Quantifiable

(Continued)

Table H-2 Nature and level of uncertainty introduced during each step of a typical activated sludge modelling project (*Continued*).

Typical modelling project steps		Details of each step	Nature and source of uncertainty	Level of uncertainty
	Additional information	Equipment failures	Irreducible: e.g. due to unexpected equipment failures	Quantifiable, scenario, recognised ignorance
Plant Model Setup	Influent model	Influent dynamics, characteristics, influent fractions	Reducible: due to simplifications of influent dynamics (applying a generic diurnal pattern to average vs. constructing a dynamic profile of the whole sewer system), due to simplifications of influent characteristics (fixed ratios for influent fractions)	Scenario
	Biological model	Model structure: processes (conversion, separation), calculation of composite variables, type of mathematical expression used to describe processes (Monod vs. enzymatic kinetics)	Irreducible: due to the inherent variability of the real system	Recognised ignorance
			Reducible: due to simplifications in model structure for example, processes not included, processes included in simplified form (one step vs. two step nitrification), due to the choice of mathematical description of processes	Quantifiable
		Model parameters: fixed, a priori chosen, calibrated, time varying	Reducible: due to our lack of knowledge of the appropriate value	Quantifiable, scenario
	Hydraulic model	Model structure: transport and mixing processes, number of trains, number of tanks in series Model parameters: fixed, a priori chosen, calibrated, time varying	Reducible: due to the simplification of transport and mixing processes in models, inadequate spatial resolution (CSTRs vs. CFD, selection of number of trains to model, number of tanks in series)	Quantifiable, scenario
	Aeration system model	Model structure: gas transfer processes, mechanical system details	Reducible: due to the simplification of gas transfer processes and aeration system	Quantifiable, scenario

(*Continued*)

Table H-2 Nature and level of uncertainty introduced during each step of a typical activated sludge modelling project (*Continued*).

Typical modelling project steps		Details of each step	Nature and source of uncertainty	Level of uncertainty
Plant Model Setup (continued)		Model parameters: fixed, a priori chosen, calibrated, time varying		
	Clarifier model	Model structure: separation processes, calculation of composite variables and type of mathematical expression used to describe processes (1D, 2D, CFD)	Reducible: due to simplifications in model structure for example, processes not included, processes included in simplified form as well as due to the choice of mathematical description of processes	Quantifiable, scenario
		Model parameters: fixed, a priori chosen, calibrated, time varying	Irreducible: due to inherently varying biomass settling properties Reducible: due to our lack of knowledge of the appropriate value	Quantifiable, scenario
	Controllers in plant operations	Control loops, sensors, actuators, time variation of set-points	Reducible: due to the oscillation of the aeration system, time delays in control loops, non-linearity of actuators	Quantifiable, scenario
	Interfaces between models	Use of one or several sets of state variables, calculation of composite variables	Reducible: due to the aggregation of state variables	Quantifiable
	Model technical aspects	Numerics: solver selections & settings, bugs Simulators: limitations of simulation platforms	Reducible: due to numerical approximations and software bugs	Quantifiable, recognised ignorance
Calibration & Validation	Model parameter selection	Selection of model parameters for e.g. biological, separation models that need to be adjusted	Model prediction error calculations. Uncertainty analysis of calibration & validation parameters	N/A
	Model evaluation	Evaluation of model prediction error for calibration & validation data sets		N/A
Simulation	Alternatives evaluation, future "what-if" scenarios	Generation of model desired results (probability distributions, statistics)	Post-calibration uncertainty analysis of simulations (sensitivity and Monte Carlo uncertainty analysis)	N/A

REFERENCES

Belia E., Amerlinck Y., Benedetti L., Johnson B., Sin G., Vanrolleghem P. A., Gernaey K. V., Gillot S., Neumann M. B., Rieger L., Shaw A. and Villez K. (2009). Wastewater treatment modelling: dealing with uncertainties. *Water Science and Technology*, **60**(8), 1929–1941.

Belia E., Neumann M. B., Benedetti L., Johnson B., Murthy S., Weijers S., Vanrolleghem P. A. (IWA Task Group on Design and Operations Uncertainty – DOUTGroup) (in preparation). Uncertainty in wastewater treatment design and operation – Addressing current practices and future directions. IWA Publishing, London, UK, ISBN: 9781780401027.

DOUT (2011). http://www.iwahq.org/f9/networks/task-groups/task-group-on-uncertainty.html.

Appendix I

Mass balancing

Mass balances are powerful tools to detect inconsistencies within WWTP data sets (Barker & Dold, 1995; Nowak *et al.* 1999; Meijer *et al.* 2002; Thomann, 2008; Puig *et al.* 2008). Mass balances can be based on a number of different process variables, which allows the setup of *parallel mass balances* (mass balance of the same system utilising different process variables). Especially useful in identifying systematic measuring errors is the setup of *overlapping mass balances* (mass balances with different system boundaries with one common measuring point).

It should be noted however, that mass balances provide no information on the precision of a specific measurement. Due to averaging over typically longer periods, mass balances cannot be used to detect time-dependent errors like drifts.

I.1 TYPES OF MASS BALANCES

Mass balancing is an engineering tool that allows the identification and confirmation of the mass flows into and out of a specific system. Based on the principle of mass conservation, a mass balance has the following general form:

$$\text{Input} + \text{Reaction} = \text{Output} + \text{Accumulation} \tag{I.1}$$

Equation I.1 can be written for different process variables (e.g. Q, COD, N, P, TSS) considering different system boundaries (the biological system, the primary clarifier, etc.). A mass balance is considered as *closed* if the mass flows of all streams entering and leaving the investigated system are known.

A *closed mass balance* represents an over determined system and subsequently allows one to cross-check different measurements for plausibility and consistency. By contrast, *closing a mass balance* means that the underlying database of a mass balance is reconciled such that the sum of all input mass flows equals the sum of all output mass flows.

An *open mass balance* refers to a system where at least one in- or output cannot be quantified. It is therefore calculated as the residual of the mass flows of the other system in- and outputs. An example is the nitrogen load (as N_2, N_2O, etc.) in the off-gas of a WWTP, which can only be measured by applying specialised equipment generally not available at municipal WWTPs.

Results from *open mass balances* can be combined with expert knowledge in order to assess data quality; for example, the total oxygen demand calculated from a combined COD and N_{tot} mass balance can be compared to the energy consumption and/or measured air flow of the blower station. Another example is to replace the P_{tot} content of the sludge (which often is not available from routine data) with a typical ratio from the literature. Typical ratios can be found in Table 5.2.4 in Chapter 5.

I.2 APPLICATION TO SPECIFIC PROCESS VARIABLES

Suitable process variables for WWTP mass balances are COD, N, P, flow and Fe. TSS can be used for a solids balance around clarifiers. Although TSS is not a conservative process variable, it can be applied with the assumption that reactions changing the TSS mass are insignificant within the applied balancing period.

For a given variable, the general mass balance equation can be expressed as:

$$\sum_{i=1}^{n} L_{\mathrm{IN},i} + V \times r_{\mathrm{v}} = \sum_{j=1}^{m} L_{\mathrm{OUT},j} + \frac{\Delta M}{\tau} \qquad \text{(I.2)}$$

where:

L	Mass load of the balancing variable in influent (IN) or effluent (OUT)	[kg/d]
i	Indices for system influent streams	
j	Indices for system effluent streams	
V	Volume of the system	[m^3]
ΔM	Change of stored mass of variable in the system for the balancing period	[kg]
τ	Balancing period	[d]
r_{V}	Volumetric reaction rate	[kg/m^3/d]

For water streams the concentration values are typically analysis results from 24h-composite samples while for sludge streams the concentration values are derived from analyses of grab samples. For the latter special care needs to be applied so that grab sampling represents daily average conditions.

Note:

A P_{tot} mass balance is often the most effective approach for assessing the quality of a WWTP data set since phosphorus can easily be measured and is not removed in the gas phase.

I.3 BALANCING PERIOD

Typically, there is insufficient data to calculate the storage term accurately in Equation I.2. Therefore, mass balances for WWTP data should be set up for longer periods and quasi steady-state conditions. As a rule of thumb, 2–3 times the SRT should be chosen.

Periods of substantial changes of the operating conditions (e.g. shutdown of treatment lines for revision, start-up periods) should not be included if possible, since they describe transient system states.

Also, substantial changes of the MLSS concentration in the aeration tank within the balancing period need to be considered. If these transient states need to be included, the mass storage term in Equation I.2 has to be used.

I.4 UNCERTAINTY OF WWTP MASS BALANCES

As a rule of thumb, it can be expected that a mass balance based on reasonable data can be closed with a residual in a range of ± 5–10%. If a mass balance cannot be closed within the above mentioned range, special investigations should be carried out in order to determine the error sources.

I.5 OVERLAPPING MASS BALANCES

Overlapping mass balances provide a step-wise approach for detecting systematic errors within WWTP data sets (Thomann *et al.* 2008; Meijer *et al.* 2002; Puig *et al.* 2008). They can be set up for different system boundaries and for different components. Figure I.1 shows system boundary options for different mass balances of a typical WWTP.

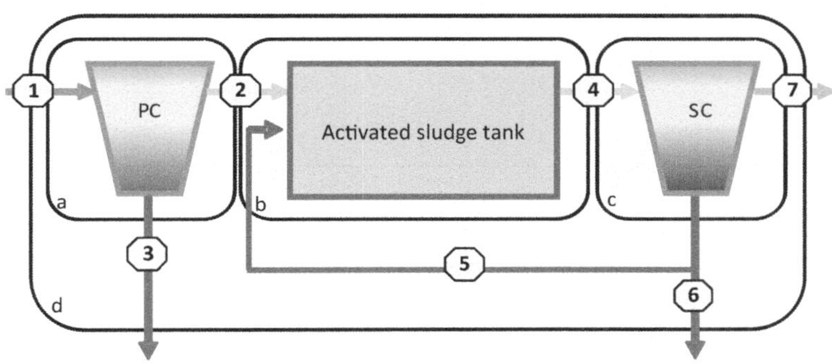

Flux	1	2	3	4	5	6	7
Description	Raw wastewater	Settled wastewater	Primary sludge	Effluent AST	Return sludge	Secondary sludge	Effluent

Figure I.1 System boundary options for mass balancing different process steps of a WWTP (adapted from Thomann, 2008).

Overlapping mass balances integrate specific measuring points in more than one mass balance. If one mass balance can be closed but another overlapping balance cannot, the error is likely to be in another variable (not the overlapping one). If both are not closing, the probability is high that the error is in one of the measured variables (flow or concentration), which overlap. Calculating a *parallel mass balance* with the same boundaries but for another component allows separating flow and concentration measurements. Introducing a correction factor for the variable containing the potential error should allow closing of all balances.

Note that the mass balances including the biological treatment steps (activated sludge tank AST) include an additional removal path for the process variables COD_{tot} and N_{tot}, which is respiration and denitrification, respectively. Since typically no measurement data are available for these outputs, they have not been explicitly included in Figure I.1. The dashed boxes define possible mass balances applicable for a WWTP (sludge treatment not included).

Table I.1 summarises the different mass balance options depicted in Figure I.1 (adapted from Thomann, 2008).

Table I.1 Mass balances and applied process variables according to Figure I.1.

System boundaries		Process variables					
Treatment step	**Ref. in Figure I.1**	**Q**	**COD**	**N_{tot}**	**P_{tot}**	**Solids**	**Application**
Primary clarifier (PC)	a	1 3	1 2	1 2	1 2 3		Removal efficiency P-balance
Aeration tank	b	2 6	AST 2	2 4 6	AST	AST 5 6	Sludge production Oxygen consumption
Secondary clarifier	c	4 5 6 7	4 5 6 7			AST 5 6	Solid balance
Whole plant	d	1 3 6 7	1 7	1 7	1 3 6 7		P-balance Sludge production

I.6 TYPICAL PITFALLS AND RECOMMENDATION

- BOD_5 cannot be used as a balancing variable, since it describes the partial oxygen consumption in biological degradation processes within an arbitrarily defined time period of five days. Since not all of the degradation processes are completed within five days, the BOD_5 does not characterise the complete oxygen load needed for a full degradation of the constituents of the sample. In order to use BOD_5 for mass balancing it must be converted to BOD_∞ or COD.
- Large disturbances such as peak loads, storms, and so on, can have a disproportionate impact on averages if they fall within the balancing period. This data 1) should be removed from the balancing period or 2) the period extended so the effect of disturbances gets minimised. Such dynamic periods – if sampled frequently enough – are useful for separate dynamic modelling investigations.
- Causes for high uncertainty of specific mass flows should be investigated in detail before assigning residual errors to specific measurements. Communication with WWTP personnel is essential in the data reconciliation process. Plant visits should be planned to discuss and review potential error sources.

REFERENCES

Barker P. S. and Dold P. L. (1995). COD and nitrogen mass balances in activated sludge systems. *Water Research*, **29**(2), 633–643.

Meijer S. C. F., van der Spoel H., Susanti S., Heijne J. J. and van Loosdrecht M. C. M. (2002). Error diagnostics and data reconciliation for activated sludge modelling using mass balances. *Water Science and Technology*, **45**(6), 145–156.

Nowak O., Franz A., Svardal K., Müller V. and Kühn V. (1999). Parameter estimation for activated sludge models with the help of mass balances. *Water Science and Technology,* **39**(4), 113–120.

Puig S., van Loosdrecht M. C. M., Colprim J. and Meijer S. C. F. (2008). Data evaluation of full-scale wastewater treatment plants by mass balance. *Water Research*, **42**(18), 4645–4655.

Thomann M. (2008). Quality evaluation methods for wastewater treatment plant data. *Water Science and Technology*, **57**(10), 1601–1609.

References

Allison J. D., Brown D. S. and Novo-Gradac K. J. (1991). MINTEQA2/PRODEFA2, A geochemical assessment model for environmental systems: Version 3.0 User's manual. EPA/600/3–91/021, United States Environmental Protection Agency, Office of Research and Development, Washington, DC, USA.

Ardern E. and Lockett W. T. (1914). Experiments on the oxidation of sewage without the aid of filters. *Journal of the Society of Chemical Industry*, **33**(10), 523–539.

ATV-DVWK-A 131E (2000). Dimensioning of Single-Stage Activated Sludge Plants. ISBN 978-3-935669-96-2, German Association for Water, Wastewater and Waste, Hennef, Germany.

Baker A. J. (1994). Modelling activated sludge treatment of petroleum and petrochemical wastes. PhD thesis. McMaster University, Hamilton, Ontario, Canada. http://digitalcommons.mcmaster.ca/dissertations/AAINN98180/.

Barker P. S. and Dold P. L. (1995). COD and nitrogen mass balances in activated sludge systems. *Water Research*, **29** (2), 633–643.

Barker P. S. and Dold P. L. (1997). General model for biological nutrient removal activated sludge systems: model presentation. *Water Environment Research*, **69**(5), 969–984.

Batstone D. J., Keller J., Angelidaki I., Kalyuzhnyi S. V., Pavlostathis S. G., Rozzi A., Sanders W. T. M., Siegrist H. and Vavilin V. A. (2002). Anaerobic Digestion Model No. 1 (ADM1), IWA Task Group for Mathematical Modelling of Anaerobic Digestion Processes. IWA Publishing, London, UK.

Belia E., Amerlinck Y., Benedetti L., Johnson B., Sin G., Vanrolleghem P. A., Gernaey K. V., Gillot S., Neumann M. B., Rieger L., Shaw A. and Villez K. (2009). Wastewater treatment modelling: dealing with uncertainties. *Water Science and Technology*, **60**(8), 1929–1941.

Belia E., Neumann M. B., Benedetti L., Johnson B., Murthy S., Weijers S., Vanrolleghem P. A. (IWA Task Group on Design and Operations Uncertainty –DOUTGroup) (in preparation). Uncertainty in wastewater treatment design and operation – Addressing current practices and future directions. IWA Publishing, London, UK, ISBN: 9781780401027.

Boero V. J., Eckenfelder W. W., Jr. and Bowers A. R. (1990). Soluble microbial product formation in biological systems. *Water Science and Technology*, **23**(4–6), 1067–1076.

Brdjanovic D., van Loosdrecht M. C. M., Versteeg P., Hooijmans C. M., Alaerts G. J. and Heijnen J. J. (2000). Modeling COD, N and P removal in a full-scale wwtp Haarlem Waarderpolder. *Water Research*, **34**(3), 846–858.

Buisman C., Ijspeert P., Janssen A. and Lettinga G. (1990). Kinetics of chemical and biological sulfide oxidation in aqueous solutions. *Water Research*, **24**(5), 667–671.

Bury S. J., Groot C. K., Huth C. and Hardt N. (2002). Dynamic simulation of chemical industry wastewater treatment plants. *Water Science and Technology*, **45**(4–5), 355–363.

Busby J. B. and Andrews J. F. (1975). Dynamic modeling and control strategies for the activated sludge process. *Journal Water Pollution Control Federation*, **47**(5), 1055–1080.

Choubert J. M., Racault Y., Grasmick A., Beck C. and Heduit A. (2005). Nitrogen removal from urban wastewater by activated sludge process operated over the conventional carbon loading rate limit at low temperature. *Water South Africa*, **31**(4), 503–510.

Choubert J. M., Stricker A. E., Marquot A., Racault Y., Gillot S. and Heduit A. (2009). Updated Activated Sludge Model No 1 parameter values for improved prediction of nitrogen removal in activated sludge processes: Validation at 13 full-scale plants. *Water Environment Research*, **81**(9), 858–865.

Corominas L. (2006). Control and optimization of an SBR for nitrogen removal: From model calibration to plant operation. PhD thesis, University of Girona, ISBN 84-690-0241-4.

Corominas L., Rieger L., Takács I., Ekama G., Hauduc H., Vanrolleghem P. A., Oehmen A., Gernaey K. V. and Comeau Y. (2010). New framework for standardized notation in wastewater treatment modelling. *Water Science and Technology*, **61**(4), 841–857.

de Haas D. W., Wentzel M. C. and Ekama G. A. (2001). The use of simultaneous chemical precipitation in modified activated sludge systems exhibiting biological excess phosphate removal Part 6: Modelling of simultaneous chemical-biological P removal – Review of existing models. *Water South Africa*, **27**(2), 135–150.

De Pauw D. J. W. and Vanrolleghem P. A. (2006). Practical aspects of sensitivity function approximation for dynamic models. *Math. Comp. Modell. of Dyn. Sys.*, **12**, 395–414.

Dold P. L., Ekama G. A. and Marais G. V. R. (1980). A general model for the activated sludge process. *Progress in Water Technology*, **12**, 44–77.

Dold P., Takács I., Mokhayeri Y., Nichols A., Hinojosa J., Riffat R., Bailey W. and Murthy S. (2007). Denitrification with Carbon Addition – Kinetic Considerations. Proceedings of the WEF/IWA Nutrient Removal Conference 2007, Baltimore, Maryland, USA, 4–7 March 2007, 218–238.

DOUT (2011). http://www.iwahq.org/f9/networks/task-groups/task-group-on-uncertainty.html.

Downing A. L., Painter H. A. and Knowles G. (1964). Nitrification in the activated sludge process. *J. Proc. Inst. Sewage Purif.*, **2**, 130–153.

Eckenfelder W. W. (1956). Studies on the oxidation kinetics of biological sludges. *Sewage and Industrial Wastes*, **28**(8), 983–990.

Ekama G. A., Dold P. L. and Marais G. V. (1986). Procedures for determining influent COD fractions and the maximum specific growth-rate of heterotrophs in activated-sludge systems. *Water Science and Technology*, **18** (6), 91–114.

Eremektar G., Karahan-Gul O., Germirli-Babuna F., Ovez S., Uner H. and Orhon D. (2002). Biological treatability of a corn wet mill effluent. *Water Science and Technology*, **45**(12), 339–346.

Frank K. (2006). The application and evaluation of a practical stepwise approach to activated sludge modelling. In: Proceedings PENNTEC 2006, Annual Technical Conference and Exhibition, State College, July 2006, Pennsylvania, USA.

Fujie K., Sekizawa T. and Kubota H. (1983). Liquid mixing in activated sludge aeration tank. *Journal of Fermentation Technology*, **61**(3), 295–304.

Garrett M. T., Jr. and Sawyer C. N. (1952). Kinetics of removal of soluble BOD by activated sludge. Proc. 7th Industrial Waste Conference, Purdue University, Lafayette, Indiana, USA, 51–77.

Germirli F., Orhon D. and Artan N. (1991). Assessment of the initial inert soluble COD in industrial wastewaters. *Water Science and Technology*, **23**(4–6), 1077–1086.

Gernaey K. V., Rosen C. and Jeppsson U. (2006). WWTP dynamic disturbance modelling – an essential module for long-term benchmarking development. *Water Science and Technology*, **53**(4–5), 225–234.

Gillot S. and Héduit A. (2008). Prediction of alpha factor values for fine pore aeration systems. *Water Science and Technology*, **58**(8), 1265–1269.

Gillot S. and Choubert J. M. (2010). Biodegradable organic matter in domestic wastewaters: comparison of selected fractionation techniques. *Water Science and Technology*, **62**(3), 630–639.

Goel R., Mino T., Satoh H. and Matsuo T. (1998). Comparison of hydrolytic enzyme systems in pure culture and activated sludge under different electron acceptor conditions. *Water Science and Technology*, **37**(4–5), 335–343.

Grady C. P. L., Jr., Gujer W., Henze M., Marais G. V. R. and Matsuo T. (1986). A model for single sludge wastewater treatment systems. *Water Science and Technology (WST)*, **18**(6), 47–61.

Gujer W., Henze M., Mino T. and van Loosdrecht M. C. M. (1999). Activated Sludge Model No. 3. *Water Science and Technology*, **39**(1), 183–193.

Gujer W., Henze M., Mino T. and van Loosdrecht M. C. M. (2000). Activated Sludge Model No. 3. In: M. Henze, W. Gujer, T. Mino and M. C. M. van Loosdrecht (eds), Activated sludge models ASM1, ASM2, ASM2d and ASM3. Scientific and Technical Report No. 9, IWA Publishing, London, UK.

Gujer W. (2006). Activated sludge modeling: past, present and future. *Water Science and Technology*, **53**(3), 111–119.

Gujer W. (2008). Systems analysis for water technology. Springer, Berlin, Germany, ISBN: 978-3-540-77277-4.

Gujer W. (2010). Nitrification and me – A subjective review. *Water Research*, **44**(1), 1–19.

Gujer W. (2011). Is modeling of biological wastewater treatment a mature technology? *Water Science and Technology*, **63**(8), 1739–1743.

Hamer G. (1997). Microbial consortia for multiple pollutant biodegradation. *Pure and Applied Chemistry*, **69**(11), 2343–2356.

Hauduc H., Gillot S., Rieger L., Ohtsuki T., Shaw A., Takács I. and Winkler S. (2009). Activated sludge modelling in practice – An international survey. *Water Science and Technology*, **60**(8), 1943–1951.

Hauduc H., Rieger L., Takács I., Héduit A., Vanrolleghem P. A. and Gillot S. (2010). Systematic approach for model verification – Application on seven published Activated Sludge Models. *Water Science and Technology*, **61**(4), 825–839.

Hauduc H. (2010). Modéles biocinétiques de boues actives de type ASM: Analyse théorique et fonctionnelle, vers un jeu de parameters par defaut. PhD thesis Université Laval, Quebec, Canada/AgroParisTech, Paris France (in English).

Hauduc H., Rieger L., Ohtsuki T., Shaw A., Takács I., Winkler S., Héduit A., Vanrolleghem P. A. and Gillot S. (2011). Activated sludge modelling: Development and potential use of a practical applications database. *Water Science and Technology*, **63**(10), 2164–2182.

Hellinga C., Schellen A. A. J. C., Mulder J. W., van Loosdrecht M. C. M. and Heijnen J. J. (1998). The SHARON process: an innovative method for nitrogen removal from ammonium-rich waste water. *Water Science and Technology*, **37**(9), 135–142.

Henze M., Grady C. P. L., Jr., Gujer W., Marais G. V. R. and Matsuo T. (1987). Activated Sludge Model No. 1. IAWPRC Scientific and Technical Report No. 1, IAWPRC, London, UK.

Henze M., Gujer W., Mino T., Matsuo T., Wentzel M. C. and Marais G. V. R. (1995). Activated Sludge Model No. 2. IAWQ Scientific and Technical Report No. 3, IAWQ, London, UK.

Henze M., Gujer W., Mino T., Matsuo T., Wentzel M. C., Marais G. V. R. and van Loosdrecht M. C. M. (1999). Activated Sludge Model No.2d, ASM2d. *Water Science and Technology*, **39**(1), 165–182.

Henze M., Grady C. P. L., Jr., Gujer W., Marais G. V. R. and Matsuo T. (2000a). Activated sludge model No. 1. In: M. Henze, W. Gujer, T. Mino and M. C. M. van Loosdrecht (eds), Activated sludge models ASM1, ASM2, ASM2d and ASM3. Scientific and Technical Report No. 9, IWA Publishing, London, UK.

Henze M., Gujer W., Mino T. and van Loosdrecht M. C. M. (2000b). Activated Sludge Models ASM1, ASM2, ASM2d and ASM3. Scientific and Technical Report No. 9, IWA Publishing, London, UK.

Henze M., Gujer W., Mino T., Matsuo T., Wentzel M. C., Marais G. V. R. and van Loosdrecht M. C. M. (2000c). Activated sludge model No. 2d. In: M. Henze, W. Gujer, T. Mino and M. C. M. van Loosdrecht (eds), Activated sludge models ASM1, ASM2, ASM2d and ASM3. Scientific and Technical Report No. 9, IWA Publishing, London, UK.

Henze M., van Loosdrecht M. C. M., Ekama G. A. and Brdjanovic D. (2008). Biological wastewater treatment: principles, modelling and design. IWA Publishing, London, UK, ISBN: 9781843391883.

Herbert D. (1958). Some principles of continuous culture. In: G. Tunevall, Almquist and Wiksel (eds) Recent progress in microbiology. 7th International Congress for Microbiology. Stockholm, Sweden, 381–396.

Hu Z. R., Wentzel M. C. and Ekama G. A. (2007). A general kinetic model for biological nutrient removal activated sludge systems: model development. *Biotechnology and Bioengineering*. **98**(6), 1242–1258.

Hug T. (2007). Lecture: Advanced wastewater treatment – Introduction to mathematical modeling. Department of Civil Engineering, University of British Columbia, Vancouver, Canada.

Hug T., Benedetti L., Hall E. R., Johnson B. R., Morgenroth E. F., Nopens I., Rieger L., Shaw A. R. and Vanrolleghem P. A. (2009). Mathematical models in teaching and training: mismatch between education and requirements for jobs. *Water Science and Technology*, **59**(4), 745–753.

Hulsbeek J. J. W., Kruit J., Roeleveld P. J. and van Loosdrecht M. C. M. (2002). A practical protocol for dynamic modelling of activated sludge systems. *Water Science Technology*, **45**(6), 127–136.

Insel G., Orhon D. and Vanrolleghem P. A. (2003). Identification and modeling of aerobic hydrolysis – application of optimal experimental design. *J Chem Technol Biotechnol,* **78**(4), 437–445.

Isermann R. and Ballé P. (1997). Trends in the application of model-based fault detection and diagnosis of technical processes. *Control Engineering Practice*, **5**(5), 709–719.

ISO 15839 (2003). Water quality – on-line sensors/analysing equipment for water – Specifications and performance tests. ISO, Geneva, Switzerland.

Itokawa H., Inoki H. and Murakami T. (2008). JS Protocol: A practical guidance for the use of activated sludge modelling in Japan. Proceedings IWA World Water Congress and Exhibition, 7–12 September 2008, Vienna.

Japan Sewage Works Agency (2006). Technical Evaluation of the Practical Use of Activated Sludge Models, Report No. 05-004, Research and Technology Development Division, Japan Sewage Works Agency, Toda (in Japanese). http://www.sbmc.or.jp/.

Johnson T. L. (1993). Design concepts for activated sludge diffused aeration systems. Ph.D. Thesis, University of Kansas, USA.

Jones G. L., Jansen F. and Mckay A. J. (1973). Substrate inhibition of the growth of bacterium NCIB 8250 by Phenol. *Journal of General Microbiology*, **74**, 139–148.

Karahan-Gül Ö., Artan N., Orhon D., Henze M. and van Loosdrecht M. C. M. (2002). Respirometric assessment of storage yield for different substrates. *Water Science and Technology*, **46**(1–2), 345–352.

Kappeler J. and Gujer W. (1992). Estimation of kinetic parameters of heterotrophic biomass under aerobic conditions and characterization of wastewater for activated sludge modelling. *Water Science and Technology*, **25**(6), 125–139.

Ky R. C., Comeau Y., Perrier M. and Takács I. (2001). Modelling biological phosphorus removal from a cheese factory effluent by an SBR. *Water Science and Technology*, **43**(3), 257–264.

Langergraber G., Rieger L., Winkler S., Alex J., Wiese J., Owerdieck C., Ahnert M., Simon J. and Maurer M. (2004). A guideline for simulation studies of wastewater treatment plants. *Water Science Technology*, **50**(7), 131–138.

Langergraber G., Alex J., Weissenbacher N., Woerner D., Ahnert M., Frehmann T., Halft N., Hobus I., Plattes M., Spering V. and Winkler S. (2008). Generation of diurnal variation for influent data for dynamic simulation. *Water Science and Technology*, **57**(9), 1483–1486.

Lesouef A., Payraudeau M., Rogalla F. and Kleiber B. (1992). Optimizing nitrogen removal reactor configuration by onsite calibration of the IAWQ activated sludge model. *Water Science and Technology*, **25**(6), 105–123.

Makinia J. and Wells S. A. (2005). Evaluation of empirical formulae for estimation of the longitudinal dispersion in activated sludge reactors. *Water Research*, **39**(8), 1533–1542.

Makinia J. (2010). Mathematical modelling and computer simulation of activated sludge systems. IWA Publishing, London, UK, ISBN: 9781843392385.

Mamais D., Jenkins D. and Pitt P. (1993). A rapid physical-chemical method for the determination of readily biodegradable soluble COD in municipal wastewater. *Water Research*, **27**(1), 195–197.

Marquot A. (2006). Modelling nitrogen removal by activated sludge on full-scale plants: Calibration and evaluation of ASM1. PhD thesis, Université de Pau et des Pays de l'Adour, France.

Martin C., Shaw A. R., Phillips H. M., Gilley A. and Ayesa E. (2010). Comparison of methods for dealing with uncertainty in wastewater treatment modelling and design. Proc. 2nd IWA/WEF Wastewater Treatment Modelling Seminar, March 2010, Mont-Sainte-Anne, Quebec, Canada.

Maurer M., Abramovich D., Siegrist H. and Gujer W. (1999). Kinetics of biologically induced phosphorus precipitation in waste-water treatment. *Water Research*, **33**(2), 484–493.

McCorquodale J. A., Griborio A. and Georgiou I. (2005). A public domain settling tank model. Proceedings Water Environment Federation 78th Annual Conference & Exposition, Washington DC, USA, 2546–2561.

McKinney R. E. (1962). Mathematics of complete mixing activated sludge. *Journal of Sanitary Engineering Division, Proc. American Society of Civil Engineering*, **88**(3), 87–92.

Meijer S. C. F., van der Spoel H., Susanti S., Heijnen J. J. and van Loosdrecht M. C. M. (2002). Error diagnostics and data reconciliation for activated sludge modelling using mass balances. *Water Science and Technology*, **45**(6), 145–156.

Meijer S. C. F. (2004). Theoretical and practical aspects of modelling activated sludge processes. PhD Thesis, Department of Biotechnological Engineering. Delft University of Technology, The Netherlands.

Melcer H., Bewley J., Witherspoon J. and Caballero R. (1993). Estimating VOC emissions. *Journal of the Air and Waste Management Association*, **43**(10), 1402–1403.

Melcer H., Bell J. P., Thompson D. J., Yendt C. M., Kemp J. and Steel P. (1994). Modeling volatile organic contaminants' fate in wastewater treatment plants. *Journal of Environmental Engineering – ASCE*, **120**(3), 588–609.

Melcer H., Dold P. L., Jones R. M., Bye C. M., Takács I., Stensel H. D., Wilson A. W., Sun P. and Bury S. (2003). Methods for wastewater characterisation in activated sludge modeling. Water Environment Research Foundation (WERF), Alexandria, VA, USA.

Monod J. (1942). Recherches sur la croissance des cultures bactériennes. Thèse Doctorat ès Sciences Naturelles, Hermann & Cie, Paris, France (In French).

Montgomery D. C. and Runger G. C. (2010). Applied statistics and probability for engineers. 5th Edition, ISBN 978-0-470-05304-1, John Wiley & Sons, Hoboken, NJ, USA.

MoST (2006). Modelling Support Tool vers. 3.1.5. Developed in the frame of the EU project HarmoniQuA. Contract no: EVK1-CT-2001-00097, Duration: 2002–2005. http://harmoniqua.wau.nl/public/Products/most.htm.

Naidoo V., Urbain V. and Buckley C. A. (1998). Characterization of wastewater and activated sludge from European municipal wastewater treatment plants using the NUR test. *Water Science and Technology*, **38**(1), 303–310.

Nowak O., Svardal K. and Schweighofer P. (1995). The dynamic behaviour of nitrifying activated sludge systems influenced by inhibiting wastewater compounds. *Water Science and Technology*, **31**(2), 115–124.

Nowak O., Svardal K. and Kroiss H. (1996). The impact of phosphorus deficiency on nitrification – case study of a biological pretreatment plant for rendering plant effluent. *Water Science and Technology*, **34**(1–2), 229–236.

Nowak O., Franz A., Svardal K., Müller V. and Kühn V. (1999). Parameter estimation for activated sludge models with the help of mass balances. *Water Science and Technology*, **39**(4), 113–120.

Orhon D., Ateş E., Sözen S. and Çokgör E. U. (1997). Characterization and COD fractionation of domestic wastewaters. *Environmental Pollution*, **95**(2), 191–204.

Orhon D., Ubay Çokgör E. and Sözen S. (1999). Experimental basis for the hydrolysis of slowly biodegradable substrate in different wastewaters. *Water Science and Technology*, **39**(1), 87–95.

Orhon D., Germirli Babuna F. and Karahan O. (2009). Industrial wastewater treatment by activated sludge. IWA Publishing, London, U.K.

Penfold W. J. and Norris D. (1912). The relation of concentration of food supply to the generation time of bacteria. *Journal of Hygiene*, **12**, 527–531.

Petersen B., Gernaey K., Henze M. and Vanrolleghem P. A. (2002). Evaluation of an ASM1 model calibration procedure on a municipal-industrial wastewater treatment plant. *Journal of Hydroinformatics*, **4**(1), 15–38.

Phillips H. M., Anderson W., Kuosman J., Pier D. and Rogowski S. (2010). Long-term operations modeling: When does validation become recalibration? Proceedings WEFTEC.10, New Orleans, Louisiana, USA.

Phillips H. M. (2011). Personal communication.

Refsgaard J. C., Henriksen H. J., Harrar W. G., Scholten H. and Kassahun A. (2005). Quality assurance in model based water management – review of existing practice and outline of new approaches. *Environmental Modelling and Software*, **20**(10), 1201–1215.

Rieger L., Koch G., Kühni M., Gujer W. and Siegrist H. (2001). The EAWAG Bio-P module for activated sludge model No. 3. *Water Research*, **35**(16), 3887–3903.

Rieger L., Thomann M., Gujer W. and Siegrist H. (2005). Quantifying the Uncertainty of On-line Sensors at WWTPs during Field Operation. *Water Research*, **39**(20), 5162–5174.

Rieger L., Takács I., Villez K., Siegrist H., Lessard P., Vanrolleghem P. A. and Comeau Y. (2010). Data reconciliation for WWTP simulation studies – Planning for high quality data and typical sources of errors. *Water Environment Research*, **82**(5), 426–433.

Roeleveld P. J. and van Loosdrecht M. C. M. (2002). Experience with guidelines for wastewater characterisation in The Netherlands. *Water Science and Technology*, **45**(6), 77–87.

Rosso D., Libra J. A., Wiehe W. and Stenstrom M. K. (2008). Membrane properties change in fine-pore aeration diffusers: Full-scale variations of transfer efficiency and headloss. *Water Research*, **42**(10–11), 2640–2648.

Rozich A. F. and Gaudy A. F., Jr. (1985). Response of phenol-acclimated activated sludge process to quantitative shock loading. *Journal Water Pollution Control Federation*, **57**(7), 795–804.

Scholten H., Van Waveren R. H., Groot S., Van Geer F., Wösten H., Koeze R. D. and Noort J. J. (2000). Good modelling practice in water management. In: Proceedings HydroInformatics2000. International Association for Hydraulic Research, Cedar Rapids, Iowa, USA.

Shaw A., Rieger L., Takács I., Winkler S., Ohtsuki T., Langergraber G. and Gillot S. (2011). Realizing the benefits of good process modeling. Proceedings WEFTEC 2011, 84th Annual Technical Exhibition and Conference. October 15–19, 2011, Los Angeles, CA, USA.

Siegrist H., Krebs P., Bühler R., Purtschert I., Rock C. and Rufer R. (1995). Denitrification in secondary clarifiers. *Water Science and Technology*, **31**(2), 205–214.

Sin G., Van Hulle S. W. H., De Pauw D. J. W., van Griensven A. and Vanrolleghem P. A. (2005). A critical comparison of systematic calibration protocols for activated sludge models: A SWOT analysis. *Water Research*, **39**(12), 2459–2474.

Smolders G. J. F., Van-Der-Meij J., van Loosdrecht M. C. M. and Heijnen J. J. (1995). A structured metabolic model for anaerobic and aerobic stoichiometry and kinetics of the biological phosphorus removal process. *Biotechnology and Bioengineering*, **47**(3), 277–287.

Sperandio M., Urbain V., Ginestet P., Audic J. M. and Paul E. (2001). Application of COD fractionation by a new combined technique: comparison of various wastewaters and sources of variability. *Water Science and Technology*, **43**(1), 181–190.

Stricker A.-E., Lessard P., Heduit A. and Chatellier P. (2003). Observed and simulated effect of rain events on the behaviour of an activated sludge plant removing nitrogen. *Journal of Environmental Engineering and Science*, **2**(6), 429–440.

Takács I., Patry G. G. and Nolasco D. (1991). A dynamic model of the clarification thickening process. *Water Research*, **25**(10),1263–1271.

Third K. A., Shaw A. R. and Ng L. (2007). Application of the Good Modelling Practice Unified Protocol to a plant wide process model for Beenyup WWTP design upgrade. 10th IWA Specialised Conference on "Design, Operation and Economics of Large Wastewater Treatment Plants", 9–13 September 2007, Vienna, Austria, 251–258.

Thomann M., Rieger L., Frommhold S., Siegrist H. and Gujer W. (2002). An efficient monitoring concept with control charts for on-line sensors. *Water Science and Technology*, **46**(4–5), 107–116.

Thomann M. (2008). Quality evaluation methods for wastewater treatment plant data. *Water Science and Technology*, **57**(10), 1601–1609.

US EPA (2009). Guidance on the development, evaluation, and application of environmental models. EPA/100/K-09/003, Council for Regulatory Environmental Modeling, U.S. Environmental Protection Agency, Washington, DC, USA.

Van Impe J. F. M., Vanrolleghem P. A. and Iserentant D. M. (1998). Advanced instrumentation, data interpretation and control of biotechnological processes. Kluwer Academic Publishers, Dordrecht, The Netherlands. ISBN 0-7923-4860-5.

Van Waveren R. H., Groot S., Scholten H., Van Geer F. C., Wösten J. H. M., Koeze R. D. and Noort J. J. (2000). Good Modelling Practice Handbook. STOWA Report 99-05, Utrecht, RWS-RIZA, Lelystad, The Netherlands. http://www.info.wau.nl/research%20projects/pub343pdf/gmp.pdf.

Vanrolleghem P. A., Insel G., Petersen B., Sin G., De Pauw D., Nopens I., Weijers S. and Gernaey K. (2003). A comprehensive model calibration procedure for activated sludge models. In: Proceedings WEFTEC 2003, 76th Annual Technical Exhibition and Conference. October 11–15, 2003, Los Angeles, CA, USA.

Villez K. (2012). Personal communication.

WEF MOP 8 (2010). Design of municipal wastewater treatment plants. Manual of practice No. 8.Water Environment Federation, Alexandria, VA, USA.

Wentzel M. C., Ekama G. A., Loewenthal R. E., Dold P. L. and Marais G. V. R. (1989). Enhanced polyphosphate organism cultures in activated sludge systems. Part II: Experimental behaviour. *Water South Africa*, **15**(2), 71–88.

Wentzel M. C., Mbewe A. and Ekama G. A. (1995). Batch test for measurement of readily biodegradable COD and active organism concentration in municipal waste-waters. *Water South Africa*, **21**(2), 117–124.

Wilkinson T. G. and Hamer G. (1979). The microbial oxidation of mixtures of methanol, phenol, acetone and isopropanol with reference to effluent purification. *Journal of Chemical Technology and Biotechnology*, **29**(1), 56–67.

Wood L. B., Hurley B. J. E. and Matthews P. J. (1981). Some observations on the biochemistry and inhibition of nitrification. *Water Research*, **15**(5), 543–551.

Index

IWA Publishing's authorised EU representative for General Product Safety Regulations is Diane D'Arras, 15 rue Duret, 75116 Paris, France, e-mail: safety@iwap.co.uk.

Printed and bound by CPI Group (UK) Ltd, Croydon, CR0 4YY
12/05/2026
02108863-0010